"Fascinating, creative, readable, rigorous and compassionate, this is one of the most important books I've ever read. *The Nature of Pandemics* is an essential book for our time, and beautifully told. Robinson incisively analyses what is causing the problem, and offers hope and a remedy to solve it. Everyone – especially anyone in a position of power – should read this book."

—Lucy Jones, author of *Foxes Unearthed*,
Losing Eden and *Matrescence*

"I loved this book! Timely and terrifying, if humans are to have any kind of future, it is imperative that we learn from our mistakes – and this book shows us exactly how."

—George McGavin, entomologist, broadcaster
and President of the Dorset Wildlife Trust

"Our relationship with nature has become warped. We cram sentient animals into industrial farms and wet markets, destroy habitats and misuse antibiotics and other chemicals. Clear-eyed and evidence-rich, *The Nature of Pandemics* lays bare the dangerous path we're on, and how restoring nature, and our place within it, is key to making the world safer and healthier."

—Ben Goldsmith, environmentalist
and author of *God Is an Octopus*

"A riveting read! Robinson has provided an enthralling analysis concerning the relationship between nature, disease and human history. Ultimately, while nature has provided the pandemics, biodiversity will also provide the solutions and recovery."

—Prof. Chris Daniels DSc, author of *Koala*
and *A Guide to Urban Wildlife*

"Robinson elegantly weaves together science and history to explore how pandemics unfold and why our society is so vulnerable. But we are not helpless in the face of the inevitable – *The Nature of Pandemics* reveals powerful opportunities for change. I sincerely hope we listen."

—Dr Rebecca Nesbit, author of *Tickets for the Ark*

"From the Midlands of England to Micronesia and from bats to bumblebees, Robinson takes the reader on a wild journey through time and space. He expertly weaves a web of gripping stories and brings to life the bigger picture of why pandemics are born, grow and how they can be stopped. With rigour, empathy and plenty of humour, Robinson untangles the complex relationship between ourselves and the ecosystems we inhabit, explaining what happens when things go out of kilter. This is interdisciplinary popular science writing at its finest!"

—Dr Alexia Barrable, social science researcher
and author of *Growing Up Wild*

"Jake is a genius at communicating complexity. His work has fundamentally changed my worldview on health, humans, and nature and improved my clinical practice. Always engaging, *The Nature of Pandemics* highlights both the root causes of pandemics and the many positive and practical ways we can meet this challenge."

—Dr Aarti Bansal, GP and Co-Founder of Greener Practice

"*The Nature of Pandemics* highlights more than ever that the longevity and health of humankind are inextricably linked to restoring and rewilding our ecosystems, to safeguarding and replenishing our biodiversity, to simply communing with nature's greenspaces, and embracing the ideals of One Health. This book is highly recommended. Robinson's gift is a persuasive, easy-to-read, science-based storytelling which is both enlightening and educational."

—Associate Professor Wayne Boardman, Wildlife Veterinarian
and One Health expert

"Robinson's writing style is so engaging and captures complex ideas in a way that feels both clear and inspiring. *The Nature of Pandemics* is a powerful reminder that the health of people, wildlife, and ecosystems is inseparably intertwined. It reveals why protecting biodiversity is vital for planetary health, also serving as a preventive public health strategy."

—Dr Marja Roslund, Environmental Scientist

THE NATURE
OF PANDEMICS

JAKE M. ROBINSON is a British microbial and restoration ecologist currently residing in Australia. His research interests span microbial ecology, systems thinking, biodiversity–health linkages, and strategies to conserve and restore nature.

Sign up for Jake's newsletter at:
www.jakemrobinson.com

Subscribe to Jake's YouTube channel: @naturegutbrain

Check out Jake's podcasts: *Interconnected* and *Naked Thinking*

Find Jake on Twitter: @_jake_robinson; Bluesky: @jake1; and Insta: @treewilding.

Jake's first book, *Invisible Friends*. It's all about how microbes (bacteria, fungi, viruses, etc.) shape our lives and the world around us – in surprisingly beneficial ways!

His second book, *Treewilding: Our Past, Present and Future Relationship with Forests*, looks at restoring global forests – the science, controversies, solutions and hope.

THE NATURE
OF PANDEMICS

Why Protecting Biodiversity is Key to Human Survival

JAKE M. ROBINSON

PELAGIC PUBLISHING

First published in 2025 by
Pelagic Publishing
20–22 Wenlock Road
London N1 7GU

www.pelagicpublishing.com

The Nature of Pandemics: Why Protecting Biodiversity is Key to Human Survival

https://doi.org/10.53061/LDYB5219

A CIP record for this book is available from the British Library

ISBN 978-1-78427-599-0 Hardback
ISBN 978-1-78427-600-3 ePub
ISBN 978-1-78427-601-0 PDF
ISBN 978-1-78427-620-1 Audio

EU Authorised Representative: Easy Access System Europe – Mustamäe tee 50, 10621 Tallinn, Estonia, gpsr.requests@easproject.com

Cover illustration by Jennifer Smith

Typeset in Minion Pro by S4Carlisle Publishing Services, Chennai, India

5 4 3 2 1

Contents

PART IV. HOW DO WE ADDRESS THESE LOOMING THREATS?

Acknowledgements

First and foremost, immense gratitude to everyone who supported this book, whether by offering encouragement, inspiration or reading early drafts. To my beautiful wife, Kate – thanks for enduring my early writing hours and scattered thoughts. Listening to my constant babble has helped transform a collection of tangled thoughts into a coherent narrative. To my readers – past, present and future – thank you for your curiosity and willingness to explore difficult questions with open minds. And to the ecologists, epidemiologists, public health workers and historians whose research and fieldwork informed this book: your tireless dedication to your craft has provided the foundation upon which these pages stand.

A big thank you to the following people, who contributed thoughts or inspiration to these pages, directly or indirectly: Prof. Graham Rook, the late Prof. Oliver Rackham, Prof. Richard Yarnell, Dr Stephen Harrison, Will Storr, Prof. Vibhu Prakash, Tim Jarvis, Dr Peggy Eby, Dr Toni Bunnell, Prof. Richard Ostfeld, Prof. Felicia Keesing, Prof. Rob Edwards, Dr Richard Beckett, Dr Sarah Kidd, Dr Tara Garrard, David Quammen, Olivia Judson, Dr Alan Hicks, Prof. David Blehert, Dr Melissa Behr, Dr Travis Livieri, Dr Erica Rosenblum, Dr Jodi Rowley, Dr Danielle Wallace, Prof. Menna Jones, Prof. Diana Bell, Dr Tim Hauck, Dr Ashleigh Blackwood, Dr Andrew Breed, A/Prof. Wayne Boardman, Dr Gillian Orrow, Dr Robin Taylor, Prof. Raina Plowright, Dr Alison Peel, A/Prof. Martin Breed, Dr Tari Haahtela, Dr Marja Roslund, Dr Aarti Bansal, Prof. John Cryan, Klaus Lotz, Dr Glynn Percival.

I'd also like to thank and pay my respect to the Kaurna People and their elders of all generations, and acknowledge that I wrote the majority of *The Nature of Pandemics* on Kaurna land (Tarntanya/Adelaide, South Australia).

A big thanks to my publisher, Pelagic Publishing and Nigel Massen, for believing in me and this book, and to my editors, David Hawkins and Charlie Wilson, for your sharp eye and honest feedback. Thanks also to Sarah Stott for your marketing prowess and to Jennifer Smith for the beautiful cover art!

Thanks to Foxtrot the cat, my paperweight and hot water bottle, for keeping me warm during those early morning writing sessions. You walked across the keyboard exactly when it mattered the most and probably contributed at least 37 random characters to the first draft.

This book wouldn't have been the same without any of you.

Finally, to those whose lives have been touched or lost to pandemics, human or otherwise: this book is, in part, a tribute to you. May understanding the nature of pandemics help shape a future where knowledge, reciprocity, compassion and preparedness prevail.

With deepest gratitude,
Jake

Invisible Foes, Planetary Woes

You cannot escape the responsibility of tomorrow by evading it today.
—Abraham Lincoln

Eyam, Derbyshire, England

They nestled their coins in vinegar-filled crevices. In exchange, the residents of my home village of Eyam received eggs, bread and vegetables from neighbouring hamlets. The people of Eyam sealed themselves off from the outside world. They successfully thwarted the spread of infectious 'plague seeds' or the invisible 'miasma' in the air. Still, during the fourteen months of quarantine, the valley echoed with grief as families endured immense suffering, losing multiple members to the bubonic plague. I cherished my time living in Eyam and was fortunate; I wasn't even a twinkle in the universe's eye when this story unfolded. The year was 1665.

In 1665, Eyam was just like any other rural village that lined the muddy trade routes connecting London to the rest of England. Yet, the collective action by its 700 residents transformed Eyam from 'just another village' to one of the most significant places in England at the time. The villagers stemmed the spread and progression of the bubonic plague to nearby towns, including Bakewell and Sheffield. The significance of their actions resonated through space and time.

Eyam has long been a quintessential destination for school trips, offering a valuable lesson in history, ecology and collective action for many years. Indeed, before I moved to Eyam, a school visit to the village may have sparked my interest in 'disease ecology' – the science of how interactions in nature influence the spread, dynamics and impact of diseases. Memories of that trip linger vividly in my mind. A decade or so later, I found myself writing about plague aetiology (the cause of a disease) at university. For my thesis, I studied how tick-borne diseases affected hedgehog behaviour. In a way, the tiny Derbyshire village and its entrenched tales of woes and resilience inspired me to 'pen' these pages.

The infamous story of Eyam (aka 'the Plague Village') goes like this: Someone sent a parcel of cloth from London to the village tailor, Alexander Hadfield. As George Viccars, Hadfield's assistant, unfurled the damp cloth, he inadvertently opened a Pandora's box. Viccars released a bundle of fleas carrying *Yersinia pestis* – the plague-causing bacterium.

The name *Yersinia* comes from Alexandre Yersin, a French bacteriologist who 'discovered' the bacterium during an 1894 plague outbreak in Hong Kong. The species name *pestis* is Latin for 'plague', and it's no coincidence that people of the time would call it the *Great Pestilence*.

Shortly after handling the cloth, Viccars fell ill, and he died within a week. This marked the beginning of the Eyam outbreak. His death was swiftly followed by several others in the village. The disease spread rapidly. In the autumn of 1665, 42 villagers died. By spring, many considered fleeing their homes to save themselves.

Cue William Mompesson, the newly appointed rector, and his predecessor, Thomas Stanley. The two clergymen devised a plan to stop the disease from spreading to other towns. On the 24th of June 1666, Mompesson told the locals that the village must be quarantined, with nobody allowed to enter or leave – a 'cordon sanitaire'. This was the original 'lockdown'. He had arranged for food and supplies to be sent in exchange for coins left in vinegar-filled holes in the village boundary stones. They had no understanding of microbes back then, though it would only be a decade or so until the unseen world was 'discovered' by Dutch microscopist Antonie van Leeuwenhoek. However, they believed that vinegar acted as a disinfectant of sorts – a potent elixir capable of warding off malevolent spirits or plague 'seeds'.

The summer of 1666 saw high temperatures, high flea activity and a high death toll in Eyam. Local farmer Elizabeth Hancock buried six of her children and her husband, all perishing in just eight days. Of the 700 residents, 257 died during the 14-month outbreak (37% fatality).

Many historians say that Eyam's quarantine saved countless lives in surrounding towns and villages. In a recent article, a local historian, Thompson, said:

> They knew they were risking life and limb, but they still agreed to do it. If it means anything at all, you almost feel responsible for doing something to remember it. There's an onus on the people in the village that you can't just turn your back on what the people did.[1]

Zoonotic Times

Let's set the stage a little more. A term that will weave through the fabric of this book is 'zoonotic'. Zoonotic diseases are transmitted between non-human animals (such as bats, birds and horses) and humans. Whether it's through a bite, a sip of tainted water, a breath of contaminated air or the pinprick of a mosquito's proboscis, these diseases leap across species boundaries. If unchecked, they ignite public health crises like outbreaks and pandemics.

Examples of zoonotic diseases include the plague (responsible for the Eyam outbreak), rabies, Ebola, Lyme disease, influenza and the one we know all too well: COVID-19 (though the debate about its origin rages on). Around 70% of emerging infectious diseases in humans are now zoonotic.[2] The remaining 30% come from food and waterborne sources, healthcare facilities and other environments.

Pandemics are rare. This is despite hundreds of thousands of potentially zoonotic species circulating in nature. The 'species' I talk of are invisible to our eyes – think bacteria, viruses, fungi, protozoa and others. They circulate 'behind the scenes'. They circulate behind a wall of dense foliage. They circulate in the 'wild'. The rarity of pandemics lies in the taxing journey a microbe must undertake to leap from a wild animal to a human. They face a gauntlet of barriers before they can successfully make this jump. If they do make the jump, they often face further hurdles to spreading efficiently among humans. Understanding the hurdles that keep most microbes in check is a good thing. It can help us predict the next pandemic-causing pathogen. I'll describe some of these challenges in the early chapters of this book.

In the seventeenth century – at the time of our Eyam story – emerging infectious diseases were hard-hitting. Yet it may come as a surprise that despite the absence of modern health services and knowledge of microbes and hygiene practices, the *diversity* of emerging infectious diseases then was probably lower than today. Emerging infectious disease events have certainly increased in the last century,[3] for reasons I'll uncover in this book. Furthermore, we've detected thirty novel human pathogens in the last three decades.[4] Research suggests that the frequency of novel diseases is poised to rise. Consequently, predicting the next pandemic can be complex.

Emerging threats are ever evolving... They're also hidden in plain sight.

Epidemiological Alphabet Soup

Let's quickly explore a few more terms that appear in this book. 'Pandemic', 'outbreak', 'epidemic' and 'endemic' are words used to describe the prevalence and spread of diseases.

A *pandemic* refers to the global spread of a new disease – the ultimate tidal wave of diseases. It affects a considerable portion of the global population – think the COVID-19 pandemic caused by the novel coronavirus or the bubonic plague that took Europe (including Eyam) by storm a few centuries ago.

An *outbreak* is a sudden surge in disease cases that can catch a community off guard, like a flare-up of measles or foodborne illnesses in localised settings. It's unexpected but usually contained.

Step up a level, and you have an *epidemic* – when a disease spreads rapidly and affects far more people than expected in a particular area or population. It can be local, regional or national – larger than a small outbreak, but not yet global.

Finally, when a disease is *endemic*, it's constantly present or prevalent in an area or population, like malaria in tropical regions or Lyme disease, where ticks roam.

The Eyam event was an outbreak, but part of a bigger centuries-long pandemic.

The word 'pandemic' comes from the ancient Greek adjective *pàndemos*, which means 'of' or 'belonging to' all the people (*pan*, 'all', and *demos*, 'people'). *Pàndemos* appeared in the *Odyssey* in the eighth century BCE. In the fifth century BCE, Plato used the term 'pandemic' to describe popular or 'pandemic' love instead of heavenly love.[5]

It wasn't until 1666 (when, incidentally, the last victim of the Eyam outbreak died) that the word 'pandemic' was used in English to describe a disease commonly occurring in a region or country—also known then as an *epidemic* (just to confuse things even more!).

Previous Pandemics and the Notion of 'Crowd Infections'

Exploring the annals of past pandemics can help illuminate the path ahead, casting light on future outbreaks. We've touched on Eyam's 1665 outbreak. But to truly grasp the pandemic's origins, we must travel back more than three centuries. In 1347, the plague first entered the Mediterranean via trade vessels transporting goods from the territories in the Black Sea. It spread rapidly throughout Europe, the Middle East and northern Africa and resulted in extensive fatalities. Estimates suggest up to 60% of the population perished in the Black Death. This was a long-lasting pandemic, known as the 'Second Plague Pandemic', which persisted for over 300 years.

But if this was the Second Plague Pandemic, when was the first? Has there been a third? And when was the first ever pandemic?

To unravel these threads, we must first discuss the concept of *crowd infections*.

Crowd infections happen when a contagious disease sweeps through densely populated areas or large gatherings. It turns them into hotbeds of transmission. Such infections are relatively new in human evolutionary terms. But why might this be? Well, it's because until a few thousand years ago, there were no 'crowds'.

Imagine living 10,000 years ago. The thoughts 'What will trigger the next pandemic?' or 'When will the next pandemic strike?' would have been utterly foreign concepts – ideas that wouldn't even cross your mind. Firstly, humanity's awareness of microbes didn't arise until the late seventeenth century. Secondly, the absence of densely populated areas meant that large-scale human pandemics simply didn't happen.

One of the earliest known cities (or 'proto-cities') in the world was Çatalhöyük. It was a settlement of around 10,000 people in southern Anatolia, Türkiye (7400–5200 BCE). As bands of humans began living near one another in urban settlements, opportunities arose for pathogens to jump from host to host and thereby spread. The earliest recorded instances of major crowd infections can be traced back to antiquity, with outbreaks such as the Plague of Athens in 430 BCE and the Antonine Plague in 165–180 CE. These epidemics, boosted by increased urban living and trade networks, resulted in considerable mortality.

Over the centuries, crowd infections have continued to shape human history, with pandemics like the Justinian Plague in the sixth century (the First Plague Pandemic). This caused 25–100 million deaths, up to half of the world's population. Globalisation and trade – the modern alchemy of commerce and connection – have further accelerated the spread of infectious diseases.

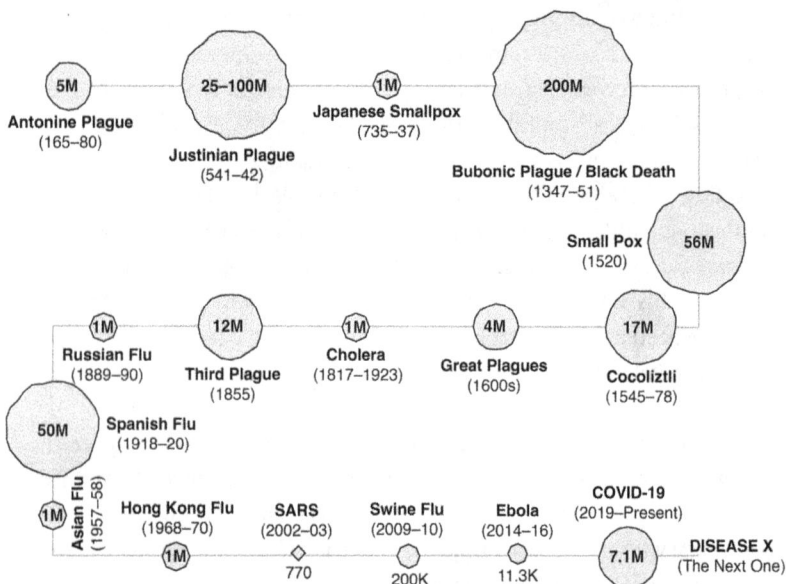

Timeline of pandemics and some major outbreaks in the current era

The timeline in the nearby figure does not include non-infectious disease events. It also doesn't include malaria. These are often considered 'endemic' rather than pandemics, despite killing millions of people worldwide. AIDS has been responsible for around 35 million deaths since 1981, but is transitioning from an epidemic to an endemic disease.[6] Malaria took 150–300 million lives in the twentieth century alone.[7]

It is remarkable that malaria, a disease often overlooked in discussions of historical pandemics, has claimed more human lives than any other affliction in recorded history. A *Nature* news article suggested that malaria "may have killed half of all the people who ever lived".[8] This assertion implies that malaria has caused around 55 billion deaths, which seems somewhat incredible. A professor of medical statistics once proposed a more conservative estimate of 4–5% of the human population throughout history.[9] Even at this reduced estimate, it still amounts to roughly five billion deaths. This number is so significant that if I drew a circle for it in the figure, it would envelop the entire timeline.

Take another look at the timeline. You may have also noticed 'Disease X' in the bottom-right corner. Disease X is a placeholder term used by the World Health Organization (WHO). It refers to a hypothetical new disease that could soon emerge and pose a serious global health threat. We'll explore this later in the book, but it's worth keeping in mind as we navigate through this historical context. Many scientists think that the diversity and frequency of major (as-yet-unknown) diseases will increase. So, many Disease X scenarios may await us on the horizon while scientists beckon us to heed their calls.

Back to Eyam, but Three and a Half Centuries Later

Having lived in Stoney Middleton (one of the villages saved by the Eyam heroics) for a little while, my wife and I decided to move to Eyam. It's a village now coloured by traditions that honour the bravery of its seventeenth-century residents. Fascinating, albeit rather morbid, commemorative plaques stand outside the original stone cottages. One, the 'Plague Cottage', was where Alexander Hadfield's assistant, George Viccars, unwittingly unfurled the cloth infested with fleas on that fateful

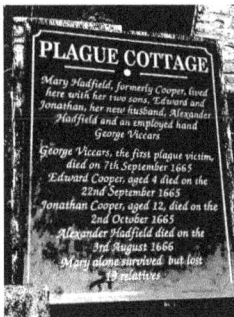

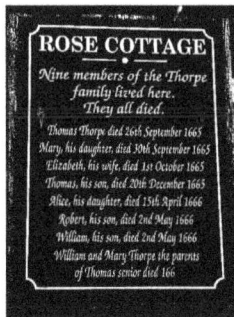

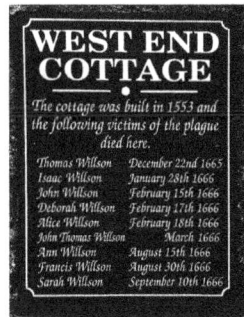

Eyam cottage plaques remembering those who perished

day in 1665. His wife, Mary, was the only surviving member of the household. She lost 19 relatives.

There's also an annual 'Burn the Rat' ceremony. No animals are harmed, except for perhaps reinforcing the rat's already unfortunate image. Hundreds of local residents and visitors from afar parade through the streets at night, carrying fiery torches and accompanied by a colossal artificial rat crafted by the local schoolchildren. The procession culminates in the rat's symbolic immolation atop a bonfire.

Eyam villagers carrying a giant artificial rat to a bonfire

We Need to Talk About Rats

Rats have long been blamed for spreading the Black Death in Europe and other diseases across the world. The main reason is that rats can be reservoir hosts for the plague bacterium *Y. pestis*. Being a reservoir means the rats carry the bacteria. The bacteria persist and reproduce in their bodies. Rats serve as a source of infection for other individuals or species. Importantly, they often harbour a few bouncy companions: fleas.

When a flea feeds on the blood of an infected animal, such as a rodent, it ingests the *Y. pestis* along with the blood. Inside the flea's digestive system, the bacteria multiply and form a blockage in its foregut. This tiny traffic jam stops the flea from drinking more blood. It leaves it thirsty and desperate – a microscopic case of indigestion that spells big trouble. When the flea bites a human or another host to feed, it regurgitates some infected blood and bacteria into the bite wound. This process introduces the plague bacterium into the host's bloodstream, leading to infection.

Back in the seventeenth century, Eyam's rat population likely continued to serve as a critical reservoir for the plague. During the winter months, the disease might have ceased without their involvement, as there were occasions when no individuals were infected. Rodent transmission remained significant. However, it only contributed to around 25% of the cases.[10] Research suggests that rats might not have been the leading carriers of the fleas and lice that caused the initial Black Death pandemic between 1347 and 1351. The culprits? Well, some say it was that large, wanderlust-filled, hairy and promiscuous mammal *Homo sapiens*.

In a 2018 study published in the *Proceedings of the National Academy of Sciences*, researchers conducted simulations of Black Death outbreaks in European cities to unravel the mechanisms of plague transmission.[11] Their simulations explored three potential models of infection. The first was rats as carriers. The second was airborne transmission. The third was the role of fleas and lice transported by humans on their bodies and garments. In most cities, the model of fleas and lice on humans emerged as the most precise way to elucidate the disease's spread. Earlier research has supported these conclusions, suggesting that the plague spread too fast for rats to be the main culprit.

Scientists have uncovered DNA evidence of the *Y. pestis* bacteria in human bones dating back four millennia. However, in the fourteenth century, a virulent strain of the bacteria surged into Europe. It caused the 'Black Death' (so-called because of the dark-purplish or blackish skin lesions that were symptomatic of the disease). It's believed to have originated in villages near the Chui Valley, now part of Kyrgyzstan. From here, some scientists speculate that the pathogen traversed to humans through fleas from infected marmots. Humans then spread it along the Silk Road to Europe.

The Silk Road was a network of trade routes that connected the East and West. It sparked the exchange of goods, ideas, cultures and technologies between different regions of Asia, Europe and Africa. This ancient thoroughfare was crucial in developing civilisations and expanding trade and commerce during antiquity and the medieval period. But it also had a knack for spreading surprises of the infectious kind. As traders, travellers, soldiers and pilgrims journeyed along these routes, they unintentionally became carriers of pathogens. They spread diseases rapidly to new populations and far-flung regions. And so began the rise of crowd infections.

The significance of the last few paragraphs lies in highlighting the human tendency to assign blame to others, whether it's rats, bats, pigeons or pangolins. Social norms and cognitive mirages of rightness can drive the notion of blame. Yet, pointing the finger at wildlife for pandemics is off track. It's a misguided heuristic that overlooks the complexity of diseases, ecosystems and human antics. It's also a way to keep *solutions* buried in a grave of misconception.

A Quick Note on Germaphobia

Having briefly discussed misconceptions, it would be remiss of me not to mention how indispensable *microbes* are in keeping us alive. Fewer than 0.01% cause human diseases, leaving 99.9% harmless or beneficial for our health and survival. I often say, "All the things we *can* see in nature intimately depend on all the things we *can't* see." This is to say that tarnishing all microbes with the same brush as those that cause maladies can spread 'germaphobia' and the fear of 'dirt', with adverse impacts on our ecosystems and health.[12] So, while I write about the 0.01% in this book, I must here express my admiration for the 99.9% of our invisible 'friends'.

Book Objective One

This book has four objectives. The first is to challenge and shift the prevailing narrative that places the blame for pandemics squarely on wildlife. This view ignores the intricacies driving the emergence of infectious diseases (Part 1 and throughout). It wrongly paints non-human animals as villains. In truth, pandemics often stem from the tangled impacts of human activities – habitat destruction, pollution, wildlife trafficking and encroachment into wilder spaces. These narratives risk spawning policies that target wildlife instead of addressing the deeper, often human-induced causes. It's time to reframe the story with solutions. I think the solutions should be rooted in the One Health approach.

In its simplest form, One Health acknowledges that the health of humans, non-human animals and our environment are inseparably linked. It's the understanding that a disease in wildlife can leap to humans and that damaging our environment can increase the likelihood of that happening. Imagine it as a triangle, where each point supports the others. If you remove one point, the whole structure collapses.

Book Objective Two

My second objective is to traverse the landscape of 'disease ecology' and discover the likely source and timing of the next pandemic (Part 2). Will it be a virus like SARS-CoV-2? Will it be a bacterium like the plague-causing *Y. pestis*? Or perhaps it'll be a fungus that turns us into cannibalistic beings like in the hit HBO series *The Last of Us*? This might seem like a jest, and indeed, the notion of zombie-like cannibalism verges on the hyperbolic. But get this: in the tumultuous year of 2020, around three million people succumbed to COVID-19. Yet, a recent study reveals that global fatalities due to *fungal* diseases are estimated to be nearly four million annually.[13] Let that sink in for a moment. Even more disconcerting is that fungus-related deaths have doubled within the last ten years.

What about protozoa ('proto' meaning first, 'zoa' meaning animal)? These are single-celled microbes. They're diverse in shape and size. They live in various environments, including freshwater, marine environments, soil and even within other organisms as parasites. The protozoan parasite *Plasmodium*, which causes malaria, is transmitted to humans through an infected mosquito bite (or 'proboscis puncture'). Malaria is a significant global health problem, particularly in tropical and subtropical regions. But could a *Plasmodium*-like organism cause the next pandemic?

Book Objective Three

The third objective is to steer us away from human-centric narratives and into the realms of wildlife 'pandemics'. In a recent article, conservation biologist Diana Bell said, "When people ask me what I think the next pandemic will be, I often say that we are in the midst of one – it's just afflicting a great many species more than ours."[14]

We'll learn from expert virologists, mycologists and veterinarians about the elephant in the room: how non-human pandemics and outbreaks reshape the natural world. From the flu to white-nose syndrome and spongey brains, diseases are exacting a heavy toll on wildlife. Again, our actions stand at the forefront of the emergence and spread of these diseases. Hence, the mantle of responsibility rests upon us to embark on resolute measures to confront these urgent plights. The imperative remains clear. Even if you align solely with the perspective that biodiversity holds *instrumental* rather than *intrinsic* value, we must staunch the tide of wildlife disease outbreaks. After all, it's from wild animals that most emerging infectious diseases in humans originate.

If you're in the 'Guess what? Life's not all about humans' camp, please warmly accept this objective (Part 3) as a hat-tip to you.

We'll also explore plant and fungi outbreaks. As the late naturalist Oliver Rackham once said, "Trees are wildlife just as deer or primroses are wildlife. Each species has its own agenda and its own interactions with human activities." Has there been or could there be a plant pandemic?

Book Objective Four

The fourth and final objective of this book is to explore *solutions* (Part 4). We'll look at the case for safeguarding biodiversity, including rats, bats, pigeons and pangolins – and their homes. Additionally, I'll discuss the evidence that healing nature can be a potent tool for preventing major disease outbreaks.

We'll discover how shifts in land use can increase the risk of spillover events. How we use the land induces two critical changes in the hosts carrying the disease. The first change is in host behaviour – how they move around and use habitats. The second change is in host energy and the burden of stress. Chronic stress affects viral infections in animals. It also affects viral shedding (the release of virus particles). This can lead to increased spillover risks. I'll aim to convey that spillovers are *ecological* issues that require *ecological* solutions – again emphasising the One Health perspective.

In the concluding chapters, we'll explore how 'green prescriptions' and immersing ourselves in nature can enrich our immunity and overall wellbeing. This could strengthen our defence against infectious maladies. Moreover, we'll navigate the realm of vaccinations for humans and wildlife, the call for another agricultural revolution, and the strides made to 'stay ahead of the curve' by predicting and responding to outbreaks.

In short, this book is about One Health on our One Planet.

I hope you enjoy reading *The Nature of Pandemics*.

PS: Throughout the book, you'll notice a few QR codes like the one below. Scan these with your smartphone to view 3D models of different organisms and pathogens. Click and drag on the model to scroll and double-click (two-finger pinch on a trackpad) to zoom.

Flea

A Trip Down Spillover Lane

*To succeed, jump as quickly at opportunities as you
do at conclusions.*
—Benjamin Franklin

"Your body is like a medieval castle," my immunology professor
said with a fervour that was contagious in its own right.

In my student days, I immersed myself in a world where
invisible forces shaped animal health. My professor would often regale
us with tales of microbial warfare and the ways that pathogens navigate
the complex terrain of the body.

"Think about it: a castle is fortified with various defences to
ward off invaders, and our bodies are equipped with a complex array
of defence mechanisms to protect against harmful pathogens. The
skin serves as the sturdy walls of the fortress. It provides a physical
barrier that prevents invaders from breaching the body's defences.
Meanwhile, the mucous membranes lining the respiratory, digestive
and reproductive tracts act as the metaphorical moat. They serve
as an additional line of defence by trapping and expelling potential
threats. If people knew back then what we know about the immune
system today, they'd probably have had far more advanced warfare
and military strategies!" Nevertheless, we can draw some uncanny
parallels.

Beyond the structural defences, the body deploys a complex
network of immune cells and molecules to detect and neutralise
invading pathogens. We can think of these immune cells as vigilant
guards stationed throughout the body. They constantly scan for signs
of trouble. When a threat is detected, these cells spring into action,
mounting a coordinated defence to eliminate the invaders. Additionally, specialised immune cells known as macrophages are like the castle's

knights in shining armour. They patrol the body's tissues and engulf any foreign invaders they encounter.

Meanwhile, antibodies, produced by immune cells known as B cells, act as the castle's archers. Antibodies precisely target and neutralise pathogens. The body's defence mechanisms work synergistically. They form an imposing barrier against infection and disease, much like the layers of defence in a fortified castle.

All animals have immune systems to defend their bodies against incoming threats. But there's also a concept called the *species barrier*. This natural barrier prevents diseases from easily jumping from one species to another. The jump involves several critical steps. First, the disease agent must find a surface it can infect and to which it can attach. Often, it must then multiply on that surface, colonise it and invade the new host. It then has to multiply inside the host. Finally, it must evade the host's immune defences. Each step depends on specific interactions between molecules in the host and the pathogen. Crucially, for the disease to successfully infect a new species, the entire sequence of events must be 'reprogrammed' – it must undergo changes, often through mutations or genetic exchanges.[1]

When a pathogen successfully crosses the species barrier and infects a new host species, it may lead to a disease *spillover* event. In addition to genetic mutations, changes in host behaviour or ecology and environmental disturbances can increase the likelihood of a spillover. Over 30% of new diseases reported since 1960 can be attributed to deforestation and land-use changes.[2] Closeness is an issue. By changing the land, we increase human–wildlife contact. As mentioned in the Introduction, scientists think that around 70% of recently emerging infectious diseases originated from non-human animals.[3] Again, closeness is an issue. The closer the two target host species are genetically, the more easily microbes can overcome biological barriers to produce successful spillovers. This genetic closeness is one reason mammals are often the primary source of zoonotic diseases.

It's worth noting here that I won't discuss bats in this chapter, as I've dedicated Chapter 3 to our flying furry relatives. However, I do recognise that bats have played an important role in the history and ecology of various spillover events.

Okay, now I'd like you to imagine stepping into a travel agency unlike any other – let's call it *Zoonotic Holidays*. You've booked a tour of a few of the world's most intriguing disease spillover events. I'll be your guide, so let's begin.

Zika

First, we're taking a trip to Uganda. In 1947, deep in the country's Zika Forest, a virus circulated amongst the resident rhesus monkeys. The virus was named Zika after the forest. 'Zika' means 'overgrown' in the local Luganda language. The forest is home to many organisms that eat, buzz, sleep and breed. Among them are 50–70 species of mosquitoes harbouring various diseases.[4]

Zika was first noted as a mild, almost insignificant pathogen in a clinical setting. For decades, it stealthily circulated among the forests of Africa and Southeast Asia. It only caused occasional and mild illnesses. It was overshadowed by more notorious viral cousins like dengue and yellow fever.

Fast-forward to 2007, and the story of Zika was far from over. This was the first time I heard about it. I remember listening to a podcast that mentioned the virus had staged a dramatic appearance on the island of Yap in the Federated States of Micronesia. This outbreak marked Zika's first significant foray beyond its traditional confines of Africa and Asia. But news of the virus was fleeting. It also lacked widespread attention.

A more significant outbreak occurred in French Polynesia in 2013, where officials reported thousands of cases.[5] This outbreak brought a little more international attention to the virus and revealed new complications associated with Zika. These include neurological disorders such as Guillain-Barré syndrome, which can cause paralysis. But the Zika virus didn't stop there. Millions of people cross oceans and borders each year. As we do, we provide a comfy vehicle for pathogens. They're the ultimate hitchhikers.

Zika soon arrived in the Americas, with devastating consequences. In 2015–2016, Brazil was at the epicentre of a public health crisis.[6] Zika infections exploded across the country. The once-dismissed virus had reached new heights on the menace scale, threatening life. It was linked to severe congenital abnormalities in newborns, most notably microcephaly. With this condition, babies are born with abnormally small heads, leading to severe developmental issues.

How did Zika manage such a feat? The primary vectors were the ubiquitous *Aedes* mosquitoes, especially *A. aegypti* and *A. albopictus*. These mosquitoes are often urban dwellers. They thrive in human environments. They made the perfect vectors, biting infected individuals and then passing the virus on to others. But unlike many other mosquito-borne diseases, Zika could also be transmitted sexually, through blood transfusions and from mother to foetus.[7] This broadened its reach and complicated efforts to control its spread. The impact was profound.

The link between Zika and congenital Zika syndrome, with its highly distressing images of affected newborns, brought a sense of urgency to the global response. Public health officials coalesced to implement mosquito control measures, from insecticide spraying to eliminating breeding sites like standing water in urban areas. Educational campaigns urged communities to protect themselves from mosquito bites and practise safe sex to curb the virus's spread.

Most of us know the ensuing chaos and headlines that come with a growing infectious disease. But as we scramble to understand a novel pathogen and find a cure, we often forget how human actions can turn a local issue into a global crisis. As with many zoonotic diseases, the Zika story wasn't just about a virus jumping from monkeys to humans – a classic spillover. There's more to the story.

This will sound a little weird, but think of the spillover as a very toxic cake. We must consider the recipe, how the ingredients are combined and how those combined ingredients are baked. To me, it sometimes feels like as a society we throw a load of toxic ingredients into the mixing bowl and then the oven. We get on with our daily lives. And suddenly, we're startled to see these bad cakes everywhere.

But what are the ingredients? At the top of the list, we have wildlife habitat encroachment and sprawling urbanisation. Then,

Ugandan forest

we have international travel at the click of our fingers. Mix these together and add a pinch of cognitive dissonance and a dash of political weakness. Then, we have an oven that's on too high, representing the effects of climate change. We've created the perfect recipe for vectors like *Aedes* mosquitoes to thrive and diseases to spill over into human populations. Once they do, they spread like cracks on shattered glass.

This is where the narrative needs to shift. It's where we need to change. Recognising our role in inadvertently helping to drive disease dynamics can allow us to shift from passive observers to active participants in prevention. But as you'll see later, it's probably far easier for some to change than others.

West Nile Virus

As we're already in Africa, we might as well stay here for the next disease on our tour. We only need to wander up the road in Uganda to the West Nile district. Here, scientists first isolated another virus in 1937. Its natural hosts were birds and mosquitoes. The West Nile virus was named after this district (after the Nile River). Once again, the virus lived in relative obscurity. It only caused sporadic outbreaks among birds and occasional mild illness in humans.

Its potential for global impact remained largely unrecognised in the 1930s. However, as the world became increasingly interconnected through trade, travel and urbanisation, the virus found new avenues for dispersal. Migratory birds continued traversing vast distances. But humans ventured into previously untouched territories, probably affecting the behaviour and ecology of migratory birds and other wild animals. The virus seized the opportunity to expand its reach beyond African borders.

The first inklings of the virus's potential for global spread came in the late 1990s. Reports emerged of unusual outbreaks of encephalitis (brain inflammation) among humans in Israel and Romania. Scientists later attributed these outbreaks to the West Nile virus. Then, in 1999, the virus dramatically appeared in the bustling metropolis of New York City.[8] Now, the Western media began paying attention. Unsurprisingly – I say with a pinch of cynicism – the arrival of the West Nile virus on North American shores marked a turning point in the virus's history. It thrust this pathogen into the global spotlight and ignited a flurry of scientific inquiry and public health responses.

West Nile Virus

Cases of human illness mounted. The virus spread across the continent. Researchers strived to understand its origins and how it was transmitted. According to the US Center for Disease Control and Prevention (CDC):

> West Nile virus is the leading cause of mosquito-borne disease in the continental United States. There are no vaccines to prevent or medicines to treat West Nile in people. Fortunately, most people infected with West Nile virus do not feel sick. About 1 in 5 people infected develop a fever and other symptoms. About 1 out of 150 infected people develop a serious, sometimes fatal, illness.[9]

The discovery that the virus could cause severe neurological disease sent shockwaves through communities and health systems.

Mosquito control efforts were intensified in the United States. Communities implemented targeted spraying campaigns, and residents were urged to eliminate standing water sources where mosquitoes bred. The United States bolstered its surveillance systems and expanded diagnostic testing to detect and track the virus's spread. Despite these efforts, the West Nile virus remains a persistent threat. It's capable of causing seasonal outbreaks with varying degrees of severity. The virus can adapt to different environments and exploit various hosts, which is critical to its success.

Once again, it shows that as globalisation ushered in an era of increased travel and trade, and as we change the behaviour of migratory animals, viruses can hitch a ride on unsuspecting travellers (human or otherwise), finding new horizons and sowing disorder far and wide.

Migratory Birds

It's August, and I'm at the Adelaide International Bird Sanctuary National Park. I love to write while out in nature; it helps me feel at one with the story I seek to unravel. This particular story is about migratory birds. At the sanctuary, I'm lucky to see a bar-tailed godwit (*Limosa lapponica*). They're large wading birds with the world record for the longest non-stop flight. They've been known to journey 13,500 kilometres from Alaska to Tasmania in just 11 days, maintaining an average speed of over 50 kilometres per hour and shedding nearly half their body weight during the flight.[10] Once here, they fatten up on molluscs, worms and insects. Several months later, they'll embark on the return journey. These sanctuaries are becoming increasingly vital for conserving migratory birds and their health.

I continue watching these elegant birds; I can't help but reflect on the broader implications of our impact on their lives. I've mentioned a couple of times that our actions have altered the behaviour and ecology of migratory birds (and vectors such as mosquitoes), potentially making it easier for West Nile virus and other pathogens to 'spill over'. Let me qualify this with some examples:

1. The expansion of cities and land development for infrastructure and agriculture has destroyed natural habitats. This 'development' forces birds to alter their migratory routes and stopover sites, often bringing them closer to human populations and domesticated animals.

2. Human-induced changes in climate patterns can affect the timing and routes of bird migration. For instance, warmer temperatures can lead to longer breeding seasons and alter the distribution of insect vectors like mosquitoes, carriers of diseases like West Nile. Birds that once migrated to specific regions may change their destinations or stop migrating altogether. This can lead to new interactions with different species and environments. It can also enhance the chances of a spillover.

3. Intensive farming practices and monoculture crops also change the environment for large populations of birds and insects. For instance, irrigation systems and standing water in agricultural fields can provide breeding grounds for mosquitoes. Birds feeding in these areas can become infected with viruses, which they can then carry to new locations during migration.

4. Human activities such as creating bird feeders and birdbaths that aren't properly cleaned can also spread disease in birds. The lack of hygiene creates a breeding ground where pathogens have more chances to leap from birds to humans or other animals. This also applies to non-migratory birds.

5. Stripping natural habitats away drives birds to gather in unnaturally high densities, often close to people. This crowding, especially in mosquito-prone areas, can create a perfect storm for spreading pathogens among birds and potentially to humans.

6. The globalisation of trade and travel influences the movement of birds and mosquitoes across regions and continents. We introduce exotic birds into new areas through the (often illegal) pet trade (a form of forced migration), and shipping containers can transport mosquitoes to different parts of the world. As we move animals around, we introduce new pathogens into ecosystems where birds and humans may have little immunity, leading to outbreaks.

While our bar-tailed godwit is happy sticking to the coastal flats and migrating from Alaska to Australia in one go, other birds need to make pitstops. However, some pitstop locations are disappearing due to damaged (or destroyed) ecosystems and climate change. In Chapter 10, we'll discuss why this affects zoonotic disease risks.

Q Fever

Let's stick with Australia. We're heading to Queensland, where our next spillover pathogen emerged from livestock rather than wild animals. Known as 'Q fever' – named after Queensland – the disease is caused by the bacterium *Coxiella burnetii*. It was first identified in 1935 when an outbreak of a mysterious fever among abattoir workers led to its discovery.[11] The bacteria are highly resilient. They can survive in harsh environments for long periods. This resilience and the fact that airborne particles can spread the bacteria make Q fever exceptionally infectious. Humans typically contract the disease by inhaling contaminated particles from infected animals' placenta, urine, faeces or milk – most commonly from sheep, goats and cattle.

Q fever often flies under the radar because its symptoms – fever, chills and muscle aches – are easily mistaken for the flu (although I'm sure we say that about most bacterial and viral infections!). Some might struggle to catch their breath or battle a persistent cough as the bacteria work their way into the lungs. It can sometimes lead to pneumonia. For others, the infection digs even deeper, reaching the liver or the heart. It can cause inflammation in places you'd least want it and a bone-deep fatigue that won't quit. The *chronic* form of Q fever is a different beast – rare but relentless. It can take months or even years to rear its head. It often targets those with existing heart conditions or weakened immune systems and can leave behind a legacy of chronic fatigue or severe heart and liver issues.

A Dutch Affair

I remember attending the annual European Scientific Conference on Applied Infectious Disease Epidemiology (ESCAIDE) in Edinburgh in 2012. I hadn't heard about Q fever at the time, but 13 years on, I vividly recall the talk introducing it.

The talk was about a major Q fever outbreak in the Netherlands. In 2005, scientists recorded Q fever on two dairy goat farms. By 2007, it had spread to the human population in the south of the Netherlands.[12] Between 2007 and 2010, more than 4,000 human cases were reported, with a noticeable annual seasonal peak. The outbreaks were primarily confined to the southern region of the country, an area characterised by intensive dairy goat farming. Up to 15% of the population may have been infected in the hardest-hit areas.

Strange to say, the highest risk for contracting Q fever didn't come from working directly with animals. It was living close to an infected dairy goat farm. Those living within five kilometres were at higher risk. The sweltering and dry weather in 2007 caused contaminated dust particles to float around. People living downwind from the farms were at a higher risk of contracting the disease.

The bacteria stick to airborne particles and travel considerable distances like invisible kitesurfers. When infected pregnant small ruminants miscarry, they release billions of *C. burnetii* bacteria into the environment.[13] That might sound alarming. But even more unsettling is that exposure to fewer than *ten* of these bacteria is enough to start an infection.

Vegetation and Wet Soils

When scientists dug deeper into the risk factors of Q fever, they stumbled upon a curious finding. Outbreaks were strongly linked to dairy goat farms grappling with abortion issues caused by the disease. However, some of these farms reported no human cases at all. The scientists studied the environments within a five-kilometre radius around twenty-seven farms experiencing Q fever-induced abortions.[14] They found that vegetation density and groundwater conditions seemed to play a role in disease transmission. Satellite imagery revealed a fascinating pattern: areas with higher vegetation densities – think trees, crops and other green spaces – had fewer human cases of Q fever. Similarly, regions with shallower groundwater levels also recorded fewer instances of the disease.

This evidence suggests that dense vegetation and soil moisture play essential roles in preventing the transmission of *C. burnetii* from infected ruminant farms to humans. But why might this be? Well, dense vegetation can act as a physical barrier, reducing the spread of contaminated aerosols. It's also known to minimise dust production from surfaces and can remove dust particles and their microbial hitchhikers from the air.

The following may sound counterintuitive, but shallow groundwater levels typically produce higher soil moisture content.[15] This is due to capillary action (where liquids move against gravity in narrow spaces due to adhesive and cohesive forces). This is particularly true in soils with fine textures. When soils are wetter, they hold together better. When they hold together better, less dust gets swept into the air. Moist soils can also act like sponges, catching dust particles that drift down and land on their surface, helping to keep the air cleaner and dust levels down.

Once again, these findings show that we have the power to act to prevent Q fever outbreaks. It's too simple to say, "Farm animals spread this disease." Farm hygiene and conditions are essential in determining the pathogen's presence and their ability to spread. Moreover, as we'll see in the later chapters, caring for the environment is vital to stopping the spread.

Bumblebees

The final stop on our spillover tour is a greenhouse complex in North America. This time, we're not looking at diseases in humans. Instead, let's consider bumblebees and a special kind of spillover.

Commercially bred bumblebees are used to pollinate greenhouses. They can serve as reservoirs for several pollinator parasites, including protozoans, invisible fungi, viruses and mites. When they escape the greenhouse environment, these commercial bees can infect wild bee populations. Infection can occur through direct mingling between commercial and wild bees or via shared flower use.

One study found that commercial bumblebees infected nearly half of all wild bees near greenhouses.[16] Infection rates and incidence declined significantly further away from the greenhouses. Instances of spillover between bumblebees are well documented globally, particularly in Japan, North America and the UK. The study's authors suggest that, given the available evidence, pathogen spillover from commercial bees likely contributes to the ongoing decline of wild *Bombus* populations in North America.

So, spillovers occur between closely related wild animal species, too, not just in humans. Remember, closeness is an issue. We're commercially breeding bumblebees that host deadly disease agents that spill over into wild populations. Shifting the narrative that wildlife is to blame for spillovers is crucial because it obscures humans' primary role in creating these situations. By focusing on human responsibility, we can better address and mitigate the root causes of these ecological disruptions and promote healthier ecosystems.

Pangolins and the Psychology of Scapegoating

It was a couple of years into the COVID-19 pandemic. Stories of pangolins (*Manis spp*) being blamed for the spillover started circulating. Some researchers thought pangolins may have spread the virus to humans. A Chinese university stated on its website, "This discovery will be of great significance for the prevention and control of the origin."[17] It turns out this was false. But that didn't stop it from causing widespread exaggerated concerns in the international media.

Scapegoating is dangerous. Poor messaging, particularly when conveying unconfirmed assumptions, is also dangerous. I've since read various posts on social media containing messages like 'Never forget the cause of COVID: the pangolin' and 'Pangolins might be the coronavirus culprits'. We should pause here and recall the early erroneous implication of palm civets (*Paradoxurus hermaphroditus*) as the primary reservoir for the SARS-CoV-1 in 2002. This led to the culling of some 10,000 captive palm civets.[18]

Pangolins are not only endangered; they are probably the most trafficked mammal in the world. They're trafficked primarily for their scales and meat, driven by cultural beliefs, traditional medicine and luxury consumption. There's no reliable evidence that pangolin scales have special medicinal value.[19] And the scales' main constituent? It's keratin – the same stuff as our nails. We might as well chew our fingernails in the hope it'll cure diseases.

Pangolins do, however, play critical ecological roles. As predators of ants and termites, they help maintain a healthy equilibrium by regulating insect populations. Their foraging behaviour, which involves digging burrows, promotes the turnover of organic matter and enhances soil aeration. Furthermore, the burrows created by pangolins provide essential shelter and thermal refugia for various other species. They support diverse commensal organisms and contribute to the ecosystem's overall health.

If pangolins were the source of diseases (and like us, they do harbour viruses and other pathogens), you know what's not going to prevent spillovers? Hunting them, trafficking them and eating them in hotpots.

The mesmerising and otherworldly pangolin

So, we know that poor communication is dangerous – it can harm wildlife and divert attention away from meaningful agendas that benefit humans. But something lurks deep in the human psyche that often drives this poor communication: *scapegoating*. It's the phenomenon whereby humans seek to place blame on specific entities or circumstances, rather than taking responsibility for their broader, systemic behaviours that actually cause an issue.

The psychological foundations of scapegoating are multifaceted. They're rooted in various cognitive, social and emotional processes. For instance, *displacement* is a defence mechanism where individuals shift their negative emotions or impulses from the source of distress to a safer or more convenient target. In the case of scapegoating, people may blame innocent parties (e.g., wildlife) for their problems to avoid confronting the actual, often more complex causes.

Cognitive dissonance occurs when individuals experience discomfort due to conflicting beliefs or attitudes. Scapegoating can help reduce this discomfort by providing a clear, albeit incorrect, explanation for a problem. Thus, it simplifies the situation and reduces the mental strain of dealing with ambiguity or complexity.

We also naturally tend to categorise others into in-groups ('those like us') and out-groups ('those unlike us'). Think of sports teams and their fans. Fans of a particular sports team often feel a strong sense of camaraderie and belonging with fellow supporters (in-group), while fans of rival teams are viewed as outsiders (out-group). Scapegoating often involves blaming an out-group. This can strengthen in-group cohesion and identity while providing a sense of control over the perceived threat posed by the out-group. Humans learn scapegoating through socialisation and cultural narratives that perpetuate blame-shifting behaviours. Societies and groups may have historical or cultural precedents for blaming certain groups or entities for collective problems. This reinforces scapegoating as an acceptable response. I think social media is changing the pace and influence of such socialisation in quite dramatic ways. Think back to those social media posts on pangolins and COVID-19 – with a few clicks of a 'repost' button, this perception can become (socially) viral.

We also often prefer simple explanations over complex ones due to cognitive ease. It's called heuristics. As Will Storr explains:

> The brain defends our flawed model of the world with an armoury of crafty biases. When we come across any new fact or opinion, we immediately judge it. If it's consistent with our model of reality, our brain gives a subconscious feeling of *yes*. If it's not, it gives a subconscious feeling of *no*. These emotional responses happen before we go through any process of conscious reasoning.[20]

Scapegoating provides a straightforward narrative that is easier to understand and accept than the nuanced reality of systemic issues and complex causes. We should also watch out for political leaders and authority figures. They can exploit scapegoating to unite followers, deflect criticism or distract from their own failures. Without naming names, these leaders can influence people to adopt scapegoating behaviours, especially in times of crisis or uncertainty, such as during pandemics!

Social Equity Issues

But isn't it too simplistic to attribute the increase in zoonotic disease outbreaks to 'human activities'? I think it's important to finish this chapter by emphasising that although we need to change our collective behaviour to reduce the risks of spillover events, it's far easier for some people to change than others. Social equity plays a significant role in this dynamic. People facing socioeconomic challenges often find it more difficult to alter their behaviour. These challenges are sometimes exacerbated by the actions of those in more privileged positions. In fact, it's often not the behaviour of the people facing social and economic challenges that needs to change.

Let's illustrate this with a couple of hypothetical stories. The first is the story of Gabriela, a farmer in rural Brazil. Gabriela lives on the edge of a rapidly shrinking rainforest. Her livelihood depends on small-scale agriculture, but as large corporations clear the forest for cattle ranching, she finds herself in closer contact with wildlife. These changes increase the risk of zoonotic diseases, such as spreading viruses from bats and rodents to humans. Gabriela's situation highlights the issue of social equity. While humans need to change their behaviour to reduce spillover risks, it's far easier for some people than others. For Gabriela, avoiding the forest is not an option; it's her backyard and her source of income. Meanwhile, the corporations driving deforestation and increasing the risk of spillovers are far removed from these immediate dangers and are not the ones directly suffering the consequences.

Dakarai is the protagonist in the second story. In the markets of a small village in West Africa, the sale of bushmeat is a daily necessity for many residents. Here, the livelihood of countless families depends on hunting and selling wild animals despite the known risks of zoonotic diseases. For Dakarai, a father of four, hunting is the only way to feed his family and pay for his children's education. "We know it's dangerous, but what choice do we have?" he says, his hands stained with the day's catch. Contrast this with the leafy suburbs of a major Western city, where the well-heeled residents (much like myself) shop at organic and artisanal markets, far removed from the risks of zoonotic spillovers. These individuals have the luxury of choice. They're able to opt for imported meats or diverse plant-based diets with ease. In one of my other books, *Treewilding: Our Past, Present and Future Relationship with Forests*,[21] I spoke about how the patterns of consumerism in wealthy nations drive deforestation in lower-income countries. In fact, G7 countries (Canada, France, Germany, Italy, Japan, the United Kingdom and the United States) alone drive an annual loss of more than three billion tropical trees (or four rainforest trees per person per year) just owing to imports.[22]

Meanwhile, Dakarai and his community face various compounded challenges. Deforestation driven by global demand for palm oil, a product found in many Western households, forces wildlife closer to human settlements, increasing the risk of disease transmission.

The behavioural changes required to prevent spillovers are complex and challenging, crossing ecological and cultural boundaries. But one thing is certain: the bats, rats, birds, bees, pangolins and forests can't make the necessary changes for us. And remember, spillovers don't just happen. They're a consequence of fractured ecosystems and close interactions between humans and other organisms. When we push nature to its limits, nature pushes back.

CHAPTER 2

What Have Vultures Got to Do with It?

*Vultures are the most righteous of birds: they do not
attack even the smallest living creature.*
—Plutarch

L et's now journey to India, the 'land of diversity', where myriad
silhouettes once soared in the expansive skies. They belonged to
one of my favourite birds of prey: vultures. Just a few decades
ago, their sheer numbers were such that attempting even a rough tally
seemed insurmountable. However, a survey spanning 18 protected
regions across India provided the basis for an intriguing extrapolation.
In the early 1990s, India's vulture population was estimated to be over
40 million – an astonishing number.[1] For comparison, this was more
than the human population in Spain at the time. Yet, by the mid-1990s,
their numbers had dwindled to the brink of oblivion, leaving an
ecological void with unfathomable impacts.[2]

This is the story of the vulture crisis in India. However, it's a challenge
that extends far beyond India's borders. In fact, it's a little-spoken *global*
crisis. Of the 23 species of vultures in the world, 16 are now threatened
with extinction. In Southeast Asia, vultures have faced a devastating
decline, with the loss of tens of millions of birds.[3] It's worth pausing here
to let that sink in.

Vultures play a vital role in maintaining the health of ecosystems
as one of nature's most efficient scavengers. They perform the essential
task of cleaning up carrion. By doing so, they prevent the spread of
diseases and reduce pollution. The pathogens they neutralise would
otherwise pose significant threats to the health of humans and other
animals. Vultures also play a vital role in nutrient cycling; they turn

decaying matter back into the earth and enrich the soil.[4] This fuels the growth of thriving vegetation. Furthermore, the presence of vultures helps regulate the population sizes of other scavenger species, thereby indirectly influencing the dynamics of entire ecosystems.

However, to say that vultures have a bad reputation would be like saying a tornado is just a gentle breeze. Their unusual appearance and diet of decaying carrion have earned them quite a terrible reputation. 'Omens of death' and 'bad luck symbols' spring to mind.[5] These perceptions have contributed to the mass vilification of vultures in some societies. And let's face it, images of vultures with their bills covered with offal, blood and ragged sinew tend to evoke less sympathy and garner less support than those of cute and cuddly species such as the giant panda (*Ailuropoda melanoleuca*), kakapo (*Strigops hagroptilus*), tiger (*Panthera tigris*) or koala (*Phascolarctos cinereus*) – the so-called charismatic animals.

With this unfavourable reputation comes a spectrum of responses, from disdainful shudders and exaggerated tales of alleged malevolence to proactive poisoning. As scavengers, vultures are also prone to being secondary victims; they feast on other animals poisoned by humans.

A Silent Spring

At this point, it's worth reflecting on a certain polysyllabic 'D' word, or more accurately, a 'DDT' word: 'Dichlorodiphenyltrichloroethane'. (Don't worry, you don't need to retain this.)

Scientists developed DDT in the 1940s.[6] Governments initially hailed DDT as a miracle chemical for its effectiveness in controlling vectors like mosquitoes, which transmit diseases like malaria and typhus. However, it was later revealed that DDT had unintended consequences. It accumulated in the food chain and thinned eggshells in birds like eagles and falcons. This led to population declines in raptor birds. The 1962 book *Silent Spring* by Rachel Carson[7] is a seminal work that brought attention to the detrimental effects of DDT on the environment. The title 'Silent Spring' refers to the eerie absence of bird songs resulting from the decline in bird populations due to pesticide use. Carson's book sparked public outcry and increased awareness of environmental issues and the need for pesticide regulation. As a result, some countries eventually banned DDT. Others restricted its use.

Similarly, vultures can also become victims when they consume substances that seem innocuous to us, such as veterinary medicines. This is what happened to vultures in India during the 1990s.

Cue another tongue-twisting 'D' word: 'diclofenac'.

Widely used as a veterinary drug to treat livestock, diclofenac proved fatal to vultures. It's suspected that the vultures ingested the drug from the carcasses of cattle treated with diclofenac shortly before their demise. It led to kidney failure in the birds. Data modelling unveiled that even a tiny fraction (less than 1%) of livestock carcasses contaminated with diclofenac could precipitate a drastic decline (more than 60%) in vulture populations. A study of carcasses showed that about 10% were contaminated.[8] Within two decades, vulture populations plummeted by more than 95%, a catastrophic decline that sent shockwaves through India's ecosystems.

The drug has now been banned in several countries, at least for veterinary usage, because of the acute danger it poses to vultures.

How does the Decline of Vultures Relate to Zoonotic Diseases?

In the absence of vultures, vast numbers of animal carcasses remained unattended in India. This build-up of carcasses posed a significant threat to public health. They served as hotbeds for infectious pathogens and attracted 'pests'. The absence of vultures also led to a notable surge in the population of feral dogs. A vulture's metabolism is a definitive endpoint for various potentially deadly pathogens, effectively halting their spread. However, in the absence of vultures, other scavengers, including mammals, assumed the role of carriers for these pathogens, perpetuating their transmission within ecosystems.

Moreover, feral dogs are primary carriers of rabies. Following the decline of vultures, India's feral dog population skyrocketed by at least five million, and a catastrophic rabies outbreak followed.[9] It's thought that the dog population boom resulted in over 38 million additional dog bites and more than 47,000 extra fatalities due to rabies, with an estimated economic toll of $34 billion.[10] Indian districts afflicted by the disappearance of vultures experienced a notable rise in overall human mortality rates, averaging at least 4.2% from 2000 to 2005.[11] Again, I want to be careful not to *blame* wildlife (i.e., feral dogs) for this outbreak. After all, it was human actions that triggered the cascade.

Rabies 101

Rabies is a viral disease that affects the central nervous system and is usually transmitted through the saliva of infected animals. A bullet-shaped rabies virus causes the disease, which belongs to the genus *Lyssavirus* (from the Greek 'Lyssa' – goddess of rage and madness) in the family Rhabdoviridae. Rabies is a zoonotic disease as it can be transmitted from wild animals to humans. The RNA viruses such as rabies have extremely high mutation rates,[12] which means they can often evolve faster and have various levels of virulence. Think of the COVID-19 virus (also an RNA virus), which has mutated and evolved into many different strains. This ability to rapidly mutate and evolve can make RNA viruses challenging to treat.

The rabies virus primarily infects mammals, with dogs being the most common transmission source to humans in many parts of the world. Other animals that can carry the virus include raccoons, bats, foxes and skunks – all reservoir hosts. But as mentioned, it's often human activities, such as encroachment into habitats or poisoning keystone animals (like vultures) that promotes transmission. Rabies typically spreads through the bite or scratch of an infected animal. However, it can also be transmitted if infected saliva encounters mucous membranes or broken skin.

Once the rabies virus enters the body, it travels to the brain and spinal cord, where it causes inflammation and ultimately leads to severe neurological symptoms. The incubation period for rabies can vary widely, ranging from a few days to several years, but once symptoms appear, the disease is almost always fatal. Early symptoms in humans may include fever, headache and fatigue, which can progress to more severe symptoms such as confusion, hallucinations, paralysis and eventually coma and death.

Rapid post-exposure prophylaxis is recommended for individuals who have been bitten or scratched by a potentially rabid animal. This consists of a series of rabies vaccinations, and in some cases 'rabies immune globulin' – a solution of antibodies that provides immediate, passive immunity against the rabies virus. It contains a high concentration of rabies antibodies derived from the plasma of human donors vaccinated against rabies.

Rabies Virus

Who Detected the Vulture Decline?

In the early 1990s, Indian conservationist Vibhu Prakash was entrusted with leading a project to discover how many vultures were alive and where they lived.[13] He gathered data on various resident species between 1990 and 1993. From 1996 to 1999, Vibhu extensively studied raptors at the Velavadar National Park in Gujarat and the Rollapadu Wildlife Sanctuary in Andhra Pradesh. During these years, Vibhu meticulously documented the alarming decline in the *Gyps* species of vulture populations across the country. Vultures became the focus of his professional life.

I decided to call Vibhu, who is still based in India, to learn more about this story. He told me that when animals died in the wild pre-1990s, hundreds of vultures would descend on the carcasses within minutes. This meant the dead bodies were rapidly recycled, and feral dogs didn't get a chance to feed on them. But that all changed in the 1990s. Dog food (dead bodies) was available everywhere for the stray canines. Vibhu said, "Female dogs would also litter their pups in the rib cages of the large carcasses." Presumably, the rib cage offers protection from predators and other threats. The carcass may also provide the newborn pups with warmth and shelter.

I asked Vibhu, "What did local people think when vultures started dying?"

He said that when the vulture populations crashed, people would ask him what had happened. Animal carcasses accumulated across the landscape, but local people hardly ever saw the carcasses of the dead vultures themselves.

He said, "And do you know what people kept saying to me?"

I wasn't expecting what was to come.

"The Americans are stealing our vultures; they're vacuuming them out of the sky!"

He said people would repeatedly blame it on the Americans. And as funny as it sounds, they were deadly serious. They found it baffling that thousands of vultures were vanishing without a trace. The most popular explanation was that another country, specifically the United States, had been secretly stealing them.

However, after doing some tree surveys, Vibhu realised that the vultures were returning to their treetop nests to die. They were so high in the canopy that nobody saw them.

How incredibly saddening.

In the early 2000s, there was a rush to determine the cause of the decline in the vulture population. Some thought it could be viral in origin. There's some irony hidden in these circumstances. The vulture decline indirectly caused a zoonotic disease outbreak. However, scientists were once worried that the cause of the vulture decline itself was a zoonotic disease.

This brings me to the late Lindsay Oaks and his team at the Peregrine Fund. Numerous teams were entrenched in the challenge of identifying the cause of the decline. Frustration and concern mounted as vulture mortalities persisted despite rigorous fieldwork and exhaustive diagnostic analyses. They investigated potential pathogens and environmental toxins present in recently deceased vultures from breeding colonies in Pakistan and Nepal. To achieve this, they collected tissue samples and shipped them to laboratories in the United States for thorough analysis.

Vultures affected by the decline exhibited widespread kidney damage. However, despite exhaustive investigations, Oaks' team found no evidence of disease-causing pathogens or environmental toxins. But there was a clue. Uric acid crystals permeated the bodies of the vultures. Conversely, vultures known to have succumbed to pesticide poisoning or collisions showed no signs of uric acid accumulation.

The researchers began considering whether the vultures could be exposed to veterinary drugs through the carcasses of dead livestock they consumed. And then, finally, the team unearthed evidence that diclofenac – the newly introduced pharmaceutical drug intended for domestic livestock – was the primary culprit behind the vultures' demise. It was known to induce kidney damage, and local veterinarians extensively prescribed it. Furthermore, its usage had surged during the same period as the vulture population had declined. Diclofenac was inexpensive. Vets and landowners routinely administered the drug to treat everyday ailments like lameness and injuries in cows before their demise.

The majestic cinereous vulture (Aegypius monachus)

To confirm their suspicions, the researchers conducted experiments. Diclofenac and meat from animals treated with diclofenac were administered to 20 non-releasable vultures rescued from nesting colonies. Oaks reflected on this aspect of their work in a *New Scientist* article: "We hated to do it."[14] The outcome was stark: even small doses of diclofenac proved fatal to these vultures. They exhibited identical symptoms to those observed in the wild vultures in India. Moreover, the likelihood of vulture mortality increased with higher doses of the drug.

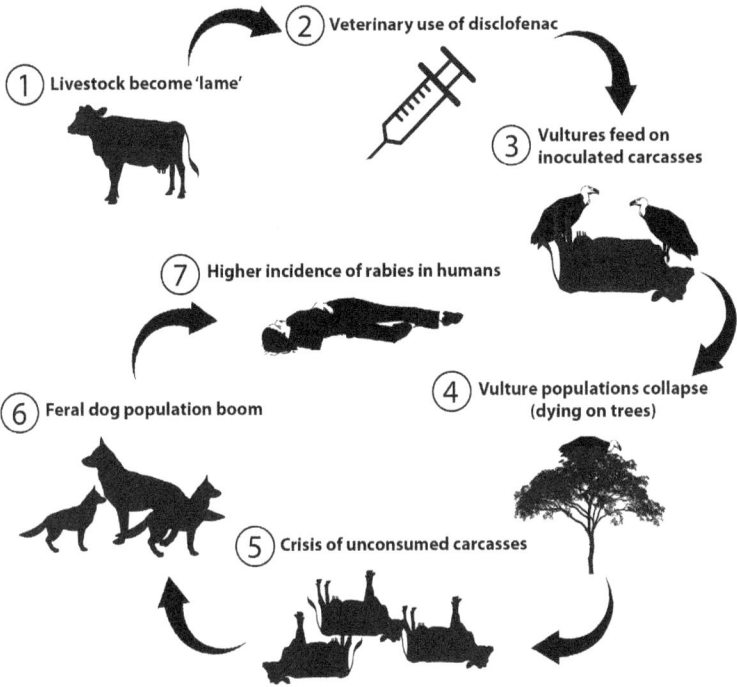

The links between the vulture decline and rabies outbreaks

Despite India's 2006 ban on veterinary diclofenac, vultures have continued to be poisoned. One study scrutinised carcasses gathered from 2011 to 2014 and revealed elevated diclofenac levels in the kidney tissue of 14 out of 29 white-backed vultures and 9 out of 12 Himalayan griffons.[15] This provided compelling evidence that diclofenac poisoning was still an issue. Some say the drug is still sold and used illegally.

Controversially, over in Spain, diclofenac was approved for veterinary use in 2013 (i.e., after it was known to kill vultures). It remains accessible despite Spain hosting approximately 90% of the European vulture population.[16] An independent simulation study underscored the peril, indicating that the drug could annually reduce vulture populations by 1–8%. Furthermore, new sanitary regulations governing animal carcass disposal in Spain diminish the vultures' food supply and escalate costs and greenhouse gas emissions.[17]

A fundamental shift in global cultural perceptions of vultures is imperative; but there is a ray of hope. Vibhu tells me that although vulture populations are not making a rapid recovery, they have been stable since 2016. However, due to their low breeding rate, the population is projected to (only) double in the next ten years. This projection hinges on factors such as conserving and restoring healthy habitats and, of course, the absence of poisoning. I cross my fingers.

Anthrax

Remember in the Introduction when I said that some, actually most, microbial species are either harmless or vital for our survival? Well, in my day job as a microbial ecologist, I once reviewed the scientific literature on these beneficial species. I created a list of more than a hundred bacterial species known to have human health benefits. Chief among them was a genus called *Bacillus*. Within this genus, I identified at least nine different species that confer various beneficial effects on humans, including reducing diarrhoea, enhancing intestinal health and promoting a robust immune system. Because of their 'friendly' status, several species of *Bacillus* are used in probiotics – a concoction of live microbes, typically bacteria or yeast, that confer health benefits when consumed in adequate amounts ('probiotic' is derived from the Greek for 'in favour of life').

However, like in many communities, there's often one bad egg, and in this group (genus), it's *Bacillus anthracis*. *B. anthracis* is responsible for causing anthrax, a severe infectious disease. 'Anthrax' is derived from the Greek word for 'charcoal', likely referring to the black eschars or skin lesions that can occur in some forms of the disease. The bacteria are transmitted to humans and other animals through contaminated soil, water or animal products. Anthrax can appear in several forms – cutaneous (skin), inhalation and gastrointestinal – and can be severe or even fatal if not promptly treated with antibiotics.

Some people have suggested that vultures act as vectors of the anthrax bacterium.[18] They can indeed carry it once they've fed on infected carcasses. However, researchers now believe that vultures – and other highly efficient scavengers – play a vital role in preventing anthrax outbreaks and reducing their impact when they do occur.[19]

One reason is related to the bacterium's ability to produce spores. Under stressful conditions, such as high temperatures or low nutrients, the bacteria form spores as a survival mechanism. These spores are highly resistant and can endure unfavourable conditions until the environment becomes suitable for growth again.

The spores have a tough, protective outer coat. This allows them to withstand extreme conditions and remain dormant for extended periods. Think of the bacterial spore as akin to the seed of a plant lying dormant in the soil during harsh winters. It's simply awaiting warmer temperatures and sufficient moisture to germinate and grow. Anthrax is clearly more nefarious than this, but you get my point.

The ability of the bacterium to form spores is crucial for several reasons:

1. The spores are highly resistant to heat, desiccation and chemical disinfectants, allowing them to persist in the environment for years. This long-term survival enables the bacteria to maintain reservoirs in soil, water and animal products, which then serve as a source of infection for susceptible hosts.

2. Spores can be ingested or inhaled. They can also enter the body through breaks in the skin. Once they do this, they can germinate and cause infection. Inhalation of spores can lead to severe respiratory anthrax, which is often fatal without prompt treatment. The persistence of spores in the environment aids their transmission to humans and other animals.

3. Environmental factors such as soil disturbance, drought or floods can trigger the release of spores from contaminated sites, increasing the risk of exposure to humans and other animals.

So, the quick and efficient way in which anthrax cells turn into spores – and how long these spores can survive – plays a role in how outbreaks happen and spread.

If carcasses are left out in the open for too long, flesh flies will lay their eggs on them. The hatched larvae (maggots) feed on the decomposing flesh. The maggots mechanically disrupt the carcass tissues during this feeding process, creating openings or wounds. This can trigger the conditions (e.g., exposure to high temperature and dehydration) that induce spores to form. When flies infest a carcass, their larvae feed on the tissues and can inadvertently encounter anthrax spores in the decomposing flesh. If the larvae or flies move to another location or encounter other animals, they may carry and transmit anthrax spores to new hosts.

Therefore, efficient carcass scavenging is vital! Failure to scavenge or remove infected carcasses often leads to prolonged anthrax outbreaks. Vultures are the efficient scavengers the world needs. If their populations plummet, more carcasses are left to rot. This means the conditions for *B. anthracis* sporulation become more likely, and so do anthrax outbreaks.

There's a pattern here. While this is an utterly morbid topic, there's something quite beautiful in the interconnectedness of it all.

At least 200 cases of anthrax were recorded in the southern states of India between the 1950s and 2001. However, three clustered outbreaks occurred in the late 1990s, including twelve cases of fatal anthrax meningitis. And what did these outbreaks coincide with? You guessed it: the decline of the vultures.[20] The village folk handled infected carcasses that the vultures would otherwise have cleaned up.

Towers of Silence (Dakhmas)

The decline of vultures was not merely an ecological crisis; it also struck the heart of the culture and traditions of some Indian communities. And it wasn't just the pile-up of rotting livestock carcasses that came with disease risk. Human bodies were also posing a danger. When animals (including humans) die, if we're not cremated, our bodies go through a natural decomposition process, where quite remarkable alchemy and ecological succession take place. If you're squeamish, you might want to skip the next two paragraphs.

At first, the body's cells break down through 'autolysis'. At this stage, a suite of enzymes digests cell membranes and organelles. This leads to the release of cellular contents and the breakdown of bodily tissues. Then comes putrefaction. Bacteria and fungi multiply, feeding on the decaying tissues and producing gases such as methane, hydrogen sulphide and ammonia. This leads to a characteristic odour associated with decomposition. At each stage of the decomposition, the body will support a different community of microbes – much like the natural succession when a shrubby habitat is destined to be a forest, its vegetation, animal and microbial communities change along the way.

During putrefaction, the body turns greenish purple due to the formation of haemoglobin breakdown products. Gases accumulate within the body cavity, and the abdomen starts to swell, causing bloating and distension. Eventually, the skin may rupture, releasing gases and fluids. Then, during the 'active decay' stage, the body undergoes rapid decomposition as microbes break down tissues and organs. Notably, bodily fluids, gases and decomposing tissues provide an ideal breeding ground for bacteria, viruses and parasites, many of which are potential human pathogens. Additionally, bodily barriers such as skin and organs breaking down during active decay increase the likelihood of contact between infectious agents and external surfaces or animal disease vectors. This can further spread pathogens to other organisms or environments.

All these gory details are to say that rotting bodies can be hotbeds for potentially dangerous pathogens. Therefore, without efficient scavengers like vultures, rotting human bodies pose a disease risk.

You may be thinking now, "But when people die, they're either cremated or buried six feet below the ground, so this is not a problem." And you'd be correct, in general. However, things are slightly different in the Parsi community – the people who came to India from Persia thousands of years ago with the Zoroastrian faith.

For millennia, the Parsis have practised unique funeral rituals. A Parsi funeral begins in a way that is relatively familiar to many people. When someone in the community dies, people gather to pay their respects and chant prayers. However, for me, at least, that's about as familiar as it gets.

Once the chants have finished, the Parsis carry the corpse to the top of a circular, raised structure known as a *dakhma* or a 'Tower of Silence'. The corpse is then left to the elements to decompose, with vultures playing the starring role. The Parsis believe that upon death, Nasu, the 'corpse demon', rushes into the body and contaminates everything it encounters. They also believe that earth, water and fire are sacred, and therefore a rotting and demon-filled corpse must not come into contact with these elements. In a way, burying a body in the ground is seen as polluting nature.

Traditionally, vultures consumed human remains placed in Towers of Silence, facilitating a swift and ecologically sound decomposition process. However, the drastic reduction in vulture numbers has disrupted this ritual, leading to partially decomposed bodies and raising concerns about public health and religious continuity.[21]

Now, the Parsi community has turned to technology to help decompose bodies. They use solar concentrators.[22] These massive structures comprise giant mirrors on top of the Towers of Silence. The mirrors direct sunlight towards the bodies and generate heat of up to 125 degrees Celsius to speed up the decomposition. Ravens and kites also help, but it takes days rather than the hours it would take vultures.

So, not only do vultures play a role in neutralising pathogens and preventing disease outbreaks, but they also have cultural importance.

Please consider supporting vulture conservation efforts!

CHAPTER 3

Why Bats Get Bad Press

*We see then that the bat is a very wonderful
creature, one of Nature's triumphs and masterpieces.*
—W.H. Hudson, *The Book of a Naturalist*

As twilight descends and the world transitions from light to dark, bats emerge from their crevices and soar covertly through the air. These disease-ridden creatures with their mysterious flight patterns and ability to navigate in the dark evoke unease in the human psyche...

Hold on a moment – is this even true? How universal is this negative perception of bats?[1] And does the perception match reality?

I love bats. They're like tiny bears with wings; they play inordinately valuable roles in the ecosystem, and many use a biological sonar to navigate. Echolocation evolved in bats due to natural selection favouring adaptations for efficient hunting and navigation in low-light environments. As a child, I considered this a bona fide superpower. Now I understand the less-mystical reality, my awe remains undiminished.

Why I Love Bats, Continued...

The sun was beginning its slow descent as we walked out of the village – our destination, the local cave. My guide brought me to the cave entrance to see a wildlife spectacle: bats leaving in their thousands to find food. We waited a few minutes, head torches in place, and then it began, right on cue.

Whoosh! Swathes of giant leaf-nosed bats (*Hipposideros gigas*) came flooding out and took to the skies. Their echolocation guided them as they left the cave entrance, narrowly but skilfully avoiding the trees (and us). Then, something unexpected happened. A colony of larger bats flew out: Egyptian fruit bats (*Rousettus aegyptiacus*). As their name suggests, they sought a nectar-rich food source in a different part of the forest.

And then came a torrent of another species – Hildegarde's tomb bats (*Taphozous hildegardeae*). Three huge colonies of different bat species shared this same cave and seemingly took turns leaving for the night.

I was in Africa doing rainforest research, and this (in addition to a giant fish eagle (*Haliaeetus vocifer*) nearly landing on my friend's head while he was snorkelling) was one of my favourite wildlife encounters. There's something mysterious about bats, but in my eyes, this only adds to their appeal. Their sheer abundance is also astounding. In terms of numbers of individuals, bats are by far the most abundant mammal on the planet – around fifty-six billion (compared to eight billion of us). Yet, they only account for less than one-tenth of the total biomass, whereas humans now weigh six times as much as all wild mammals combined.[2] Bats are also the most diverse group of mammals after rodents, with around 1,400 species.[3] Some species can live for more than 30 years, making them some of the longest-living mammals for their size.[4]

In the UK, we have 18 species; in Australia, where I live, there are around 80 species. Many of them echolocate. This means they have a sensory system that enables them to navigate and hunt prey in darkness, emitting high-frequency sounds that bounce off objects. The bats detect these bouncing sound waves with their sensitive ears. Echolocation is a biological sonar that allows them to create a mental map of their surroundings, even in poor light.

Some species feed on fruit rather than insects. They're aptly called 'fruit bats'. Scientists once believed that fruit bats didn't echolocate. Why would they have evolved this? After all, they didn't need the precision targeting of insects in flight. However, studies show that some fruit bats use a form of echolocation. Rather than using their larynx and making high-frequency screeching sounds like many microbats, some fruit bats click their tongue and produce signals like dolphin clicks.[5] Other research suggests they may also use their wings to produce a rudimentary form of echolocation.[6]

Fruit bats can also see quite well – they often have huge eyes. They probably switch and combine sensory modes between bright and dark environments, giving them flexibility.

Grey-headed flying fox (Pteropus poliocephalus) *hanging in a tree in Adelaide, South Australia*

Tequila!

Some bats function as natural pest controllers, consuming insects that are harmful to crops such as corn and pecans. Others pollinate plant species that give rise to our favourite foods, like bananas, coconuts, avocados and agave – an ecological role that people often associate with bees. Fruit-eating bats also aid in wild plant growth. They disperse seeds of mangoes, cashews, figs and almonds.[7] Their droppings (guano) are also rich in natural fertiliser. Without bats, many plants would struggle to reproduce, and the populations of insects (some of which carry diseases) could surge unchecked.

If you asked me, "What are your favourite cocktails?" I would put margaritas in my top three. If you're a fellow margarita or tequila enthusiast, you might not immediately think of bats as you sip your poison. Yet tequila, derived from the agave plant, has long relied on bats for pollination and seed dispersal. The Mexican long-nosed bat (*Leptonycteris nivalis*) has coevolved with the agave plant for millennia.

The bat has an impressive 7.5-centimetre-long tongue, which means its tongue is around 80% of its body length.[8] (If the human tongue were proportionally as long, it would be over 1.5 metres!) The bat uses its tongue to delicately slurp nectar from agave flowers under the cover of night. It undertakes annual migrations along a 'nectar corridor' of blooming agave and cactus plants, from west Texas to Mexico, playing a crucial role in pollination.

Bat Houses to Control Agricultural 'Pests'

Several years ago, I embarked on a mission to install a suite of timber bat boxes on the trunks of trees. It was an act of hope. It was also a tangible effort toward supporting our endangered nocturnal friends. My fellow ecologists and I returned to the same spot a year later. We were filled with anticipation. Did our efforts bear fruit? As we inspected the boxes, a heartening sight greeted us: several soprano pipistrelle bats (*Pipistrellus pygmaeus*) occupied the boxes. It was deeply satisfying to see these tiny houses providing safe havens for the bats.

I believe bats are intrinsically valuable – they have value in their own right. However, as we've discussed, they're also instrumentally valuable – they play various roles that benefit humans – and it turns out that building little bat houses could yield unexpected benefits for agricultural pest control.

Troy Swift has been a pecan farmer in Texas since 1998. He never considered incorporating bat houses near his orchards until Merlin Tuttle, a renowned bat conservationist, suggested that he use bats as part of his pest-control programme.[9] Inspired, Swift promptly began constructing bat houses. Picture a wooden box mounted on a tall pole. The box has multiple chambers and narrow entrances to mimic natural roosting sites such as tree hollows or building crevices.

Within six months, bats had moved into Swift's bat houses. Now, his farm has 17 bat houses, and Swift collaborates with Tuttle's organisation to gauge the bats' impact on his crops. They deployed bat acoustic detectors and collected DNA samples from droppings. The findings were impressive. At least seven bat species were identified on Swift's farm. They discovered that these bats consumed over a hundred insect species within six weeks. They're still gathering data, but it seems that simply having bats around helps to suppress the activity of insects that consume and parasitise crops.

An exciting research question would be: Do bats help reduce the prevalence of zoonotic diseases by controlling disease vectors (such as mosquitoes and tsetse flies)?

As I finished writing this section, I saw a post on social media from a researcher, Eyal Frank, about his latest paper.[10] In his study, Frank found that farmers in the United States had compensated for bat declines in recent years by increasing their insecticide use by 31.1%. He linked this increased insecticide use to a 7.9% increase in human infant mortality in affected areas.

Although it's not an infectious disease link, this illustrates the One Health concept beautifully:

1. Our actions have led to a huge decline in the number of bats, which would ordinarily keep agricultural 'pests' in check.
2. Because the bats are no longer around, these pests have increased.
3. Farmers have turned to chemical pesticides to keep them in check.
4. This has increased infant mortality (pesticides are linked to maladies in children).
5. The solution? Protect bats and their habitats!

In Asia

Good sex. Good fortune. Good death. For many, these are probably not the first words that spring to mind when thinking about bats. In China, however, the bat motif carries a distinctly different connotation from the familiar Western perception. The Chinese characters for bat (蝙蝠) are pronounced 'biānfú'. Now, it just so happens that the second character (蝠) bears a close resemblance to the character for fortune (福), pronounced 'fú'. Because of this resemblance, uttering the word 'bat' invokes a blessing. In Chinese symbolism, the sight of a bat in flight signifies the arrival of good fortune.

Bats are renowned for their ability to hang upside down in perfect stillness for extended periods. This trait is revered in some Eastern cultures. It symbolises patience, meditation and a serene stillness. It's believed to bring longevity and a peaceful transition into the afterlife.

Farmers in South Sulawesi, Indonesia, believe that the presence of flying foxes near their rice fields ensures a bountiful harvest.[11] Likewise, fishermen in the Philippines regard mangrove-roosting flying foxes as protectors of their fishing areas that help them catch more fish and shellfish. Across Malaysian Borneo, it is culturally inappropriate to disturb or harm fruit bats. What happens if you harm a bat? Well, it's believed to bring misfortune to unborn babies, particularly in cases where a woman is nearing childbirth.

In Sarawak, Borneo, some believe a bat flying into the house signifies a shaman bringing good vibes.[12] And sadly, in some Asian countries, bat body parts are consumed as medicine.

Still, elsewhere, bats have a bad reputation.

The 'Harbinger of Doom' Perspective

In some cultures, bats have historically been distrusted. Superstitions, myths, and folklore often link them with darkness, death and evil spirits. Over centuries, these negative perceptions have been perpetuated through literature, art and media. It's safe to say that bats are misunderstood. Their nocturnal habits and cryptic behaviour compound the misconceptions.

The depiction of bats as sinister, vampire-like creatures in movies and other forms of media has long fuelled public fear and misunderstanding.[13] Such portrayals usually distort reality and spread misleading information about bats. I remember watching the movie *Indiana Jones and the Temple of Doom* as a child. In one scene, Jones and his nightclub singer companion Willie Scott are riding through an Indian forest on elephants when a colony of flying foxes (large fruit bats) flies overhead.

Willie says, "Wow, what big birds." But suddenly, the theatrical music turns sinister, and Jones replies in an equally sinister voice, "Those aren't big birds, sweetheart; they're giant vampire bats!"

For a start, there are no *giant* vampire bats, and certainly not in India (although I like to think Indy was just pulling Willie's leg, so to speak). Secondly, these visual, auditory and dialogical cues strengthen the perception of bats as fiendish creatures. Many millions of people watch movies. They have a major role in shaping people's perception of animals, for better or worse.

There's a video on YouTube called 'Bats evolution in cinema: Scary bats in 60 movies by size'.[14] It shows how bats, particularly in horror movies, are invariably portrayed in an evil light. You may say, "But what about Batman? And what about Robin Williams' rapping bat in *Fern Gully*? My, perhaps feeble, response would be: these aren't horror movies.

There's been a call to associate bats with more positive emotions to prevent children from developing unhealthy negative perceptions of these animals.[15]

Bats were often linked to the underworld in ancient Greek and Roman mythology.[16] They were considered symbols of evil or nocturnal creatures associated with darkness and death. During the Middle Ages in Europe, people frequently associated bats with witches, black magic and vampires.[17] Stories of bats transforming into vampires or being familiars of witches were prevalent in folklore and contributed to their negative portrayal. Gothic literature frequently used bats as symbols of darkness and mystery.

A Bat

There's also an association between bats and Vlad the Impaler. Vlad is a historical figure who inspired the character of Count Dracula in Bram Stoker's novel.[18] He was also known as Vlad Dracula. In the fifteenth century, he ruled Wallachia, a region in present-day Romania. But how did Vlad pick up his epithet 'the Impaler'? If you studied European history at school, you'll probably know that Vlad had a reputation for cruelty. In particular, he would impale his enemies on stakes while they were still alive. This contributed to the dark imagery surrounding the legend of Dracula. In some interpretations of the Dracula story, bats are portrayed as one of the forms Dracula can take, adding to the menacing image. However, Vlad's connection to bats is fictionalised and based on literary interpretations rather than historical facts.

The Disease Perspective

Now, there's no denying that bats carry diseases. Just like all animals (including *Homo sapiens*), bats possess their own suite of pathogens. They can be bacterial, viral, fungal and others. However, they've garnered attention for their ability to harbour various viruses. Some have the potential to spill over into human populations. For instance, bats are known reservoirs for several coronaviruses. These include those closely related to the viruses responsible for severe acute respiratory syndrome (SARS) and Middle East respiratory syndrome (MERS) in humans.

Many scientists believe the recent COVID-19 pandemic, caused by the SARS-CoV-2 virus, originated in bats before spilling over to humans, possibly through an intermediate host – although this theory is hotly contested.[19]

Evidence suggests that fruit bats also transmit the Ebola and Nipah viruses.[20] Studying these viruses is important because they pose significant public health risks.

Bats serve as proficient hosts for a range of viruses due to factors such as their long lifespans, high metabolic rates and unique immune systems. Bats are also social creatures. The way bats hang out together (pun intended) helps viruses spread easily among them. This might lead to more types of viruses emerging in bat groups than in other animals.

But Why don't Bats Get Sick?

Despite the prevalence of numerous perilous viruses jumping between them, bats seemingly defy the odds. They rarely succumb to these ongoing infections themselves. Bats can strike a delicate balance between managing viral infections and avoiding the deadly consequences of an overactive inflammatory response in their body. This is a physiological balancing act. It eludes many other hosts, including us.

I mentioned earlier that bats have a high metabolic rate. Well, the thing that drives this heightened metabolic rate is *flight*. Flying elevates their core body temperature to around 38°C, akin to a persistent fever in humans. Researchers have proposed that this metabolic adaptation could serve as a protective mechanism against viral infections, aiding bats in their survival.[21] The idea is that the higher body temperatures may serve as a powerful selective force against virulence and promote virus diversification (i.e., selecting for different variants that can thrive in these conditions).

Viruses can unleash destruction on hosts by hijacking cellular machinery to replicate – they're often called 'obligate parasites'. They can also trigger an excessive inflammatory response known as a 'cytokine storm'.

Let's use a sci-fi analogy to explore this further. Imagine a peaceful garden where each flower represents a cell in your body. Now, picture a mischievous alien intruder sneaking in. The intruder starts to raise hell. It hijacks the flowers to reproduce itself (I did say it was sci-fi!). In response, the flowers send out distress signals, like alarm bells ringing throughout the garden. These signals, known as cytokines, are meant to rally the gardeners to fight off the intruder. However, the alien intruder is cunning. It tricks some of the flowers into sending out too many cytokines. It's like a false alarm triggering a disproportionate response from the gardeners. Suddenly, the garden is flooded with an overwhelming number of gardeners rushing in, causing chaos, bumping into each other and trampling over the flower beds. This flood of gardeners (immune cells) overwhelms the garden. It destroys the intruders but damages the surrounding flowers and structures. In our bodies, this exaggerated immune response is the 'cytokine storm'.

So, bats' evolutionary adjustment to flight may equip them with enhanced heat tolerance. This adaptation may also allow them to withstand the detrimental effects of inflammation better than other mammals. It turns out that some bats may also have a gene mutation that reduces the production of inflammation-causing proteins. This mutation could also help them to avoid damage caused by cytokine storms. The gene, which all mammals have, is called the stimulator of an interferon gene (or 'STING').[22] Interestingly, when this gene is faulty in humans, they are more likely to have autoimmune diseases (where the body's immune system attacks the body's own cells). They're also more likely to have other inflammatory disorders.

This means that the physiology, genes and behaviour of bats allow some dangerous viruses to thrive and diversify.

How Dangerous are Bats to Us?

Some viruses carried by bats can indeed harm humans. However, there are important nuances to consider. For instance, we shouldn't be closely interacting with bats in the first place. We know that direct interaction with bats, such as handling or consuming them, increases the risk of transmission of bat-borne viruses to humans. For example, many emerging infectious diseases, including Ebola and Nipah virus infections, have been linked to contact with bats or their excrement. Avoiding behaviours that increase the risk of exposure to viruses is key. What do these behavioural changes look like? Avoiding the consumption of bats ('bushmeat') is a good start. We must reduce the risk of bat–human contact in areas where bats and humans coexist – more on this later in the book.

But there's another crucial nuance to consider: how human activities increase stress in bat populations, which in turn increases the likelihood of viral shedding. And what exactly is *viral shedding*?

Viral shedding is the release of virus particles from an infected host into the environment or bodily fluids. When a virus infects an animal, it replicates within their cells, producing new virus particles. The body then sheds these virus particles through various routes, such as respiratory secretions (e.g., coughing, sneezing), saliva, faeces, urine, blood or genital secretions. Viral shedding is a critical aspect of viral transmission. It essentially allows the virus to spread from one host to another. The duration and magnitude of viral shedding can vary. It depends on factors such as the type of virus, the stage of infection and the individual's immune response – but stress also exacerbates the issue.

Wildlife ecologist Peggy Eby, who specialises in flying fox behaviour and disease in eastern Australia, recently said to me that bats in urban and agricultural areas rely on suboptimal food, which may lead to nutritional stress that facilitates viral shedding.

Flying foxes are the reservoir hosts of the Hendra virus, a deadly zoonotic disease that primarily affects horses and humans in Australia. Around 70% of people infected with Hendra have died.[23] To put this into perspective, the mortality rate for SARS-CoV-2 is around 1%, and for the Ebola virus, it's about 50% (ranging from 25–90% in previous outbreaks).[24] The Hendra virus has probably circulated in bats for far longer than Europeans have occupied Australia. However, we first identified a Hendra virus spillover in 1994.[25] This means that despite its danger to humans and despite bats being the reservoir hosts for centuries, no confirmed cases in humans had ever been recorded before 1994.

What does this tell us?

Firstly, it highlights the importance of understanding zoonotic diseases and their potential for spillover into human populations. We've recognised bats as the reservoir hosts of the Hendra virus for a long time. However, the absence of human cases before 1994 indicates that spillover events were either:

1. Extremely rare.
2. Not so rare but went unnoticed (very unlikely, given the high mortality rate).
3. Non-existent – the conditions were such that the spillover simply didn't happen.

Secondly, it suggests that the conditions that prevented a spillover event from bats into humans either gradually or abruptly changed in the early 1990s.

Historically, flying foxes in Australia were known for their nomadic lifestyle, traversing vast distances to follow fleeting bursts of Eucalyptus trees that flower and provide their favourite foods – nectar, pollen and fruit. It was uncommon for the flying foxes to occupy roosts continuously. During the summer, numerous tree species across the landscape provided abundant food sources for bats. However, food availability dwindled in the winter, with only a few tree species offering sustenance. Unfortunately, human activities like urbanisation and agriculture restricted the distribution and abundance of these trees. We clear vast swathes of forests without considering the ecological and disease consequences. As a result, brief periods of mass flowering in remaining forests attracted a significant proportion of the bat populations to concentrated areas, particularly in coastal lowlands. The loss of these winter food sources could devastate bat populations.

Sometimes, trees fail to flower during winter or spring. This phenomenon, which happens every one to four years, is due to fluctuations in temperature and rainfall. When this occurs, bats experience short-term food shortages that usually last between three and twelve weeks. In response, the bats adjust their behaviour. They assemble in small groups near dependable but often subpar food sources found in urban gardens and agricultural lands (such as fruits from ornamental, commercial or weed species). These behavioural adaptations were once transient. They only lasted as long as the acute food stress persisted. Bats typically reverted to their nomadic lifestyle and nectar-feeding habits when food shortages abated.

However, the more forests we chop down, the less transient these urban visits become, the more likely we are to come into regular contact with the bats and the more likely they are to become stressed. This combination of factors is a recipe for disaster.

How did the Hendra virus story unfold? In 1994, a mysterious illness struck horses in the Brisbane suburb of Hendra, Australia. The disease was characterised by respiratory distress, neurological symptoms and high mortality rates among infected horses. Veterinary pathologists discovered a novel disease. They later named the Hendra virus the causative agent.

Further investigation revealed that the virus had likely originated from fruit bats, particularly the black flying fox (*Pteropus alecto*) and the spectacled flying fox (*P. conspicillatus*), which are native to Australia.[26] These bats are natural reservoirs of the virus and can shed it in their urine, saliva and birthing fluids without showing symptoms of illness.

The spillover of Hendra virus from bats to horses occurred primarily through close contact when horses grazed under bat-inhabited trees or consumed contaminated fodder. Once infected, horses can transmit the virus to humans through respiratory secretions or bodily fluids. This can lead to severe and often fatal illnesses. Since its discovery, several human cases of Hendra virus infection have been reported, mainly among horse handlers and veterinarians. Incidentally, as I write this (in July 2025), another case of Hendra has been reported in Queensland, and a horse has sadly died.

In response to the Hendra virus threat, public health authorities in Australia have put in place a range of safety measures. These include vaccinating horses and educating the public on avoiding contact with bats.

Peggy and many other scientists now think human encroachment into habitats creates opportunities for viruses like Hendra to jump species barriers and infect new hosts. Therefore, it's not bats themselves that are inherently dangerous to humans. The conditions created by human activities increase the risk of disease transmission. Our view of bats as disease-causing villains – often reinforced by movies – fuels a harmful cycle of misunderstanding about the real sources of zoonotic diseases.

This mindset stems from the phenomenon of scapegoating (see Chapter 1). We can look to science to explain this behaviour. It's rooted in psychological theories such as attribution theory and cognitive dissonance. By attributing diseases solely to bats, we absolve ourselves of responsibility and discomfort. In a way, blaming bats for disease outbreaks is like blaming the messenger for bad news. It's like tutting and pointing our finger at a mirror for showing our reflection, conveniently ignoring the entity that created the image in the first place. As long as we demonise bats without addressing our own behaviours, we remain blind to the urgent need to heal nature and coexist with wildlife. To effectively reduce the risks of future pandemics, we must shift our mindset from fear and blame to understanding and accountability. But there is hope and a remedy, which we'll explore later in the book.

So, whether it's for tequila, controlling pests or easing stress in bats to reduce the likelihood of a spillover event, it's clear that we must protect our chiropteran friends and their natural habitats.

CHAPTER 4

Biodiversity and the Dilution Effect

Biodiversity is the totality of all inherited variation in the life forms of Earth, of which we are one species. We… save it to our great benefit. We ignore and degrade it to our great peril.

—E.O. Wilson, from the foreword to *The Living Planet in Crisis*

The story of this chapter begins in the 1970s in the tiny town of Lyme, Connecticut, United States. A cluster of children began developing unusual symptoms, including rash, fever and joint pain. Residents and doctors became concerned. They soon realised they were witnessing something unprecedented: a newly recognised disease that would come to be known as Lyme disease. By 1977, the first 51 cases of 'Lyme arthritis' were described, reflecting the unusual joint pain caused by the disease.[1]

At this time, the black-legged or 'deer' tick (*Ixodes scapularis*) was linked to the transmission of the disease. People realised they'd been bitten by a tick. Often, the bite resulted in a large, red, circular skin rash followed by more debilitating symptoms. Because joint pain was one of the symptoms, the Arthritis Foundation created the first brochure addressing Lyme disease.

Four years later, in 1981, the causative agent of Lyme disease was finally identified: a bacterium named *Borrelia burgdorferi*, after medical entomologist Willy Burgdorfer, who discovered it during his research on Rocky Mountain spotted fever. It's also known as a 'spirochete' bacterium due to its characteristic spiral shape. It's like a loosely coiled spring.

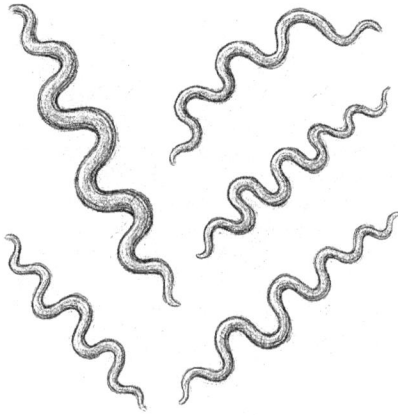

Spirochete bacteria (like B. burgdorferi*)*

B. burgdorferi makes its rounds through Ixodid ticks, hitching rides from one host to another. While it's only been recognised as the causative agent of Lyme disease relatively recently, the bacterium has been circulating in mammals, ticks and birds for millennia.

A groundbreaking study led by researchers at the Yale School of Public Health uncovered genetic evidence suggesting that *B. burgdorferi* has ancient origins in North America.[2] It circulated within forest ecosystems for at least 60,000 years. This is long before humans arrived in the region, around 24,000 years ago.

Lyme disease is one of the fastest-growing vector-borne infections in the United States. How many new cases a year? About 500,000.[3] The Yale team's discovery shattered the notion that the Lyme epidemic is a recent phenomenon driven by a new introduction of the bacterium or a genetic mutation boosting its transmission. Instead, it turns out the real driver is a familiar one. This epidemic is deeply entwined with the sweeping ecological changes we've set in motion across much of the United States and other temperate regions.

Over the last century, the breaking up of forests and suburban sprawl have crafted the perfect environment for ticks to thrive. In the twentieth century, deer populations surged across suburban areas. They thrived without natural predators like wolves. This boom in deer numbers was a catalyst for the rapid spread of deer ticks. The rise in ticks has, in turn, fuelled the persistent nature of the ongoing Lyme disease epidemic.

Climate change has also played a role. Warmer winters accelerate tick lifecycles and enable them to survive an estimated 45 kilometres further north each year.[4] Ticks have encroached into suburban landscapes, abundant with animals such as white-footed mice (*Peromyscus leucopus*) and American robins (*Turdus migratorius*). This expansion has provided ample opportunities for the bacterium to reproduce and spread.

The tick feeds on the reservoir host and ingests the bacteria and the host's blood. The bacteria then multiply within the tick's gut. It's a 'friendly' environment. Once infected, the tick can transmit the bacteria to other hosts during subsequent feeding sessions. Indeed, when the infected tick feeds on a new host, it transmits the bacteria wrapped up in its saliva. This feeding display creates the perfect conditions for the disease to spread.

The Host-Seeking Behaviour of Ticks

Many moons ago, I studied the parasitic relationship between the European hedgehog (*Erinaceus europaeus*) and the hedgehog tick (*Ixodes hexagonus*). I spent hours in the dark of night roaming agricultural fields in the UK with a powerful torch. My goal: to detect tiny, colourful reflections, like cat eyes, amongst the crops.

In the weeks before, I'd attached small reflective tags to one of the roughly 7,000 spines on the coats of various hedgehogs. This enabled me to track them at night. It was quite a buzz to see the telltale glint in the torchlight, indicating I'd found one – sometimes, I could be out there for hours without seeing one.

When I found the hedgehogs, I would look for ticks on their bodies. I'd count them and assess the hedgehogs' condition. I also *attempted* to collect ticks in the field to study their behaviour. My interest lay in how they were attracted to their hosts.

I had read about 'tick flagging'. This is where you attach a sheet of white fabric to a pole, like a giant flag. You drag the sheet over grasses or forest leaf litter and hope ticks will attach themselves to the flag. When finding a suitable host, some tick species, like the castor bean tick (*Ixodes ricinus*) prefer an ambush-style strategy.[5] They literally 'sit and wait' atop a blade of grass or sedge and hop onto the host as it walks past. This sitting and waiting makes it easy to collect them using a tick flag.

It turns out that hedgehog ticks aren't this kind of species. I didn't know this at the time. However, the realisation came swiftly when only the castor bean tick was clinging to my tick flag.[6]

Some ticks, like *Ixodes hexagonus*, exhibit 'endophilic' or 'inside-loving' behaviour. This means that at each lifecycle stage (egg, larvae, nymph and adult), the ticks live *inside* the nests or burrows of their animal hosts. This makes collecting them in the field tricky and explains why initially I only managed to collect the other species.

Other ticks are aggressively mobile and are attracted to carbon dioxide, something that animals emit in huge quantities. For these species, you can install tick pitfall traps containing dry ice (a solid form of carbon dioxide) in the field to attract the ticks.

When I was doing my surveys, British wildlife biologist Toni Bunnell published a study on how hedgehog ticks were attracted to the faecal odour of sick hedgehogs.[7] This confirmed the findings of other studies that showed healthy hedgehogs were less likely to carry ticks than unhealthy ones. The ticks essentially sniff out particular volatile compounds (including indole) in the hedgehog's faeces. This sensory prowess allows them to discern between hosts. They probably favour those with weakened immune systems over those with robust cellular weapons that may cause them issues.

Incidentally, hungry hedgehog ticks display 'questing' postures. The first pair of their legs extends outwards. Then, they essentially wave in response to stimuli, such as the faecal odours, that indicate the presence of a host. They're like picky hitchhikers with their thumbs out, waiting patiently but selectively for the perfect ride that matches their taste.

Both the hedgehog tick and castor bean tick can carry the Lyme disease pathogen in the UK. However, the castor bean tick receives more attention due to the higher prevalence of parasitism between this species and humans (and livestock).

But let's circle back to the black-legged tick – the tick in the United States that is (inadvertently) driving the Lyme disease epidemic. Recent research has suggested that these ticks are surprisingly picky.[8] They're drawn to some resident microbes on the skin of their animal hosts, yet completely turned off by others.

Why is this pickiness important?

It implies that the skin microbiome (the resident microbes and theatre of activity) may determine whether ticks will parasitise an individual. The microbiome also plays a vital role in a host's immune system. We know that damaged ecosystems and human disturbance can negatively disrupt animal microbiomes[9] and, as Fackelmann et al. (2021) said, "make animals vulnerable to infections that may become zoonotic".[10]

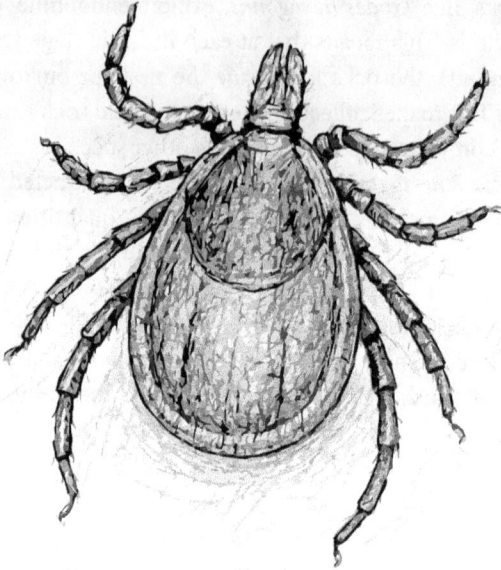

Ixodid tick

I'm now wondering if changes in the microbiome due to a damaged ecosystem could influence the likelihood of animals being parasitised by ticks and their susceptibility to infection by the Lyme disease pathogen. It seems the carbon dioxide emissions and the skin microbiome of the white-footed mouse are just right for the deer tick in the United States.

Might this happen in humans, too?

If our microbiomes are not in great shape (due to pollution, poor diets, antibiotics and so on), are we at greater risk of being parasitised? I don't know the answer, but I do know the human skin microbiome produces the scents that are most attractive to mosquitoes![11]

Competent versus Incompetent Reservoirs

I mentioned white-footed mice and robins earlier. These are relevant because they're 'highly competent' reservoir hosts for the Lyme disease pathogen. In the United States, the white-footed mouse is probably *the* most competent reservoir host for the Lyme disease agent. If you're not a disease lingo aficionado, this may sound a little confusing. So, let's unpack the term 'competent host'.

The host's immune system drives the host's competence (ability) to support the Lyme disease bacterium. The immune system response differs from species to species. For instance, one host species, the western fence lizard (*Sceloporus occidentalis*), produces a unique protein that kills the spirochetes within the guts of feeding ticks. Other species' immune systems are less hostile to the bacterium.

So, when we talk about an animal being a 'competent host' for the bacterium, we essentially mean its physiology allows the bacterium to live, feed and replicate inside the animal's body. In other words, a competent host's body provides a friendly environment, and the body's security guards (the immune system) have no major objections.

Then, along comes the tick.

Tick

The Dilution Effect

We know that the black-legged tick is attracted to the white-footed mouse, the pathogen's most competent host. We also know that Lyme disease has boomed in the last few decades. But what does the 'dilution effect' have to do with it?

Well, many ticks are out there, some sitting and waiting, some actively seeking hosts. Yet, many ticks never become infected with the Lyme disease pathogen. This is because:

a. Some hosts aren't very good at transmitting the bacterium to feeding ticks.
b. Some hosts simply don't carry the bacterium, so they can't transmit it.

In general, larval ticks hatch from their eggs free of the Lyme disease pathogen. This is because mother ticks rarely transmit the bacterium to their offspring. So, recently hatched larval ticks are incapable of infecting hosts. However, larval ticks that feed on the blood of an infected host may acquire the bacterium (40–80% of them do) and carry the infection for the duration of their lives.[12] Yet the percentage of ticks infected with the Lyme disease agent can vary dramatically from region to region, habitat to habitat.

Back in the 1990s, ecologists Josh Van Buskirk and Richard Ostfeld used computer simulations to see how different animal hosts might affect the number of infected ticks.[13] They found that tweaking the balance between competent hosts (which are good at passing the Lyme disease bacteria to ticks) and incompetent hosts (which aren't) didn't change the overall number of ticks.

Think of going to a well-stocked, reasonably priced and tasty buffet restaurant. If the dishes are delicious and nutritious, you'll likely enjoy your meal regardless of whether ten or a hundred dishes are available.

So, the tick densities didn't change in their simulations. But what about the infection prevalence? When the relative abundance of non-mouse hosts was increased (i.e., the number of different incompetent hosts), the infection prevalence among ticks reduced significantly.

Going back to our buffet analogy, it's a little like having a buffet with many dishes, but only one is your favourite. If that favourite dish gets crowded out by all the other options, you're less likely to eat as much of it. Similarly, when the landscape is filled with a variety of animals that aren't the tick's preferred host (the 'incompetent' ones), the ticks are less likely to find their ideal meal, resulting in a drop in infection prevalence.

This concept is known as the dilution effect. It suggests that by reducing the dominance of the highly competent white-footed mice

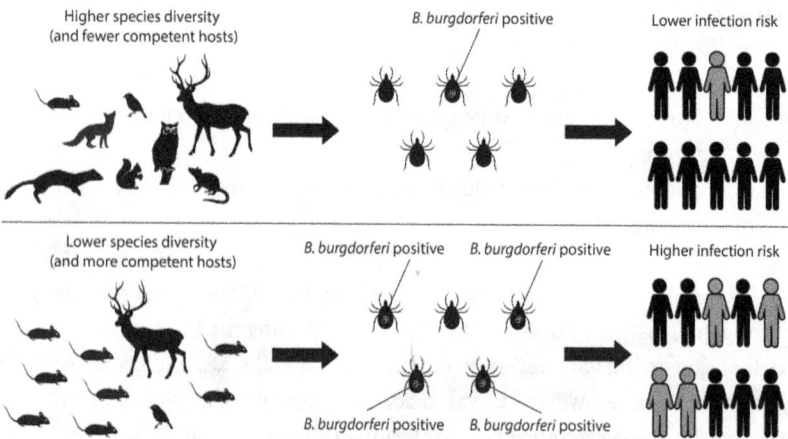

The dilution effect

relative to other hosts, the proportion of ticks carrying the Lyme disease pathogen falls. In simpler terms, higher biodiversity equals lower disease risk. Therefore, strategies to curb tick-borne infections might include reducing the population of white-footed mice while promoting the presence of other host species or boosting the number of incompetent animal hosts in the ecosystem.

Later on, Ostfeld teamed up with fellow disease ecologist Felicia Keesing to study this empirically. The team collected species diversity data, and their findings, published in 2001, further supported the idea that an increase in species diversity may reduce the ability of white-footed mice to infect the ticks with the Lyme disease bacterium.[14] This is because more ticks would feed on incompetent hosts, causing a dilution effect (see the diagram nearby).

The Effect of Habitat Fragmentation

Forests fragmented by roads, urban areas and agricultural land are common in many countries, including the United States. As we break habitats into smaller patches or fragments, the connectivity between them is disrupted. This can lead to the isolation of populations. Imagine taking a large, thriving neighbourhood of people and splitting it up with walls and other barriers. You take a walk to meet a friend, but the huge physical barrier stops you. You pick up the phone to call your relative, but the phone lines are cut off. You open your laptop to search the internet or connect with a loved one, but the internet has been cut off, too. Suddenly, what was once a community where everyone could easily interact is now divided into isolated blocks, making it impossible for residents to connect and thrive.

In natural habitats, isolation makes it hard for species to move around, find mates (and mix genes!), access resources and sustain healthy populations. Edges of these patches are often harsher – buffeted by wind or bathed in too much sunlight, which can drive away species that prefer different conditions. Smaller patches offer less space and resources. This can harm species that need larger areas. Fragmented habitats are also more vulnerable to disturbances like pollution and invasive species, further harming native species and disrupting ecosystems. In fragmented habitats, biodiversity can dwindle.

It's inordinately more complex than this, but I've explained the impacts of fragmented habitats in a nutshell.

As forests in the United States have become more fragmented, species diversity is thought to have decreased.[15] White-footed mouse populations have also boomed. So, in addition to being the most *competent* host for the Lyme disease pathogen, they become the most *abundant* host.[16]

The Dilution Effect in Other Pathogens

We've covered Lyme disease quite thoroughly in this chapter. But what about other pathogens? Is there any evidence of the dilution effect in viruses? And what about plant hosts rather than animals?

I found one study where the dilution effect occurred in plants.[17] In ecosystems with higher plant species diversity, viral and fungal disease outbreaks in plants were lower. The authors attribute this to:

a.   A reduction in the density of competent hosts
b.   The interception of spores and virus particles (also known as 'virions') by incompetent hosts

This study suggests that the magnitude of the dilution varied from ecosystem to ecosystem, so it's probably not a universal rule. However, the authors maintain that the general findings show the importance of protecting biodiversity to promote healthy ecosystems.

And then we have hantavirus.

Hantavirus is a genus of viruses that can cause various human maladies. They are primarily transmitted to humans through contact with rodents, particularly mice. Infection can occur when humans inhale airborne particles contaminated with the virus from rodent urine, droppings or saliva, or through direct contact with contaminated materials.

Researchers in Sweden investigated the zoonotic Puumala hantavirus in bank voles (*Myodes glareolus*).[18] They also examined the potential dilution effect on the disease by two competitor species – the field vole (*Microtus agrestis*) and common shrew (*Sorex araneus*) – and a predator, Tengmalm's owl (*Aegolius funereus*). Their research was based on long-term data on hantavirus infection from 2003 to 2013 in northern Sweden.

In support of the dilution effect, the researchers observed that the likelihood of infection in bank voles *decreased* as common shrew density *increased*. This indicates that common shrews played a role in reducing hantavirus transmission. Additionally, field voles suppressed bank vole density in meadows and clear-cut areas, indirectly contributing to the dilution of hantavirus infection.

But what about the Tengmalm's owl?

Well, the owl's population dramatically declined from 1980 to 2013. Scientists widely attribute this decline to the logging of old-growth forests. The male owls prefer to hunt in mature and old-growth coniferous forests and avoid hunting in open habitats and agricultural land. So, if we chop down the owls' hunting grounds, their numbers start to plummet. No surprises there, right?

The owls primarily hunt bank voles, especially during winter, as they are more accessible due to their movement patterns on snow-covered surfaces (they tend to move more on the surface than other small mammals). The voles also climb trees, making them even more accessible to the owls. Bank voles thrive in mature forests rich in bilberries, as they feed on bilberry leaves and berries. However, in nearby Finland, the coverage of bilberries in coniferous forests has significantly decreased over the years due to deforestation. It declined by 50% from the 1950s to the 1990s and continued to decrease in the 2000s.[19] This decline in bilberry coverage and ongoing clear-cutting of old-growth forests since the 1970s has likely reduced the food available to bank voles. As a result, predators like Tengmalm's owls may also be affected. It's a disruption to the food chain - fewer berries mean fewer voles, and in turn, fewer owls.

Additionally, willow and crested tits, along with other small forest birds, serve as important alternative prey for Tengmalm's owls, especially in winter. However, the population densities of these bird species have declined by approximately 70–80% in forests over the study period, further impacting the owls' food sources.

So, the owls are getting hit from all directions.

But what does this have to do with disease risks? We're removing the owl's hunting grounds and food sources, leading to a considerable decline in its population. In the same period as the owl has declined, the researchers found significantly higher rates of hantavirus infection in bank voles. They suggest the decline in the owl population may have led to this increase in disease.

This is incredibly sad, especially if you're an owl-lover. However, much like in our vultures (Chapter 2), it's also a beautiful illustration of the interconnectedness of life.

In theory, predators of reservoir hosts can reduce disease risk by (a) selectively taking infected host individuals and (b) regulating host density. Here, we see how different food web members can contribute to the dilution effect. The cascading impacts of logging are also illuminated. We remove vast swathes of trees. This reduces owl hunting grounds. This, in turn, reduces the ability of owls to pick off infected hosts, which likely increases disease risk.

It's important to note that a dilution effect is most likely to occur when there's a spectrum in host competency, with many different incompetent hosts. I can't help but think about this in evolutionary terms.

We can speculate that in some cases, evolutionary pressures may favour competent hosts, especially if they provide essential ecological functions or if the pathogen has adapted to exploit specific host species. In such scenarios, the ecosystem may exhibit a higher diversity of competent hosts, potentially leading to increased disease transmission and prevalence.

This brings me to the *amplification effect*.

The Amplification Effect

This is a phenomenon in ecology in which certain environmental factors *increase* the abundance or transmission of a particular species or pathogen.[20] Unlike the dilution effect, which reduces disease risk through increased diversity of incompetent hosts, the amplification effect enhances disease transmission.

The amplification effect can happen in ecosystems where various species shaped by evolution to play important roles are also good at hosting certain pathogens or parasites. This means that these ecosystems, despite their diversity, can have a high risk of spreading diseases because so many of their species can carry and transmit harmful organisms. This competency tips the balance in favour of the pathogens. As the hosts thrive in these environments, they may experience a boost to reproduction rates or expanded geographic ranges, leading to higher transmission of the pathogens and an elevated risk of outbreaks.

So, the amplification effect specifically relates to an increase in bio-diversity, leading to a rise in disease prevalence. As discussed in Chapter 1, environmental changes, such as deforestation or climate change, may also create the conditions for pathogens or parasites to flourish, thereby amplifying the disease risk.

This is why protecting biodiversity and treading lightly on our planet is crucial. Reducing the things that drive climate change and minimising our footprint in natural habitats gives nature a better chance to keep diseases in check.

The following isn't about amplification or dilution effects per se, but it still relates to the notion that biodiversity can reduce the negative impact of pathogens. However, this time I'm talking about our bodies, our 'walking ecosystems'.

Diverse Microbiomes and Ecology Across Scales

I stood near a large pond in a forest, watching a small brown Daubenton's bat chasing down mosquitoes freshly hatched from the stagnant water body. I imagined I was wearing space-age goggles and could magnify my vision over a million times and peek into the bat's body at a microscopic scale. At that very moment, as the bat chased down the mosquitoes in the woodland ecosystem, strange, alien, spider-like viruses chased down bacteria inside the bat's gut ecosystem. These are called bacteriophages – viruses that 'target' and kill bacteria.

The only fantasy element of this story is the space-age goggles. Indeed, principles of ecology – such as predation, competition, cooperation and energy flows – apply at myriad scales, including within the human body and the intertwined macro and microscopic ecosystems that we depend upon for survival.

We can view the human body as a 'holobiont' – a host plus trillions of microbes working symbiotically to form a functioning ecological unit. We can also view the rest of nature – the plants, deer, snakes, lichens, and so on – surrounding us as a vast collection of holobionts. We each emit a million biological particles every hour. We share the invisible constituents (the microbes) of our human holobiont with all the other holobionts (the plants and animals), whilst they share their invisible constituents with us. It's an unseen and subconscious gift economy! We must recognise this deep interconnectedness.

So, our human microbiomes are colonised by millions of microbes from our food and our surrounding environments. While this is a simplification, these invisible colonists include 'goodies' and 'baddies', and some in between that can switch from friends to foes in the right conditions. It's widely agreed that more than 99% of microbial species are either harmless or beneficial to us. And yet, a few bad ones sometimes break through the immune system's various lines of defence and wreak (often short-term) havoc on our walking ecosystems. This pathogenic assault can be worse for those with weakened immune systems and those whose microbiomes are in a state of 'dysbiosis' ('life in distress'). This could be due to overuse of antibiotics, poor diet, exposure to pollutants and so on. Dysbiosis is where your microbiome is out of kilter, often containing fewer beneficial species and genes, and typically characterised by lower biodiversity.[21]

When the microbiome is in a state of 'eubiosis' (a healthier balance in the community), it plays a crucial role in protecting the host from incoming pathogens.[22] One way the microbiome does this is by what's known as 'colonisation resistance'. This defence mechanism is not solely attributable to one or two species. Instead, it emerges from the complex interactions within a diverse microbial community.

A recent study highlights how colonisation resistance is a higher-order effect. In laboratory and animal studies, the scientists observed that a diverse microbiome can collectively outcompete incoming pathogens for essential nutrients, hindering their establishment in the host.[23]

The diverse microbiome essentially tells the pathogens, "There's no room for you at the dining table!"

This predictable colonisation resistance happens when the 'friendly' microbes in the host have proteins that look a lot like those of the invading counterparts. This match-up strengthens the host's defences and supports overall health. It creates a kind of natural shield against infections.

So, while this isn't strictly an example of *the* dilution effect, having a diverse microbiome can dilute the nutrients available to pathogens.

Diversity is the ultimate shield.

CHAPTER 5

Disease X – It's Coming

*The cloud never comes from the quarter of the
horizon from which we watch for it.*
—Elizabeth Gaskell, *North and South*

We could start this story in many ways. Perhaps we should begin by rejoining our friend Dakarai from Chapter 1. He's on his way to a bustling marketplace in a small, rural town. Stalls are lined with fresh produce, local crafts and an array of bush meat and live animals (which gives me a gut-wrenching feeling) ready for sale. The energy is palpable as vendors call out to potential buyers. However, as we know, the continued encroachment into the forest and close interaction with wild animals pose some undeniable risks.

Alternatively, we could go to a high-security infectious disease lab. Here, we might see researchers in full-body, air-supplied positive-pressure suits. They move with meticulous precision under the glow of sterile lights. They're handling Petri dishes and vials containing enhanced pathogens. In this lab, they're conducting controversial gain-of-function research – modifying viruses to better understand their lethal potential. But we're only human, and mistakes happen.

Or we could begin in an intensive agriculture complex. Here, our cochlea is invaded by the deep drone of machinery and the bleats of densely packed livestock. Rows of animals are confined to cramped spaces. Their eyes reflect a mix of exhaustion and distress. It's a grim but real scene – modern, high-intensity farming's relentless quest for efficiency. Farmers administer antibiotics en masse to stave off inevitable infections. The sheer density and proximity of the animals create a breeding ground for new, virulent pathogens.

Within any of these scenes lies a ticking time bomb that could unleash the next global health crisis. Welcome to the world of *Disease X.*

Disease X isn't an actual disease. It's a placeholder term coined by the World Health Organization (WHO) to represent the next unknown pathogen that could cause a severe pandemic.[1] It's a deliberately mysterious term. It embodies the uncertainty and unpredictability of infectious diseases in our interconnected world.

In a way, it reminds me of a concept in ecology called the *Red Queen Hypothesis*.[2] This evolutionary theory suggests that species must continuously adapt and evolve to gain a reproductive advantage and survive while competing against ever-evolving opposing species. The name is derived from a scene in Lewis Carroll's *Through the Looking-Glass* (1871)[3], the sequel to *Alice's Adventures in Wonderland* (1865).[4] In this story, the Red Queen tells Alice that in her realm, "it takes all the running you can do, to keep in the same place".

In evolutionary terms, this means that organisms must constantly evolve to keep up with the adaptations of other organisms within a changing environment. For example, predators evolve better hunting strategies while prey evolve better defence mechanisms. This creates a continuous adaptation cycle or an 'evolutionary arms race'.

Similarly, humanity must constantly adapt its defences against emerging diseases. Disease X symbolises the perpetual race against new and evolving pathogens. It's an ongoing struggle to keep pace with microbial evolution – a relentless drive for survival.

As we know, history is replete with examples of infectious diseases emerging seemingly out of nowhere, bringing turmoil to populations before medical science could catch up. Yet, we now know that various factors can contribute to the perfect storm – encroachment into wildlife habitats, globalisation and climate change, to name a few.

During a Pentagon news briefing in 2002, former US Secretary of Defense Donald Rumsfeld said:

> Reports that say that something hasn't happened are always interesting to me because, as we know, there are *known knowns*; there are things we know we know. We also know there are *known unknowns*; that is to say, we know there are some things we do not know. But there are also *unknown unknowns* – the ones we don't know, we don't know. And if one looks throughout the history of our country and other free countries, it is the latter category that tends to be the difficult ones.[5]

OPEN Things that you and everyone else knows	BLIND Things that you are unaware of but others know
HIDDEN Things that are known by you but unknown to others	UNKNOWN Things that are unknown by you and everyone else

The Johari Window

The idea of 'unknown unknowns' was developed in 1955 by two American psychologists, Joseph Luft and Harrington Ingham.[6] They created the Johari Window, a technique to help people better understand their relationship with themselves and others.

With Disease X, we're preparing for an unknown unknown. In the Johari Window, this is the fourth quadrant (bottom right). With unknown unknowns, it is, by definition, tricky to anticipate and prepare for events or information that are entirely outside our current realm of understanding. It's a cognitive blind spot that can lead to overconfidence, underestimation of risks and flawed strategic planning. The inability to recognise what we do not know creates a barrier to learning and adapting. It often requires significant shifts in perspective or new experiences to uncover these hidden gaps.

'X' stands for 'unexpected'.

The aim of having the Disease X placeholder is to spur proactive thinking about potential pandemic-causing pathogens. The idea is to broaden the focus beyond the usual suspects like influenza – or even novel but endemic ones like COVID-19 – to anticipate and prepare for the unknown threats that could emerge. It's supposed to ensure that we're not fixated on past outbreaks and get ready for future ones.

Since scientists coined the term Disease X in 2018, the world has already seen one of them – COVID-19. It was as if coining this term tempted fate! The first outbreak probably happened the following year in Wuhan, China.

I just looked through my phone messages to see when I first mentioned 'COVID' or 'Coronavirus'. It was the 31st of January 2020, in a message I sent to my wife:

> *Excellent* [said in a sarcastic tone]... *2 corona virus cases confirmed in... Yorkshire!*

The victims? Two Chinese nationals staying in a hotel in York tested positive. It turns out these were the first two confirmed cases in the UK. We were extra surprised because my wife and I were both in Yorkshire at the time.

I vividly remember planning my final PhD project at the University of Sheffield (also in Yorkshire) around this time. About a month later, a lecturer at the university tested positive for COVID-19. This event led to a pause in all teaching. It immediately hit the headlines in the city newspaper, highlighting how novel it was. One person got COVID-19, and it made the front page. That would be unimaginable now, even if it were Taylor Swift or Rihanna.

As of April 2024, there were **704,753,890** confirmed cases and **7,010,681** deaths globally.[7] The sheer scale of this is nearly incomprehensible. It's safe to say this particular Disease X took most of us by surprise. It felt like a slow-motion disaster we see in movies, but never expected to live through. Then came the surge. Numbers skyrocketed, and each day brought staggering new figures. The reality of millions infected and the relentless toll of deaths was hard to comprehend. It was a rollercoaster of emotions – fear, sadness and a sense of helplessness as we watched the world struggle to cope.

The mayhem that followed the initial months of uncertainty was surreal. Extended social lockdowns, restrictions on exercising outdoors, failed government schemes aimed at boosting the economy, such as 'Eat Out to Help Out' (a perplexing UK scheme), and missing the funerals of family and friends. These things should remind us to take the concept of Disease X deadly seriously. I was living in Eyam at the time (see the Introduction), which made it even more surreal. I was essentially quarantined in the village renowned for quarantining itself during the seventeenth-century plague.

Disorientated. I think that's the word to describe how I feel about the COVID-19 pandemic – formerly Disease X.

Adding to this disorientating mix, we now find ourselves in a strange new phase. We've almost normalised the situation, and at a rapid pace. The headlines have shifted, and life has regained a semblance of 'normalcy'. Yet, it seems the narrative of 'building back better' and using this experience to create a better world was a candle in the wind – and that's the risk of striving for 'normal'. There's an unsettling sense of déjà vu as we slip back into pre-pandemic routines, almost as if we risk forgetting the collective trauma and hard-earned lessons. The initial urgency to rethink our systems, improve public health infrastructure, protect biodiversity and address deep-seated social inequalities now seems to be waning. The comfort of the familiar overshadows them. Yet this normalcy is the antithesis of effective Disease X planning.

COVID-19 has demonstrated that pandemics can trigger a cascade of other health issues. Indeed, in one sense, we could say that Disease X accompanies Diseases Y and Z – long-term and secondary illnesses and mental health conditions. Long COVID, with its persistent symptoms, affects millions globally. Secondary diseases have emerged as healthcare systems become overwhelmed, delaying the treatment of other critical conditions. Mental health crises have surged. Anxiety, depression and other psychological issues have increased, stemming from prolonged lockdowns and social isolation. So, when we plan for Disease X, we should also prepare for Diseases Y and Z.

Remember how I said I was planning my final PhD project when the pandemic started? Well, it was supposed to involve gathering people together to create 'pocket wildlife gardens' on the grounds of various medical centres (GP surgeries). Local community members would spend time in the gardens, and I would collect biodiversity data and health data from the participants, including microbiome samples and immune markers. However, I had to cancel this entire project (which involved a six-month-long NHS ethics process!) as the coronavirus spread rapidly throughout the UK. In a later chapter, I'll talk more about the study that replaced this project, but essentially, we explored nature's role in supporting health during the pandemic.

What are the Chances?

So, what are the chances of living through another pandemic like COVID-19?

A recent study compiled a global dataset of historical epidemics from 1600 to the present, aiming to quantify the probability of future extreme pandemics.[8] By analysing 395 documented epidemics, researchers discovered that epidemic intensity – the number of deaths relative to the global population and epidemic duration – follows a power-law distribution. This essentially means that extreme epidemics, like the Spanish flu, are more likely than previously thought.

The study employed novel statistical methods to estimate the yearly probability of such events. The findings suggest that the likelihood of a pandemic with the intensity of the Spanish flu varies from 0.11% to 0.89% annually, with a mean recurrence time of 879 years today. However, the emergence of new diseases, accelerated by environmental changes, could triple this probability in the coming decades. Notably, the research also highlights a high probability of pandemics like COVID-19 occurring. The current likelihood of experiencing another COVID-19-like event in one's lifetime is 38%, a figure that may double in the coming decades due to increasing rates of disease emergence.

Indeed, the latest disease risk modelling from London-based forecasting company Airfinity suggests a 27.5% probability of a pandemic as deadly as COVID-19 occurring by 2033, driven by factors like climate change and increased travel.[9] Lower estimates indicate that the chances of another COVID-19-like pandemic occurring in a given year are over 2%.[10] It's hard to put this into perspective, and it sounds low, but remember, this is *in a given year*. The chances of being struck by lightning in a given year are around 1 in 1,600,000 or 0.0000625%. This means there's at least a 32,000 times greater chance of a COVID-19-like pandemic occurring in a given year (based on lower estimates).

I'm not sure if this is frightening or comforting!

SARS-CoV-2 Virus

How Might Disease X Play Out?

Let's do an epidemiological thought experiment. We're on day one. It might start with a handful of unexplained fevers and respiratory symptoms at a local hospital. Doctors are puzzled as patients, seemingly healthy just days ago, begin to deteriorate rapidly. Laboratory tests come back negative for known viruses like influenza or SARS-CoV-2. The medical community grows anxious as the number of cases doubles every few days. It starts to spread from the initial cluster to other parts of the city.

Fast-forward two weeks and epidemiologists are called in to investigate the outbreak. Contact tracing reveals a startling pattern. The pathogen spreads through casual contact. It also survives on surfaces for several days and thrives in crowded spaces. The reproduction number (R0) is estimated to be around 7, making it more contagious than seasonal flu and COVID-19 but less than measles. This R0 of 7 means that, on average, each infected person will transmit the disease to seven other people.

Measles has an R0 between 12 and 18. Still, our pathogen is highly contagious. Hospitals begin to overflow with patients. Healthcare workers are stretched thin. Initial estimates suggest a case fatality rate (CFR) of about 3%, indicating a significant threat. This fatality rate might seem low, but it represents a considerable threat to the population if the number of cases continues to rise. For example, if one million people are infected, a 3% CFR means thirty thousand people will die.

Fast-forward another two weeks. International travel has accelerated the spread of Disease X. Within a month, cases are reported in several major cities worldwide. The novel pathogen has an attack rate of 40%, meaning more than one in three people exposed to the pathogen become infected. Governments scramble to implement travel restrictions, quarantine and social distancing measures. However, the pathogen's stealthy transmission during the incubation period, where individuals are infectious but asymptomatic, renders these measures partially effective at best.

A few months later, the global economy is severely hit as industries shut down and supply chains are disrupted. Panic buying leads to shortages of essential goods. Public health messages emphasise the importance of hygiene and self-isolation, but misinformation and fearmongering create additional challenges. Healthcare systems are overwhelmed, with hospital bed occupancy rates reaching 150% in some regions. Vaccine research is fast-tracked, with scientists working around the clock to decode the pathogen's genome and develop a preventative treatment.

A year into the outbreak, Disease X has infected an estimated 500 million people globally, with a death toll reaching 14 million.

However, international collaboration and scientific innovation bear fruit as the first vaccines become available. Gradually, the tide begins to turn. Governments launch vaccination campaigns. The immediate threat recedes, but the scars on society are deep. All sounds scarily familiar, right?

Antimicrobial Resistance

The following chapters will explore pathogens that could cause the next human pandemic. But first, let's talk about antimicrobial resistance. Could this phenomenon make Disease X worse? Do we only need to worry about antimicrobial resistance in bacterial pandemics? Could the rising tide of antimicrobial resistance be considered a pandemic in itself?

The introduction of antibiotics into clinical use was arguably the greatest medical breakthrough of the twentieth century. Beyond treating infectious diseases, antibiotics revolutionised medicine, enabling organ transplants, cancer treatments and open surgeries. We can trace the foundation of modern anti-infective drugs back to Paul Ehrlich, who, around 1910, pioneered the use of synthetic arsenic-based pro-drugs, salvarsan and neo-salvarsan, to combat *Treponema pallidum*, a bacterium and the causative organism of syphilis. This work also laid the groundwork for chemotherapy.

The 'golden age' of antibiotic discovery began with Alexander Fleming's accidental discovery of penicillin in 1928, sparking a surge of antibiotic discoveries that peaked in the mid-1950s. In just over a century, antibiotics have dramatically transformed modern medicine. Before antibiotics, bacterial infections such as pneumonia, tuberculosis and septicaemia were often deadly. The advent of these drugs turned once-fatal illnesses into manageable conditions, dramatically reducing mortality rates. Common infections which previously claimed countless lives could now be effectively treated, saving millions.

Academics say antibiotics have extended the average human lifespan by around 23 years.[11] By effectively combating bacterial infections, antibiotics have also reduced the disease burden, allowing people to live healthier lives. The ripple effects of this improvement touch every aspect of society, from reduced healthcare costs to increased time and health for productivity and creativity.

How Antibiotics Work

We use antibiotics to kill disease-causing bacteria. It's fascinating to note that microbes themselves are the source of many antibiotics. Some microbes have naturally produced antibiotics for millions of years to defend themselves against other microbes. As I describe in another book, *Invisible Friends*,[12] it's an archaic microbial warfare played out on an unseen battlefield long before humans harnessed antibiotics for medical use. Scientists have traced the origins of antimicrobial resistance back to around 350 to 500 million years ago.[13] This means that long before humans (or any mammal, for that matter) walked the Earth, microbes were already developing ways to outsmart each other's defences. After all, some microbes consume other microbes, and some compete with other microbes for other resources. We also know that some bacteria, for instance *Actinobacteria*, evolved genes that confer resistance to the antibiotic compounds they produce themselves. In other words, they evolved to protect themselves from their own arsenal!

But how do antibiotics work? They function by targeting specific features of bacterial cells that are essential for their survival and reproduction. As you can probably imagine, after hundreds of millions of years of evolution, there are several different types of antibiotics, each with a unique mechanism of action. Some antibiotics, like penicillin, attack bacterial cell walls. Bacteria need a strong cell wall to maintain their shape and protect themselves from their environment. Penicillin weakens this wall, causing the bacteria to burst and die. Whereas other antibiotics, such as tetracyclines, interfere with protein synthesis. Like us and other organisms, bacteria rely on proteins for almost all their functions, from structural components to enzymes that drive metabolic reactions. By binding to the bacterial ribosome – a molecular machine responsible for protein production – tetracyclines prevent bacteria from making the proteins they need to grow and thrive.

Then we have antibiotics like ciprofloxacin. These target bacterial DNA. They inhibit the enzymes bacteria use to replicate their DNA, effectively halting their ability to multiply. Without the ability to replicate, bacterial populations dwindle, and the infection is controlled.

We can also classify antibiotics as either bactericidal or bacteriostatic. Bactericidal antibiotics kill the bacteria outright, while bacteriostatic antibiotics stop bacteria from growing and multiplying – and in humans, this allows the immune system to step in and finish the job. This selective targeting enables antibiotics to destroy or inhibit bacteria without harming our own cells, making them indispensable in our fight against infectious diseases. However, the evolutionary arms race mentioned earlier continues today. Bacteria and other microbes constantly evolve new mechanisms to resist the antibiotics, and now against the ones we deploy. Understanding the ancient and invisible battleground tactics is crucial for our health and survival. But through the inappropriate use of antibiotics – for instance, in us and livestock – we're making it harder to stay ahead.

How Antimicrobial Resistance Works

Microbes have short generation times, some reproducing in minutes. They are the speed demons of life. Take *Escherichia coli*, or *E. coli*, a common bacterium found in the intestines of humans and animals. Under optimal conditions, *E. coli* can double its population every 20 minutes. This doubling means a single bacterial cell can proliferate into billions within a few hours. Such speed is nearly incomprehensible in human terms, where it takes about 20 years to produce the next generation. Other microbes, such as yeast, have slightly slower but still rapid generation times. *Saccharomyces cerevisiae*, the yeast used in baking and brewing, can double every 90 minutes. Even the tuberculosis-causing *Mycobacterium tuberculosis*, which has a comparatively slow generation time of about 20–24 hours, reproduces much faster than most, if not all, animals.[14]

But why does this rapid generation time matter? The swift reproduction rate of microbes is a double-edged sword. On the one hand, it enables beneficial processes in ecosystems and applications in biotechnology and medicine. Scientists can harness these fast-growing organisms to efficiently produce pharmaceuticals, biofuels and other valuable substances. The rapid lifecycle also makes microbes ideal subjects for genetic research. It allows scientists to observe evolutionary changes over short periods.

On the other hand, the quick turnaround poses significant challenges, such as antibiotic resistance. When bacteria reproduce rapidly, the chances of genetic mutations increase, and some of these genetic mutations may confer antibiotic resistance. Unlike humans, who pass on their genes from parent to offspring through sexual reproduction – a process known as vertical gene transfer – some microbes can exchange genes directly with one another without any need for sex. This process, called horizontal gene transfer, enables microbes to acquire new traits rapidly. One of the most alarming traits they can pick up is antibiotic resistance. Bacteria are constantly swapping these resistance genes across vast landscapes, creating a major threat to human health. This relentless gene exchange means that the antibiotics we rely on to treat infections are becoming less effective, posing a severe challenge in our fight against bacterial diseases. A single mutation in a bacterial population can quickly proliferate. This single mutation can render treatments ineffective, leading to so-called 'superbugs'.

These pathogens now have two advantages: (1) they already have the swords in hand to harm you, and (2) they're charging at you in chain mail armour. However, the real fuel for dangerous mutations is how we misuse antibiotics – by overusing them, or not finishing the course.

Fleming even warned of this in his 1945 Nobel Prize acceptance speech: "It is not difficult to make microbes resistant to penicillin in the laboratory by exposing them to concentrations not sufficient to kill them, and the same thing has occasionally happened in the body."[15]

So, we promote antimicrobial resistance through a combination of overusing antibiotics for things like viruses (which aren't affected by antibiotics) and underdosing when we do use them (which allows some of the target organisms to survive, mutate and evolve resistance). Agriculture and medical waste are also huge drivers of antimicrobial resistance. For instance, livestock are often continuously given low doses of antibiotics to promote growth and prevent disease. This constant, low-level exposure creates an environment where bacteria are under selective pressure to evolve resistance mechanisms. Antibiotics in agriculture can also spill into the environment through animal waste, contaminating soil and water. This contamination can expose wild bacteria to swathes of antibiotics. This invisible chemical onslaught promotes the development and spread of resistance in nature. The microbial communities are passing around the genetic code for the chain mail armour, fortifying bacteria against our medical arsenal.

How could Antimicrobial Resistance Influence the Path of Disease X?

If Disease X turns out to be a bacterial pathogen and one that is highly resistant to antibiotics, we'll be up the proverbial creek without a paddle. Disease X would then be Superbug X. This is why antimicrobial resistance is important in the spillover narrative. The case fatality rate would skyrocket, and we'd need to rely on strict travel restrictions, quarantines and the development of alternative treatments such as phage therapy (viruses that infect and kill bacteria, as an alternative to antibiotics). These alternative therapies take time to develop.

However, Disease X doesn't have to be a bacterial pathogen for us to worry about the impact of antimicrobial resistance when it arrives. Disease X might be a viral disease. It could be a malady caused by other non-bacterial pathogens. But guess what? Many people who acquire these non-bacterial diseases die of secondary infections caused by bacteria. So, if those secondary infections are untreatable due to anti-microbial resistance, the creek without a paddle situation is back.

Researchers attribute many deaths to secondary bacterial infections during severe viral respiratory diseases like the flu and COVID-19. For instance, during the 2009 H1N1 influenza pandemic, bacterial co-infections were identified in up to 33% of cases, contributing to a higher mortality rate.[16] One study showed that around 40% of patients who died from COVID-19-related complications had bacterial infections.[17]

So, when battling a viral infection, your immune system works overtime to fight off the invader. This intense focus on the virus and the damage caused by inflammation can leave your body more vulnerable to other infections, particularly bacterial ones.

If these bacteria are the kind with both swords and chain mail, what use is a paddle anyway?

In a way, we can loosely view antimicrobial resistance as a pandemic in itself – it occurs globally, affecting many people across multiple countries and continents, and typically causes widespread illness and disruption. Purists would reject this. However, we can certainly view antimicrobial resistance as a potential facilitator of a pandemic. Moreover, the deadly synergy between antimicrobial resistance and a highly infectious pathogen could lead to catastrophic outcomes. Let's keep Superbug X at the forefront of our conversations about Disease X.

CHAPTER 6

Going Viral

*We think we are done with the pandemic, but the
pandemic is not done with us.*
—Gitanjali Pai, *Public health commentary*

"Before we [spoiler alert] jump into why viruses are the great
doom-bringers of society, tell me, what have viruses ever
done for us?" I asked Rob Edwards, a microbiologist from
Flinders University in South Australia.

As a microbial ecologist, I already knew a few things about viruses,
but the invisible world is a near-bottomless well of discovery.

Rob said, "Okay, well, in research, viruses have helped us understand
how cells work. They've taught us a lot about how DNA makes copies of
itself, how genes are turned on and off, and how cells produce proteins.
Viruses also keep bacteria in check. In fact, half of all the bacteria on the
planet are killed by viruses every forty-eight hours!"

That last fact is staggering. Rob was referring to the bacteriophages,
or 'phages', mentioned in earlier chapters. These viruses help maintain
the biological equilibrium in our ecosystems, including our walking
ecosystem: the human body. Phages blend random movement with
precise targeting. They drift through environments, carried by fluid
currents, and diffuse naturally to spread out.

In some cases, they can even sense and move toward chemical
signals emitted by bacterial cells. When a phage encounters a bacterium,
it attaches specialised proteins onto receptors on the bacterial surface,
initiating a high-stakes molecular interaction. Once attached, the
phage injects its genetic material. It then commandeers the bacterium's
machinery to produce new viruses.

There's also evidence that phage viruses enhance photosynthesis in
cyanobacteria (bacteria with green chlorophyll pigments like plants).[1]

They carry photosynthesis-related genes and transfer these genes to their cyanobacterial hosts. Given that these bacteria help produce 50% of the oxygen in the atmosphere, that's a mighty contribution. About 10^{31} individual viral particles are inhabiting the oceans alone at any given time. This is 10 billion times the estimated number of stars in the known Universe.[2] "In the soil," Rob said, "these phage viruses infect and lyse [break down] bacteria, which helps regulate population sizes and nutrient cycling. This process releases organic matter that feeds other organisms, supporting the food web. And the same thing happens in our bodies."

Indeed, each of us has 380 trillion of these viruses in our body at any moment.[3] Take a moment to let that sink in. You might be thinking, "That number is astronomically high. I can barely process it." Let me provide a comparison: each of us has a hundred times more viruses inside us right now than the number of trees on the entire planet. Here's another comparison: the total mass of viruses on Earth is equivalent to 75 million blue whales, the largest animal ever to grace the planet.

Viruses are omnipresent and clearly mega-abundant, and they play critical ecological roles. If we jump into 'social construct' mode and into the realm of 'value', many of them, particularly phages, we can probably call 'goodies'. Rob talks about phage therapy – the use of phages as a treatment to combat bacterial infections. His friend once had a urinary tract infection caused by a combination of E. coli and Staphylococcus bacteria. No matter how many antibacterial drugs they took, the bacteria resisted. The situation was getting serious. If they couldn't defeat these pathogens, they could die. They heard about phage therapy specialists in Georgia who were reporting positive results. So, off they went to Georgia. The phage specialists designed a treatment, and voilà, it worked! The viruses targeted the pathogens, and Rob's friend was cured.

Oceans and cliffs

Rob tells me how the White Cliffs of Dover were formed, in part, by the action of phage viruses that infected marine microbes called coccolithophores. Around 70 to 100 million years ago, these single-celled algae thrived in warm seas, and their calcium carbonate plates (coccoliths) accumulated on the seafloor as they died. Viruses played a crucial role in this process by infecting and lysing (breaking open) coccolithophores, releasing their chalky plates into the ocean. Over millions of years, these plates settled on the seabed and were compacted into the chalky layers that now form the cliffs. Hence, the White Cliffs of Dover (and similar cliffs) are largely composed of the fossilised remains of coccolithophores, and the role of viruses in their death and sedimentation was pivotal in creating this iconic geological feature.

As if this wasn't fascinating enough, Rob then tells me how those tangy ocean smells we breathe in when we stand on the seashore are also caused by phage viruses. Phages infect and lyse marine bacteria, releasing a chemical called DMSP, a compound that is then broken down into dimethyl sulphide (DMS), which is responsible for the ocean's distinctive smell.

Rob then said, "And have you heard of the viral placental hypothesis?"

"Well, I know that eight per cent of our DNA is viral. Is this related to the hypothesis?" I said.

"Yes! The viral placental hypothesis suggests that the evolution of the mammalian placenta was driven by endogenous retroviruses – viruses that integrated their genetic material into the host genome millions of years ago. These viral genes are thought to play a key role in placental development. The genes help form a layer of cells in the placenta that's crucial for nutrient exchange between mother and foetus, immune modulation and the fusion of placental cells."

At this point, my neurons were rapidly firing as I frantically took notes: "The hypothesis proposes that these viral elements were co-opted by the host and repurposed to enable the structure and function of the placenta. This allowed for more efficient reproduction and development in mammals. Hence, viruses are believed to have been pivotal in the evolution of viviparity [live birth] in mammals."

In other words, we wouldn't exist without viruses... for many reasons.

I think that this answers the 'What have viruses ever done for us?' question. But now, let's move on to the grim stuff. After all, awe and peril often travel together in virology. Because viruses are omnipresent, mega-abundant and masters of parasitising hosts and hijacking their genetic machinery to replicate, they are acutely dangerous to eukaryotes like us.

By hijacking the things that physically make us who we are, viruses divert the cell's resources away from its normal functions. They disrupt cellular processes. The body's immune response to the viral infection can also contribute to disease symptoms. Inflammation, fever and other immune reactions aimed at fighting the virus can cause tissue damage and symptoms like pain, swelling and fatigue. Moreover, some viruses produce toxic proteins that can directly damage tissues and organs, exacerbating diseases.

So, which virus is likely to cause the next pandemic? That's what we're all here for, right?! Let's head towards our Chapter 1 travel agency, Zoonotic Holidays. But this time, we're going next door to Deadly Virus Holidays. First on the tour is Paramyxovirus.

Paramyxovirus

A day before I started writing this chapter, I saw the news headline 'Incurable Nipah brain virus kills teenager in India'. The Nipah virus is from a family known as Paramyxoviridae (pronounced 'para-mix-o-viri-dee'). This virus has quite a scary reputation, causing severe illness in humans and other animals.

The story of the Nipah virus begins with fruit bats, which are the natural carriers. The bats can carry the virus without getting sick themselves (see Chapter 3). Sometimes, the virus jumps from bats to other animals, like pigs. From there, it can make another jump to humans. It's a classic spillover event that is increasingly likely as we encroach into 'wild' habitats.

Evidence suggests that in 1997/1998, immediately before the first Nipah virus outbreak, widespread slash-and-burn deforestation created a thick, smoky haze that covered much of Southeast Asia. As a result, many trees could not flower, depriving flying foxes of their vital food sources in their shrinking habitats. Desperate for food, these bats moved into fruit orchards near Ipoh, Malaysia, where the initial outbreak occurred. The combination of deforestation, drought and the proximity of pig farms to these orchards allowed the virus to jump from bats to pigs and eventually to humans.[4]

When humans catch the Nipah virus, it can cause severe symptoms. Imagine the worst flu you've ever had, and then add some nasty complications like encephalitis (brain inflammation), which can lead to seizures or even coma. The symptoms start with fever, headaches and muscle pain, but things can quickly get worse, leading to disorientation, drowsiness and, ultimately, severe neurological problems. The first major Nipah outbreak caused over 100 deaths.[5] This outbreak led to massive pig culls to stop the virus. Since then, outbreaks have occurred in other parts of Asia, including Bangladesh and India, often with even higher death rates.

One of the reasons Nipah is so scary is that there's no specific treatment or vaccine available yet. Doctors can only provide supportive care to help manage symptoms and complications. This makes it crucial to prevent the virus from spreading in the first place. The mortality rate of Nipah virus infections is high, ranging from 40% to 75%.[6] It's always a worry when you hear about it in the news. Before this year, the most recent outbreak of Nipah occurred in September 2023 in the Kozhikode district of Kerala, India. This outbreak involved six confirmed cases and two deaths. Scientists detected the virus in a man who presented with pneumonia and acute respiratory distress syndrome. Following his death, the virus spread to close contacts, including family members and healthcare workers. By late September 2023, more than 1,200 contacts had been traced and quarantined to prevent further spread. The most recent case, which hit the headlines as I wrote this chapter, was in Kerala, too. It's known as the most at-risk area globally for the virus, primarily due to rapid urbanisation and tree loss.

However, the likelihood of the Nipah virus becoming a pandemic is considered low, though it's not impossible. There are a few things to consider. Nipah virus primarily spreads through direct contact with infected animals (like bats and pigs) or their bodily fluids. Human-to-human transmission can occur but is typically through close contact, such as caring for infected individuals or exposure to respiratory secretions. Nipah doesn't spread as easily from person to person as highly contagious respiratory viruses like influenza or COVID-19.

Moreover, countries where Nipah outbreaks have occurred have implemented effective public health measures to control the virus. This includes quarantine, surveillance and culling infected animals. These measures help prevent the virus from spreading more broadly. Other preventive measures include avoiding contact with sick animals, not drinking raw date-palm sap (which can be contaminated by bat saliva or urine) and improving sanitation and biosecurity in animal farms.

So, while the Nipah virus might not be a household name like some other viruses, it's one to watch out for. Scientists are working hard to learn more about it and develop ways to fight it; in the meantime, being aware and taking precautions can help keep this virus at bay.

Various other paramyxoviruses are of concern. Another virus in this family with the potential to cause major global outbreaks is perhaps a more famous one: measles. As mentioned in Chapter 5, it's a highly contagious virus with a basic reproduction number (R0) of 12–18, meaning one infected person can infect between twelve and eighteen others in a susceptible population. The potential for measles to cause a pandemic is heightened due to the 'perfect storm' of conditions created by other diseases, such as COVID-19.[7]

In the first two months of 2022, global measles cases rose by around 80%.[8] Disruptions caused by the COVID-19 pandemic, conflicts and crises led to significant declines in vaccination rates, leaving millions of children vulnerable. The highly contagious nature of measles means that even a small decrease in vaccination coverage can lead to large outbreaks. The weakening of the global immunisation infrastructure increases the risk of measles outbreaks that could quickly escalate into more widespread pandemics. This is particularly pertinent in communities grappling with COVID-19, potentially leading to twin pandemics.

Therefore, while some viruses may pose a relatively low risk globally, we must consider the synergistic effects of other diseases and social events.

Influenza

Next on our virus tour is influenza or the 'flu'. Most of us have had the flu in our lifetime. It can knock you for six with achy and feverish symptoms. Each year, it affects up to 10% of the global adult population and 20–30% of children during seasonal epidemics.[9] There are four types of influenza: A, B, C and D. Influenza A and B viruses cause seasonal epidemics of disease in people; it's widely known as the 'flu season'.

Flu season hits during the autumn and winter months. This is because influenza viruses thrive in cold, dry air, making them more stable and easier to spread. Plus, when it's chilly outside, we huddle indoors, often in close quarters, making it much easier for the virus to jump from person to person. In addition, our immune systems can weaken during these months. All these phenomena create the perfect recipe for a flu outbreak. So, every year, as the leaves fall and the temperatures drop, the flu season kicks into high gear.

Influenza C is generally less severe than A or B, and influenza D mainly affects cattle and pigs, not humans. When it comes to thinking about pandemics, scientists are more concerned about influenza A. This is the type that includes avian influenza viruses. To complicate things, there's 'low pathogenic avian influenza' (LPAI), a relatively mild version, and highly pathogenic avian influenza (HPAI) – the nasty one. It spreads quickly and can make birds very sick, often leading to death. This type of flu can cause significant problems for poultry farms and can spread to humans.

Classifying flu viruses can get convoluted. It's like sorting keys by their unique shapes. Each key has a distinct pattern that fits a specific lock. The same goes for classifying flu viruses. Except that the keys are proteins on the virus's surface. You might have heard of the flu called 'H5N1' or similar. This code describes these surface proteins. Influenza A viruses are categorised based on the types of two proteins: haemagglutinin (where the 'H' comes from) and neuraminidase (where the 'N' comes from). These proteins are essentially keys that unlock different parts of the immune system. So far, scientists have identified 18 types of H's and 11 kinds of N's. In birds, there are sixteen H and nine N subtypes. These H and N subtypes can mix and match in numerous combinations, leading to various influenza A strains in avian species. This diversity is one reason that bird flu can be challenging to control and monitor. Subtypes H5 and H7 can mutate from low-pathogenic avian influenza into highly pathogenic avian influenza forms when introduced into poultry. This spells bad news.

While bird flu is rare in humans, with fewer than 900 cases in the past 20 years (mostly among poultry workers), experts are vigilant for mutations that could enable person-to-person spread.[10] If such mutations occur, bird flu could potentially become a pandemic much deadlier than COVID-19, which has killed over seven million people to date. Unlike COVID-19's roughly 1% case fatality rate, the H5N1 bird flu strain has killed over half of the people it has infected, highlighting its potency.

The H5N1 flu virus recently caused an outbreak in US dairy cows. As of the time of writing (July 2024), there have been 172 confirmed cases of infected dairy milking cows in 13 US states. This flu strain has also afflicted ten humans since April 2024. A recent *Nature* study indicates that H5N1 in cows might be more adaptable to humans, escalating concerns about the virus's spread.[11] Transmission between mammals, such as cows, might increase the risk of the virus mutating in a way that allows it to adapt to mammal-specific defences and physiology. The H5N1 strain is highly capable of infecting mammary gland cells that

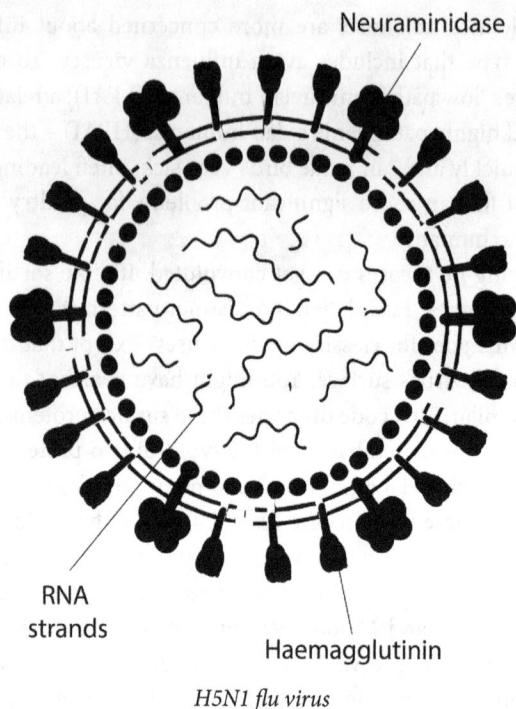

H5N1 flu virus

produce milk. Of course, humans also have this gland. Whether this poses an additional risk to us is yet to be determined.

H5N1 is a rapidly evolving virus. However, diagnostic testing and genetic sequencing are absent. This hampers our understanding of its mutations and potential human-to-human transmission. While I consider the likelihood of viruses like Nipah causing the next pandemic low, I believe a mutated flu virus is a top candidate. We must learn from past mistakes, like those made during the COVID-19 pandemic, and take the threat of H5N1 seriously. Many call for greater investment in rapid, equitable, affordable tests and enhanced surveillance. Will we learn? The only way to stop major pandemics is to acknowledge our mistakes and learn from them.

Alphavirus

It's time to imagine another virus, tiny and spherical, cloaked in a protective layer made of lipids stolen from the host cell it infects. Like the coronavirus, this outer shell bristles with spike-like proteins, perfect tools for attaching to and invading new cells. Inside, a single strand of RNA sits ready to commandeer the host's machinery. It guides the production of new virus particles and spreads the infection further. I'm talking about alphaviruses in the family of Togaviridae.

Alphavirus

Another candidate for the next pandemic is a strange-sounding alphavirus. It's called the Chikungunya virus, pronounced 'Chicken Gun Yer', although it has nothing to do with chickens. To find the origins of this disease, we need to jump in our tour van and head to East Africa. The name 'Chikungunya' comes from the Makonde language, spoken by the Makonde people of Tanzania and Mozambique; it means 'to become contorted' or 'that which bends up'. Scientists first described the disease during an outbreak in southern Tanzania in 1952.[12] Its name aptly describes the characteristic stooped posture of those affected by the illness due to the severe joint pain it causes.

Although the Chikungunya virus is a contender for causing the next pandemic, it's unlikely to be on the scale of COVID-19 or the flu. Spread by *Aedes* mosquitoes, which also transmit dengue and Zika, the virus is expanding its reach due to climate change and global travel. When people catch Chikungunya, they experience a sudden fever followed by severe joint pain, muscle aches, headache, nausea, fatigue and sometimes red rashes. Though the illness is rarely fatal, the joint pain can be relentless, lingering for months and making daily life a struggle.

Past outbreaks have affected millions, especially the 2004–2011 outbreak in the Indian Ocean and India.[13] There's no vaccine or specific treatment yet, so controlling mosquito populations and taking public health measures might be our best immediate defence. With mosquitoes spreading further and global travel rising, the risk of new outbreaks in previously unaffected areas is growing. Other alphaviruses, like the eastern equine encephalitis and Mayaro viruses, also pose threats. Still, Chikungunya's history and current impact make it a leading concern – but, in terms of the number of cases, rapidity of its spread and death toll, it ranks relatively low from a pandemic perspective.

Flavivirus

Flaviviruses are another group of viruses that have slipped into our world with the help of mosquitoes and ticks (and the help we've given to spread mosquitoes and ticks!). They're somewhat like alphaviruses in that they're spherical, cloaked in a fatty layer bristling with proteins and have a single strand of RNA that acts as the virus's blueprint. They cause a range of maladies, including dengue, Japanese encephalitis, yellow fever, Zika and West Nile disease – quite a jewellery box of doom. These diseases are often not life-threatening, typically causing fever, sometimes accompanied by rash or painful joints. However, a small proportion of those infected can develop severe or complicated infections. For instance, the Japanese encephalitis virus can lead to severe brain inflammation, and researchers have linked the Zika virus to congenital disabilities (see Chapter 1).

Some of these diseases are also candidates for a pandemic if conditions align. However, despite the threat and diversity of the diseases in this family, I think flaviviruses, like alphaviruses, rank relatively low on our pandemic likelihood and impact scoreboard. This shouldn't denigrate their importance, though. Thousands of people are affected by these viruses each year.

One with a rich history is yellow fever. Scientists don't know for sure, but yellow fever has likely affected people for at least 3,000 years.[14] Genetic evidence suggests it originated in the rainforests of Africa around 1000 BC. In the mid-twentieth century, the disease travelled on barges and sailing ships to tropical ports worldwide, following the slave trade to the Americas. The disease famously interrupted the building of the Panama Canal and left a trail of graves in its wake.

The initial attempt to build the canal was led by the French under Ferdinand de Lesseps, who had successfully constructed the Suez Canal. This phase (1881–1889) was plagued by difficulties.[15] Engineering challenges were rife, and the mortality rate among workers due to diseases was sky high. The lack of understanding about the transmission of these diseases resulted in a devastating loss of life, contributing to the failure of the French effort.

Cuban physician Carlos Finlay was the first to hypothesise that mosquitoes transmitted yellow fever. In 1881, he suggested that *Aedes aegypti* mosquitoes were responsible for spreading the disease.[16] However, his ideas were met with scepticism and were not widely accepted. The definitive proof came from the work of the US Army Yellow Fever Commission, led by Major Walter Reed, in Cuba. From 1900 to 1901, the commission conducted experiments that conclusively demonstrated that *A. aegypti* mosquitoes were the primary vectors of yellow fever.[17] The commission's experiments involved volunteers exposed to mosquitoes that had fed on yellow fever patients. Those bitten by infected mosquitoes contracted yellow fever, while those exposed to other conditions did not. These experiments provided clear evidence supporting Finlay's hypothesis.

Yellow fever starts like the flu. Symptoms include headache, fever, muscle pain, nausea and vomiting. However, about 15% of those infected develop a severe form of the disease.[18] This severe form causes high fever, jaundice, internal bleeding, seizures, shock and organ failure. Up to half of the people who progress to this severe stage do not survive. Yellow fever remains a significant threat in tropical regions, with approximately 200,000 cases annually. The disease is maintained through sylvatic and urban transmission cycles involving mosquitoes and primates. Vaccination with the 17D live attenuated vaccine is the primary preventive measure.

Due to its zoonotic nature, eradication of yellow fever is unlikely, but routine immunisation can substantially reduce the disease burden. Indeed, because of the mode of transmission, environmental restrictions, global surveillance and vector control, and effective vaccination, yellow fever is unlikely to cause a pandemic on the scale of COVID-19 anytime soon.

But wait! These things could change, right? The virus could mutate. Urbanisation and climate change could alter the vector's range. We could get complacent with surveillance, and we know vaccination programmes can be affected by global events like COVID-19. I've used the term before in this book, but 'perfect storms' do happen. Therefore, despite the current low risk, several factors could increase the likelihood of a yellow fever pandemic (and the same applies to other vector-borne diseases).

Rapid urbanisation in tropical regions creates ideal breeding grounds for mosquitoes, especially in areas with poor infrastructure where standing water accumulates. And guess what... urbanisation *is* rapidly increasing. Indeed, it's been one of the defining trends of the modern era, dramatically reshaping the global population landscape over the last several decades. Since 1950, the world's urban population has skyrocketed from 751 million to 4.2 billion by 2018.[19] This shift reflects a nearly six-fold increase in the number of people living in cities. Currently, over 56% of the global population resides in urban areas (well over 80% in some countries; Australia, where I live, is 86% urban). This trend shows no signs of slowing down. Projections indicate that by 2050, around 70% of the world's population will be urban dwellers.[20] This means the urban population is expected to reach about seven billion people. So, this risk factor for yellow fever is happening.

Climate change is also a major concern, as warmer temperatures and changing rainfall patterns allow *A. aegypti* mosquitoes to expand into new areas, potentially spreading yellow fever to previously unaffected regions. Warmer temperatures extend the mosquitoes' breeding and biting seasons, increasing the potential for disease transmission. In some areas, this has lengthened the transmission season by several months. In addition, climate change is causing more frequent and intense weather events. Floods create standing water, ideal for mosquito breeding, while droughts lead people to store water in containers, which are also perfect breeding sites for *A. aegypti*. Also, rising global temperatures allow *A. aegypti* to expand into higher latitudes and altitudes previously too cool for them. These regions include Southern Europe, North America and even parts of Australia. They're now seeing increased mosquito populations and disease outbreaks.

There's also a synthetic revolution going on, and not a good one. Many societies considered plastic a miracle invention of the twentieth century, providing a versatile material for many industries. However, it has become "the curse of the 21st century".[21] Its annual production soared from 2 million metric tons in 1950 to a staggering 381 million metric tons in 2015.[22] This production comes with massive amounts of waste. Although rarely considered in infectious disease epidemiology, plastic waste, ubiquitous in many urban areas, provides ample breeding grounds for *A. aegypti*. These mosquitoes often lay eggs in discarded plastic containers, cups, buckets, bottles and tyres, which collect rainwater. In 2019, the world produced approximately 353 million metric tonnes of plastic waste.[23] This figure reflects the widespread use of plastics in various sectors, from packaging to textiles. It also reflects our throwaway culture. Despite initiatives to reduce waste, by 2060, global plastic waste is projected to have almost tripled, reaching over one billion metric tonnes annually.[24] Another tick towards that perfect storm. But with this problem comes a solution – we must reduce plastic waste!

Increased global travel and trade can also raise the risk of outbreaks in new locations by facilitating the movement of infected individuals and mosquitoes. For example, during the 2016 outbreak in Angola, the virus spread to China via unvaccinated travellers.[25] Furthermore, disruptions in the supply or distribution of yellow fever vaccines can lead to shortages, leaving more people vulnerable to infection. In recent years, vaccine shortages during outbreaks in Angola and Brazil highlighted the fragility of global vaccination efforts.

And what about mutations? Unlike other RNA viruses, the yellow fever virus has a relatively low mutation rate. Mutations can affect a) *virulence*, making the virus more severe or deadly, and b) *transmission*, enhancing the virus's ability to spread between hosts.

The lower mutation rate of the yellow fever virus compared to many other RNA viruses is partly due to its replication mechanism. This mechanism includes a proofreading function that corrects errors during viral RNA synthesis. It's microscopic quality assurance! Unlike the error-prone replication in other RNA viruses, which results in about one mutation per ten thousand bases replicated, the yellow fever virus is significantly more stable. It averages only two mutations per ten million bases.[26] This lower mutation rate contributes to its relative genetic stability compared to other RNA viruses like influenza or coronaviruses.

Moreover, transmission depends on the *A. aegypti* mosquito, which adds a layer of complexity to the virus's spread. There's no evidence to suggest that the yellow fever virus is currently mutating in ways that significantly increase its virulence or ease of transmission. However, continuous monitoring is essential to detect and respond to any changes that might pose a higher risk.

Yellow fever is unlikely to cause the next pandemic. However, we're quickly gathering all the ingredients for the perfect storm (or the toxic cake?!) recipe. So, along with other mosquito-borne RNA viruses, it's one to watch. And the solution is to reduce and remove these ingredients.

Coronavirus

A colony of horseshoe bats flitted between the branches and tangled vines of the north-central forests in China. The bats harboured an invisible passenger: a virus.

In a wet market miles away, vendors sold their wares. Cages of exotic animals sat stacked beside stalls brimming with fresh produce. A trader reached into a cage to handle a wild animal. The animal hissed and scratched. Unbeknownst to the trader, the virus had found a new host. A brief scratch, and the pathogen crossed the species barrier. It was an invisible leap with monumental consequences.

Days later, in a crowded city street, the flow of people moved like a living river, each person brushing past another. Each passing came with a chaotic merging of microbial clouds, an accidental exchange of unseen hitchhikers. A woman coughed as she pushed through the crowd. Her hand reached for a subway rail. Her breath, carrying the virus, mingled with the air around her. Each touch and breath spread the invisible virus further.

In a sterile hospital room, a patient wheezed. The virus had multiplied, filling the lungs with fluid. Each breath was a battle. Despite layers of protection, his doctor felt the weight of the unseen enemy pressing closer.

And so, the virus jumped from one person to the next. It spread across the world like wildfire. It was aided by three phenomena: (1) the ubiquity and use of global transport systems, (2) the immense human hunger to disperse and connect, and (3) the unpreparedness of our governments and medical institutions.

We still don't know the exact details of how COVID-19 emerged, but it could have unravelled somewhat like I have described. Humanity stood still in 2019; our systems came crashing down. We were unprepared for this coronavirus, which shows in the figures – more than seven hundred million cases and seven million deaths (likely to be much higher).[27] But what exactly is a coronavirus, and could another cause the next pandemic?

Coronaviruses are a large family of viruses that can cause illnesses ranging from the common cold to more severe diseases. A few strains can cause common colds – they account for around 20% of cases of seasonal sniffles. Under the microscope, coronaviruses look like a tiny spiky ball or crown – hence the name 'corona', Latin for crown. These spikes (much like in our flu analogy) are like little keys that help the virus unlock and enter our cells, turning them into virus factories.

As with many of the viruses covered in this chapter, coronaviruses are RNA viruses, which means their genetic material is like a twisted ladder but with only one side. They're wrapped in a fatty layer that helps them stick to and invade cells. Once inside, they hijack the cell's machinery to replicate and spread. Ecologically, they thrive in various animals, including bats, camels, birds and humans. Bats are natural reservoirs for many coronaviruses. They often host the virus without getting sick, acting as mobile virus libraries.

Occasionally, these viruses jump from wild animals to humans, most often when we encroach on their habitat, handle wild animals in markets and so on. Hence, the likely beginning of the COVID-19 pandemic, as described earlier. However, it's probably worth briefly mentioning the 'lab leak' theory, too.[28] The lab leak theory suggests that the COVID-19 virus (SARS-CoV-2) may have accidentally escaped from a laboratory, specifically the Wuhan Institute of Virology in China, rather than originating from a natural spillover from wild animals to humans. Proponents of this theory point to the proximity of the lab to the initial outbreak and speculate about possible safety breaches and research activities involving corona-viruses (such as 'gain of function' research involving manipulating viruses to enhance their properties – e.g., increasing their transmission or virulence – to better understand potential threats and develop preventive measures). While this theory has been debated and investigated, most scientific evidence currently supports a natural zoonotic origin. However, some scientists have not entirely ruled out the lab leak theory.

Lab leaks aside, hundreds of coronaviruses are circulating in wild animal populations, but only seven are known to infect humans. These include 229E (alpha coronavirus), NL63 (alpha coronavirus), OC43 (beta coronavirus) and HKU1 (beta coronavirus).[29] These are all known to cause common colds. Three others, though, cause severe symptoms. These are SARS-CoV (severe acute respiratory syndrome coronavirus), MERS-CoV (Middle East respiratory syndrome coronavirus) and SARS-CoV-2 (causing COVID-19).

You may have seen these other two severe coronaviruses in the news headlines in the past couple of decades or so. SARS, caused by the SARS-CoV virus, emerged in 2002 in southern China. It's believed to have originated in bats and then jumped to humans through an intermediate host sold in live animal markets – a familiar story, right? SARS-CoV is a highly contagious coronavirus that primarily causes severe respiratory illness. The outbreak led to over 8,000 cases and nearly 800 deaths worldwide before it was contained through aggressive public health measures.[30]

The human-to-human transmission and significant mortality rate raised global concerns. SARS has been effectively contained (it's thought this lineage is now extinct).[31] However, due to the continued presence of coronaviruses in animal reservoirs and the potential for zoonotic spillover, a similar virus could emerge and cause another pandemic. Notice the high case fatality rate of SARS – around 10% compared to 1–3% for COVID-19. That's a scary figure.

So, why *didn't* SARS spread as far and wide as COVID-19?

The main reason is that SARS patients typically became highly infectious only after symptoms appeared. This made it easier to identify and isolate infected individuals before they could spread the virus widely. In contrast, COVID-19 patients can spread the virus even before showing symptoms (presymptomatic transmission) and when they have mild or no symptoms (asymptomatic transmission). This makes containment more challenging.

MERS, caused by the MERS-CoV virus, was first identified in Saudi Arabia in 2012.[32] This coronavirus is also believed to have originated in bats and then spread to humans through camels, which serve as the primary reservoir and intermediate host. MERS-CoV causes severe respiratory illness, often leading to pneumonia and multi-organ failure. It also has a higher fatality rate than SARS. The virus has caused sporadic outbreaks, primarily in the Middle East, with over 2,500 cases and around 850 deaths reported – a huge case fatality rate of 34%.[33] While MERS has not spread as widely as SARS, its high fatality rate and the presence of the virus in camels mean it could potentially cause a future pandemic. This is especially true if it mutates to become more easily transmissible between humans. Continued surveillance and research are crucial to prevent such an occurrence. It's this combination of easy and 'under the radar' transmission with a high case fatality rate that will be most devastating. No vaccine exists, highlighting the need for a One Health approach to control the spread of the virus.

Other emerging coronaviruses from wild animal reservoirs potentially threaten humans due to their ability to mutate and cross species barriers. Our tendency to encroach on 'wild' habitats and cause animal stress heightens the risk, but this also means we know how to reduce the risk. To some extent, it's a 'known known'.

The coronavirus family is large. We also don't know enough about them. What we do know is that coronaviruses can be unpredictable, highly transmissible and deadly.

Back in the forest, the cycle continues.

So, which virus is likely to cause the next pandemic? Based on the research I've done for this chapter, I would say that there's an imminent threat from highly pathogenic strains of influenza. The flu would be my top candidate if the next major pandemic were viral. However, another coronavirus comes in a close second.

CHAPTER 7

Can We ESKAPE this Bacterial Blitzkrieg?

It is the microbes who will have the last word.
—Louis Pasteur

Imagine playing a game of chess. You have a king, a queen, a couple of rooks, bishops and knights, and a row of pawns. Now, let's properly stretch your imagination. Instead of merely moving some glass, plastic or wooden pieces around the board to fight for the pride of winning a game, you're fighting for life. We can bend the rules of this chess game a little – a dash of poetic licence. The board is set. Your side, the black pieces, embody antibiotics positioned like vigilant knights, ready to strike against the oncoming host of bacterial invaders. Weird, right? Bear with me.

The white pieces – the pathogenic bacteria – are cunning and relentless. They orchestrate their moves like grandmasters. With a swift advance, you move one of your antibiotic knights, targeting a vulnerable pathogen. But the bacteria counter with dexterity, deploying mutations and biofilm pawns that fortify their defences. As the game intensifies, you launch your antibiotics in a coordinated assault; each move embodies a combination of therapies and strategic tactics. Yet, the bacteria adapt, sidestepping with their rapid evolution and horizontally transferred genes. The battle escalates with every turn as your side fights to corner your adversaries, aiming for the elusive checkmate that secures global health. It's a high-stakes duel – a true evolutionary arms race.

Currently, we're struggling in this chess game against our antimicrobial-resistant foes. In Chapter 5, I discussed how we could loosely view antimicrobial resistance as a pandemic. There's growing concern over several pathogenic bacteria being able to 'escape' the effects of antibiotics. 'Evade' might be a better word, but 'escape' has a special history and double meaning.

In 2008, Louis B. Rice published a paper titled 'Federal funding for the study of antimicrobial resistance in nosocomial pathogens: No ESKAPE'.[1] In the paper, Rice highlighted six bacterial pathogens that can 'escape' the impacts of antibiotic drugs. He emphasised the need for increased research funding and efforts to combat the threat posed by these highly resistant organisms, particularly in healthcare settings where they cause severe infections. You may notice something strange about the wording in the title. Rice used 'ESKAPE' rather than 'escape'. Of course, the 'K' and the capitalisation were deliberate.

ESKAPE is an acronym for the six pathogens of concern, namely, *Enterococcus faecium*, *Staphylococcus aureus*, *Klebsiella pneumoniae*, *Acinetobacter baumannii*, *Pseudomonas aeruginosa* and *Enterobacter* species.

Enterococcus faecium

It's quite remarkable to think we have trillions of microbes living in and on us at any given moment. As we've previously discussed, most of these microbes are harmless, and many are vital to our survival. Living in the shadows, harmless and unnoticed – this was the realm of *Enterococcus faecium*. This gram-positive bacterium coexisted in the intestinal tracts of humans and other animals, playing its part in the complex habitats of our bodies, our 'walking ecosystems'.

E. faecium was once considered a 'commensal', a benign companion that neither harmed nor helped its host. However, this unassuming bacterium has dramatically transformed over the past few decades. It has emerged from relative obscurity to become a menace in hospitals worldwide.

The story of *E. faecium*'s rise to infamy begins in the late 1980s and early 1990s. During this time, strains of the bacterium started to show resistance to vancomycin, one of the last lines of defence in the antibiotic arsenal. The appearance of vancomycin-resistant *Enterococcus* shocked the medical community. The chemical had been a reliable weapon against stubborn infections, including those caused by methicillin-resistant *Staphylococcus aureus* (MRSA). The emergence of vancomycin-resistant *Enterococcus* posed a major challenge, high-lighting *E. faecium*'s ability to adapt and survive in environments saturated with antibiotics. The bacterium's capacity for horizontal gene transfer allows it to acquire resistance elements from other microbes. This genetic agility has enabled *E. faecium* to thrive in hospital settings, where the heavy use of antibiotics creates the conditions for resistant strains to flourish.

While *E. faecium* might not be the harbinger of the next global pandemic, its resistance patterns and stealthy spread within hospitals spell trouble. It may not sweep across continents with rapid human-to-human transmission. However, its presence demands our vigilance, lest it finds new ways to exploit our weaknesses.

Staphylococcus aureus

The year was 2007. Students at Staunton River High School in Moneta, Virginia, were in shock as they gathered to honour their classmate Ashton Bonds, aged 17. Ashton had died after a week-long battle with an infection that seemingly came out of nowhere. It was MRSA, or methicillin-resistant *Staphylococcus aureus*.[2]

Determined to prevent another tragedy, students took to social media and organised a rally, insisting that the school be thoroughly cleaned before they returned. As they led school officials through the building, pointing out unsanitary conditions, it became clear that the threat was not isolated. Reports surfaced of MRSA in schools across Connecticut, Maryland, Ohio and Michigan, prompting closures and heightened hygiene measures.

The school outbreaks sparked concern nationwide. Statistics revealed that MRSA had claimed more lives in the USA than HIV/AIDS in 2005. Indeed, a study published in the *Journal of the American Medical Association* estimated that MRSA infections occurred in nearly 95,000 Americans in 2005, and an estimated 18,650 people died – a huge case fatality rate of almost 20% (20 times higher than COVID-19).[3] It's worth letting that sink in for a moment.

I found this to be surprising. Twenty per cent of people with the infection die? After reading several studies, I learned that one of the most severe forms of the disease, MRSA bacteraemia (presence of bacteria in the bloodstream), has a case fatality rate of up to 64%.[4] Elderly and immunocompromised people are most at risk.[5]

This sentence caught my eye in the summary of another study: 'Just one organism, methicillin-resistant *Staphylococcus aureus* (MRSA), kills more Americans every year (~19,000) than emphysema, HIV/ AIDS, Parkinson's disease and homicide combined.[6] Wow.

Staphylococcus aureus is arguably the most famous of the ESKAPE pathogens. It's a bacterial species that lives on the skin and in the nasal passages of many people without causing harm. However, it has been a known cause of infections for many decades. The development of penicillin in the 1940s was a breakthrough in treating *S. aureus* infections, but the bacteria quickly developed resistance, primarily due to the inappropriate use of antibiotics. By the 1960s, MRSA had emerged, becoming a major public health concern. MRSA is infamous for causing difficult-to-treat infections, especially in hospitals and other healthcare settings. It can lead to skin infections, pneumonia, bloodstream infections (the severe form of the infection mentioned earlier – bacteraemia) and surgical site infections.

As illustrated by the case of the student at Staunton River High School, MRSA has spread beyond hospitals into the community, affecting otherwise healthy individuals. The rise of MRSA has not only driven up healthcare costs but has also resulted in extended hospital stays and significantly worsened patient outcomes.

Isn't it already a pandemic? Some would argue yes.

If the primary characteristics of a pandemic include widespread geographic spread, high infection rates and significant impacts on public health and society, we could consider MRSA in the context of a pandemic. Like many pandemic diseases, it can spread from person to person. However, unlike airborne diseases, it mainly spreads via direct contact or contact with infected surfaces. It's also one of those nasty pathogens that can make more traditionally defined pandemics (such as COVID-19) far worse by causing secondary infections.

Staphylococcus aureus

Klebsiella pneumoniae

I was browsing patient stories on the website of the European Centre for Disease Prevention and Control and came across one for *Klebsiella pneumoniae*.[7] This is the story of Lill-Karin, a resilient 66-year-old retired schoolteacher from Norway. Lill-Karin has three grown-up children and five grandchildren and has been living alone since her husband passed away. She fills her days with reading, needlework, writing poems and travelling – a passion that has taken her on numerous adventures around the world. In 2010, Lill-Karin planned an exciting trip to Kerala, India, which she had never visited despite her many journeys to India. Incidentally, it's the same Kerala I mentioned in Chapter 6 – the epicentre of a Nipah virus outbreak.

Lill-Karin arranged to stay with a local family to immerse herself in the local culture. However, her trip took a dramatic turn right from the start. As her host drove her from the airport, their car collided with a lorry, leaving her with a severely broken leg. Lill-Karin spent two days in a crowded room at the hospital, surrounded by other sick patients. With limited mobility and only a bowl of water provided each day, she couldn't even wash herself or change her clothes in the stifling heat.

Eventually, she was moved to a private room and underwent an operation. In India, the family typically cares for the patient in the hospital, so she saw few doctors and nurses. Her host was her only company during this lonely period. After several challenging weeks, she was finally allowed to return home to Norway.

Klebsiella pneumoniae

Back in Norway, Lill-Karin was immediately admitted to the hospital again. She was isolated in a special room when the doctors discovered that she had contracted a highly antibiotic-resistant bacterium from the urinary catheter used during her surgery in India. It was *Klebsiella pneumoniae*. Although she didn't feel any symptoms of the infection, it could still be fatal. Everyone who visited had to wear protective clothing, and her family, cautious of the infection, kept their distance, particularly as there was a newborn baby to consider. She missed family weddings and baptisms, and the unhealed wound from her surgery only compounded her fears about her recovery and survival.

Through perseverance and medical care, Lill-Karin eventually overcame the infection and fully recovered. She was lucky, though. *Klebsiella pneumoniae* was responsible for 192,000 global deaths in 2019.[8] The rise of multidrug-resistant strains of *Klebsiella pneumoniae*, including those resistant to last-resort antibiotics like carbapenems, poses a growing challenge for healthcare systems worldwide.

Acinetobacter baumannii

When it comes to sheer number of species, soils are the densest habitats on the planet. Recent studies estimate that between 57 and more than 99.9% of all species on the planet live in soil.[9] Some bacteria have lived in the soil for thousands of years without a peep – most are harmless and play vital roles in the ecosystem; others are opportunistic pathogens. One group of bacteria, called *Acinetobacter*, are abundant in soils. *Acinetobacter* comes from Greek roots: *akinetos*, meaning 'non-motile', and *bakter*, meaning 'rod'. This etymology reflects the bacterium's non-motile nature, which distinguishes it from many other bacteria that move via flagella or other means.

One species of *Acinetobacter*, *A. baumannii*, has a hidden talent for causing trouble. Historically, *A. baumannii* wasn't typically harmful to healthy individuals. It had low virulence. This meant it didn't often cause disease. However, in the sterile, controlled environment of a hospital, it found new opportunities. In the mid- to late twentieth century, this bacterium began colonising intravenous fluids and medical equipment. It lay in wait for a chance to invade the bodies of the sick and vulnerable. Patients with weakened immune systems became prime targets for *A. baumannii*. Once inside these weakened hosts, the bacterium causes severe infections, often spreading through hospitals from one patient to another, leading to outbreaks. Once again, it's the antimicrobial-resistant strains that pose the greatest issue. Over the years, it has become increasingly resistant to the drugs designed to kill it, making infections more challenging to treat – the now 'usual story'. For patients already battling multi-organ disease, the arrival of a drug-resistant *Acinetobacter* infection can be devastating. In such cases, fatality rates have reportedly soared as high as 70%.[10]

Pseudomonas aeruginosa

My dog was restless, shaking her head repeatedly. Her ears kept flapping, and she pawed at them incessantly. I knelt, lifting her floppy ear gently, and was met with an unpleasant, sour odour and a noticeable discharge. This wasn't just a simple itch – something more sinister was at work.

Under bright lights, the vet peered into my dog's ear with a scope. "*Pseudomonas aeruginosa*," she said, with a quite serious expression. This bacterium, she explained, thrives in moist environments like a dog's ear canal. Its signature fruity smell and blue-green pigment are unmistakable signs. In fact, the latter is where part of its name comes from: *aeruginosa* is derived from the Latin word *aerugo*, meaning 'copper rust' or 'verdigris', a green pigment that forms on copper. This refers to the characteristic blue-green pigment, pyocyanin, produced by the bacterium, often visible in infected wounds or cultures.

Back home, the routine began. Ear drops twice a day, careful cleaning and a few treats to keep her still. The vet had prescribed a potent combination of antibiotics to fight this resilient pathogen. *P. aeruginosa* is notorious for its ability to form biofilms, creating a protective barrier that makes it resistant to many treatments. More on biofilms later.

Every day, as I administered the medication, I saw the stubborn bacteria fighting back. But slowly, surely, the redness faded and the discharge lessened. My dog started to shake her head less, and her energy returned. This particular chess game was turning in our favour.

P. aeruginosa is a fearsome foe, not just for pets but for humans, too. In hospitals, it typically afflicts the vulnerable – those with weakened immune systems or chronic conditions. It can cause severe infections, from pneumonia to sepsis, making its presence a serious concern. The bacterium tends to cause infections in different parts of the body, and bacteraemia (blood infection) has a case fatality rate of between 18% and 61%.[11] Once again, these are staggering numbers.

Like the other ESKAPE pathogens, *P. aeruginosa* infections could already be classed as a pandemic of sorts. They affect hundreds of thousands of people (and pets!) across the world.

Incidental Story About *Pseudomonas* and Eye Drops

In November 2022, a 72-year-old woman in Cleveland experienced an alarming decline in her vision, and when she was rushed to the emergency department, doctors discovered she had a severe corneal ulcer. The culprit? *P. aeruginosa.*

Infectious disease experts and microbiologists identified the source: contaminated EzriCare artificial tear eye drops. Treating the infection proved challenging due to the bacterium's resistance to common antibiotics. The patient was eventually treated with a powerful antibiotic called cefiderocol and two other topical antibiotics. While her condition improved, the full recovery of her vision remained uncertain.

This was not, however, an isolated case. Researchers from the Centers for Disease Control and Prevention (CDC), Food and Drug Administration (FDA) and various health departments uncovered 81 cases across 18 states, initially linked to an ophthalmology clinic in Los Angeles. The impact was severe. Nearly a third of the patients were treated at three healthcare facilities in different states. Tragically, four of the fifty-four patients with clinical cultures died within thirty days of their diagnosis. The infections were devastating for those with eye infections – four patients had their eyes removed, and an additional fourteen suffered significant vision loss.

While the CDC issued a warning about the eye drops, there is concern that people may still have them in their medicine cabinet.[12] So, there's a chance that people will still become infected in the future.

Enterobacter Species

And then there's the final 'E'. In the early twentieth century, scientists discovered a genus of bacteria named *Enterobacter*. It was initially identified in the intestines (hence the name, derived from *enteron*, the Greek word for intestine), but scientists found that *Enterobacter* species were more than just harmless gut residents. Researchers observed these bacteria in various environments, including soil, water, plants and animals. Their adaptability was noteworthy. However, it wasn't until the mid-1900s that the medical community started recognising their dual nature, both as commensal organisms and as potential pathogens, the old 'friend turned foe' scenario.

Let's imagine being in a hospital ward. An elderly patient with a catheter lies in bed, recovering from surgery. The sterile environment is meticulously maintained, yet *Enterobacter* can find its way through the tiniest breaches. It might be on a nurse's glove. It might be on a piece of equipment. It could even be in the hospital's water system. Once inside, this opportunistic pathogen can cause various infections, from urinary tract infections to life-threatening septicaemia.

Enterobacter species like *E. cloacae* are notorious in healthcare settings. These bacteria thrive in moist environments and persist on surfaces for extended periods, making infection control a significant challenge. Once again, infections are prevalent in patients with weakened immune systems, such as those in intensive care units or undergoing invasive procedures. *Enterobacter* is classed as an ESKAPE pathogen because it has evolved a way to break defences over time. It renders many antibiotics ineffective.

The battle against *Enterobacter* and other ESKAPE pathogens is ongoing. Researchers are exploring new antibiotics, alternative treatments and improved infection control measures. Understanding the behaviour and impact of these bacteria is crucial for developing strategies to combat their spread and protect vulnerable populations.

In summary, we can answer the question: What are the chances of an ESKAPE pathogen causing the next pandemic? The answer is that some scientists would argue they're already causing a pandemic. We're living through one right now. Hundreds of thousands to millions of people are dying across the world at the hands of antimicrobial-resistant bacteria. We can quibble over terminology, but there's an argument that we need to officially call this a pandemic in order to get the world to act more urgently.

By recognising the severity of antimicrobial resistance as a global crisis (and I know this is recognised in several circles), like COVID-19, we can prioritise the development of new antibiotics. We can also enforce stricter regulations on antibiotic use, invest in global healthcare infrastructure and, most importantly from my perspective, change our behaviours and the way we live with the land. Addressing this issue with the urgency it merits is crucial to preventing a future where simple infections become deadly and friends become foes.

Biofilms

As I walk through the busy Central Market in Adelaide, the high-intensity energy envelops me. Stalls are brimming with colourful fruits and fresh vegetables. Aromatic spices and diverse foods create a feast for the senses. Vendors call out to potential buyers, children laugh as they weave through the crowd, and the rhythmic clatter of knives chopping and coins clinking fills the soundscape. It's a hive of activity, teeming with life and motion. I can't help but think about the parallels of this visible aggregation of life with *biofilms* – structured conglomerates of microbes encased in a self-produced matrix that exhibits increased resistance to environmental stresses.

Each stall could represent a bacterial colony, distinct yet interconnected, working symbiotically with the wider community. The market's structure, with its organised chaos, mirrors the complex architecture of a biofilm, where microbes cling to surfaces and each other, forming protective layers.

The network of passageways is like the extracellular matrix in a biofilm, holding everything together and enabling chitchat and transport. And similar activities occur at the microscopic level. Some microbes exchange nutrients. Some microbes transport substances over great distances. Some compete, and others cooperate. Some microbes make a mess; others clean it up. Some microbes talk to each other through a process called quorum sensing.

Quorum sensing also plays a starring role in biofilm formation. Indeed, biofilms aren't just random gatherings of bacteria; they're structured, fortified colonies that pose significant challenges to human health. And much like well-organised human gatherings like rallies, biofilm formation requires good communication.

The beginning of a biofilm often goes something like this: solitary bacteria anchor themselves to a surface and start to chatter, releasing tiny chemical signals into their surroundings. As more bacteria join the gathering, this chemical conversation intensifies. The messages, carried by molecules known as autoinducers, swirl and accumulate. When the chatter reaches a crescendo, an extraordinary transformation occurs. Now aware of their numbers, the bacteria shift their behaviour in unison. They produce a sticky, protective matrix, encasing themselves in a communal fortress. This gooey shield, composed of polysaccharides, proteins and DNA, holds the colony together and shields it from external threats.

Within this biofilm, the bacteria thrive, 'talking' continuously to maintain their stronghold and ensure their collective survival. Quorum sensing is the language that turns solitary microbes into a formidable, organised community. But why is this relevant to human health and potential pandemics? Well, because biofilms shield the bacteria from external threats, they inherently boost their ability to withstand antibiotics and immune attacks.

Imagine a patient in a hospital recovering from surgery. Unbeknownst to them, a biofilm might be forming on their medical devices, such as catheters, prosthetic joints or heart valves. These biofilms are incredibly resilient. This makes infections associated with biofilms particularly difficult to treat. Biofilms are responsible for a significant percentage of chronic infections. For example, cystic fibrosis patients often suffer from lung infections caused by biofilms of one of our ESKAPE pathogens, *P. aeruginosa*.[13] In diabetic patients, biofilms can form on chronic wounds, preventing them from healing and leading to severe complications.

Now, imagine these bacterial metropolises resisting attacks and exchanging information. Bacteria within biofilms can transfer genetic material, including genes responsible for antibiotic resistance. This means that a single biofilm can become a breeding ground for superbugs – bacteria resistant to multiple antibiotics. In hospitals, where many patients are already vulnerable, the spread of such infections could be devastating. Imagine an outbreak of a biofilm-associated superbug in a crowded hospital. The bacteria could spread quickly, causing severe infections that are difficult to treat. The potential for biofilms to contribute to the next pandemic is enormous. They could increase the chances of deadly secondary infections. They could make the pandemic pathogen itself more challenging to treat.

Biofilms: Another Perspective

It's worth pointing out that biofilms aren't all bad. They play a crucial role in nature as essential components of ecosystems. These complex microbial communities form on surfaces such as rocks in rivers and roots of plants and within soil, where they contribute to nutrient cycling and plant health. In aquatic environments, biofilms act as a primary food source for various microbes and invertebrates, forming the foundation of the food web. On plant roots, biofilms facilitate the uptake of nutrients and water, enhancing plant growth and resilience against pathogens. Soil biofilms help to decompose organic matter, releasing nutrients back into the soil and maintaining soil health. Additionally, biofilms can protect plants by outcompeting harmful pathogens and promoting beneficial microbial interactions. Thus, biofilms are often integral to healthy ecosystems.

Scientists are investigating biofilms as a potential way of stopping the spread of superbugs like MRSA. I wrote about this in a recent paper called 'Probiotic cities: Microbiome-integrated design for healthy urban ecosystems'.[14] Co-author Richard Beckett has experimented with inoculating a 'friendly' bacterium, *Bacillus subtilis*, into ceramic wall tile. This has the effect of reducing the spread of MRSA.

Quorum Quenching

Remember how I said bacteria often talk to each other to form biofilms in a process called quorum sensing? Well, there's an opposite process called quorum quenching. If quorum sensing is the language of bacteria, then quorum *quenching* is the equivalent of hitting the 'mute' button. We think this is how *B. subtilis* fights against MRSA.

Quorum quenching refers to mechanisms that disrupt or inhibit quorum sensing, effectively blocking the communication between bacteria and preventing the coordinated behaviours that rely on this system.

Quorum quenching can occur in several ways. One method is to break down signal molecules using special enzymes, reducing their concentration and preventing them from reaching the threshold required to

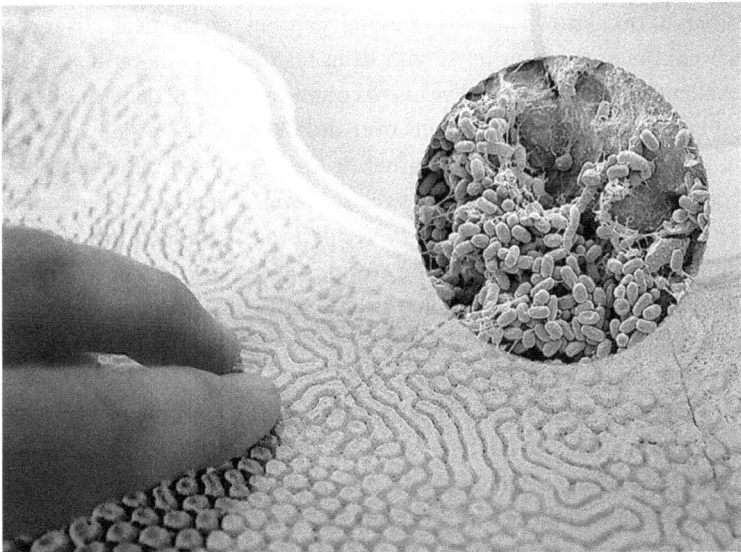

Richard Beckett's ceramic tiles inoculated with Bacillus subtilis

trigger a 'conversation' between the biofilm-forming pathogens. Other methods might involve blocking the signal receptor sites on bacteria or producing molecules that mimic the signal molecules, which could interfere with the quorum sensing process.

Many pathogenic bacteria rely on quorum sensing to initiate infection, forming biofilms or producing virulence factors. By interfering with this process through quorum quenching, we may be able to develop new strategies for preventing or treating bacterial infections.

MRSA (and other ESKAPE pathogens, like *P. aeruginosa*) use quorum sensing to form biofilms and produce toxins. A recent review suggested that *B. subtilis* produces quorum-quenching agents that can block the harmful effects of *S. aureus* in mice and its colonisation in humans.[15] Moreover, using quorum quenching instead of traditional antibiotics has an added advantage. Since quorum quenching targets communication instead of killing the bacteria or inhibiting their growth, it exerts less selective pressure for developing antibiotic resistance.

Other Bacteria

While the world should be vigilant against the ESKAPE pathogens, other bacterial pathogens could spark significant outbreaks, if not a pandemic. Take *Mycobacterium tuberculosis*, the ancient scourge behind tuberculosis, known as TB. Despite decades of medical advances, TB continues to claim lives, with drug-resistant strains posing a dire challenge to global health. I've heard countless stories of people noticing a persistent cough that deepens over the weeks. They initially dismiss it, attributing it to the cold weather or the lingering flu. But soon, the cough is accompanied by night sweats, a fever, fatigue and weight loss.

Unbeknownst to them, *M. tuberculosis* has found a home in their lungs. Like millions around the world, they're now a host to one of humanity's deadliest pathogens. Every breath they exhale carries the potential to infect those around them, perpetuating a cycle of transmission that has plagued humans for millennia.

The victims of TB often live in overcrowded apartments with poor ventilation and close quarters, allowing the bacterium to spread easily. Family members, neighbours and colleagues are all at risk, particularly older people and those with weakened immune systems. The bacterium lies dormant in some, waiting for a moment of vulnerability to strike. In others, it triggers a range of nasty respiratory symptoms. Despite the availability of treatments, *Mycobacterium* can withstand harsh conditions, evade the immune system and develop antibiotic resistance.

According to the World Health Organization, TB currently ranks as the second most deadly infectious disease, surpassed only by COVID-19 and outranking HIV/AIDS.[16] In 2022, approximately 10.6 million people were diagnosed with TB worldwide, including 5.8 million men, 3.5 million women and 1.3 million children.[17] TB affects individuals across all countries and age groups. However, it thrives in poverty, overcrowding and weakened healthcare infrastructures. Therefore, in a way, we're already at pandemic-level impacts. If the TB situation were to get considerably worse, it would probably be due to social, political and infrastructure issues that increase poverty and prevent effective treatments.

Similarly, *Vibrio cholerae*, the agent of cholera, thrives in areas with poor sanitation, causing severe dehydration and death if left untreated. Its rapid spread in low-income regions highlights a critical vulnerability. Meanwhile, one of our earlier unseen protagonists, *Yersinia pestis* (the bacterium responsible for the Black Death), still lingers in the shadows, poised to cause outbreaks under the right conditions.

On a different front, *Campylobacter jejuni*, commonly found in undercooked poultry, not only causes gastrointestinal distress but can lead to serious neurological complications like Guillain-Barré syndrome. Meanwhile, in healthcare settings, *Clostridioides difficile* lurks. It causes life-threatening colitis, particularly among those with recent antibiotic exposure. Over in water systems, *Legionella pneumophila* thrives and can cause severe pneumonia outbreaks in buildings with complex plumbing systems. These candidates are unlikely to cause a pandemic due to environmental restrictions. But remember, mutations can change this, and bacteria are adept at mutating!

A Quick Note on Zooanthroponosis

Some of the common bacterial pathogens of humans, such as *M. tuberculosis* and MRSA (and non-bacterial pathogens like influenza A virus, *Cryptosporidium parvum* and *Ascaris lumbricoides*), can be transmitted from humans to non-human animal populations. These spillover events are called 'zooanthroponoses', which translates to 'human diseases that affect animals'.

It's worth thinking about this because everything is connected to everything else. The way we live, the way we treat our environments, and the way we treat our bacterial infections can all impact wild animals. In turn, wild animal and environmental health can affect disease dynamics in nature, which can come back to bite us. This way of thinking underscores the One Health triad, something we'll discuss more later.

CHAPTER 8

A Fungal Frenzy – Could This Be 'The Last of Us'?

Fungi are the interface organisms between life and death.

—Paul Stamets, *Mycelium Running*

It's midnight, and you're lying in bed, wide awake. You're on high alert. Your sympathetic nervous system, the 'fight-flight-or-freeze' component, prepares your body to respond to threats. Neurotransmitters and hormones pulse through your bloodstream, increasing arousal and alertness. Your heart rate and blood flow are enhanced, glucose is released, and your pupils dilate to improve vision. This internal alchemy takes a toll on your body, especially when it happens night after night.

Bang! Bang! Bang! Boom! "Ahhhh!"

Gunshots, explosions and screams send a chill of unease down your spine. You pull your knees towards your chest and cup your hands over your ears, desperately trying to block out the noise. And then a moment of silence. Your brain rapidly tries to interpret this sudden change. You release your hands. A micro-moment of nostalgia washes over you as you recall peaceful times. But the feeling is short-lived as a rogue vibration invades this tranquil moment. As the vibration reaches your outer ear, your brain takes just a few milliseconds to interpret it. You know this sound all too well. "Grrraaahhh."

The gnarly growling sound is coming from downstairs. With it comes bad news. You already know what the next sound could be: crackling from footsteps crushing the broken glass you laid out on the stairs.

Crackle. Crackle. Crackle.

This event is what your body has been preparing you for. You leap into action, grab the gun next to the bed and point it at the door. You're waiting for either the crackling sound to stop and the growling to disappear, or for it to get louder as the source of the growling approaches the door from the other side.

It gets louder... and louder... and louder.

"Grrraaahhh!"

Your heart is pumping like a drum, pounding relentlessly in your chest as the adrenaline surges through your veins.

The door smashes open.

A grotesque figure lurches into the room, its eyes glazed with a malevolent hunger. You squeeze the trigger, the gunshot echoing in the confined space. The creature staggers but keeps moving forward.

You fire again and again, each shot pushing you further back against the wall.

Finally, the growling ceases, and the creature collapses at your feet.

Breathing heavily, you lower the gun. Your muscles are trembling from the adrenaline rush. The immediate threat is gone, but the noise will undoubtedly attract more of them. You know you must move quickly. Gathering your few essential supplies, you prepare to leave the relative safety of your room.

As you step over the lifeless form and head towards the window, you take one last look back. The world outside is a chaotic mess, but you're still alive. You climb out and make your way into the uncertain night; the distant growls are an ever-present reminder of the dangers that lie ahead.

The world was a different place before the outbreak. Streets were alive with the bustle of everyday life, children played in parks, and the hum of urban wildlife and human conviviality filled the air. But all of that changed when the first cases of the fungal infection appeared. Initially, scientists dismissed it as a minor health concern, something rare and unlikely to spread. However, this fungus, an insidious and aggressive variant of *Cordyceps*, quickly proved otherwise. It began with a few isolated incidents of people exhibiting strange, violent behaviour. Their eyes clouded over with a sinister glaze. News reports trickled in describing entire communities falling to the infection almost overnight. The fungus didn't spread through the air like a common virus. It had an even more horrifying method of transmission – direct contact with the infected or their spores.

As the infection spread, cities descended into chaos. Governments around the world declared states of emergency, but it was too late. The fungus moved faster than any containment effort, transforming its victims into monstrous, mindless beings driven only by the need to spread the spores further. Civilisation crumbled under the relentless advance of the infection. Those who survived did so by hiding in the shadows, scavenging for supplies and avoiding the infected at all costs. The once-thriving world was reduced to a landscape of abandoned buildings and haunted by the ever-present threat of the fungus. The last remnants of humanity struggled to hold on, facing the grim reality that they were fighting a losing battle against an unstoppable force.

Okay, let's get back to non-fiction mode!

As mentioned in the Introduction, we typically relegate the concept of a fungus causing zombie-like cannibalism and a global apocalypse to the realm of horror fiction. However, fungal diseases are responsible for nearly four million deaths annually – a figure that has nearly doubled in the last decade. So, are we already living through a fungal pandemic? And what are the chances of a fungus turning us into zombified cannibals? In this chapter, I made it my mission to find out.

The HBO Series – Spoiler Alert

The Last of Us is an HBO post-apocalyptic drama series based on a popular video game by Naughty Dog. The series is set 20 years after a global pandemic caused by a fungal infection (*Cordyceps*) devastates civilisation. This infection turns humans into aggressive, zombie-like creatures, leading to societal collapse. The story follows Joel, a hardened survivor portrayed by Pedro Pascal, tasked with smuggling Ellie, a 14-year-old girl played by Bella Ramsey, across a ravaged United States. Ellie is immune to the infection, and there's hope she might be the key to creating a vaccine. The series explores their harrowing journey as they navigate dangerous landscapes, confront hostile survivors and form a deep bond despite the grim circumstances.

Cordyceps Fungus

So, how does the fungal pandemic begin in *The Last of Us*?

As mentioned, the pandemic is caused by a mutated strain of the *Cordyceps* fungus, which originally infects insects. This real-life fungus manipulates its insect hosts to spread its spores, often resulting in the host's death. In the series, the fungus undergoes a mutation that enables it to infect humans, a dramatic leap from its natural setting and behaviour. This mutated *Cordyceps* invades human brains, taking control of their behaviour and turning them into aggressive, zombie-like creatures. Infected individuals spread the fungus through bites and airborne spores, leading to a rapid and devastating global pandemic. The series suggests that climate change might have facilitated this mutation, allowing the fungus to thrive at human body temperatures. The fungi flourish in specific environmental conditions that match the body temperature of their insect hosts. However, as global temperatures rise, the fungus must adapt to these changing conditions or die out.

A Note on Mutations and Natural Selection

Organisms, including fungi, constantly adapt to survive the ever-changing environment, driven by mutations and natural selection. Mutations are random changes in an organism's DNA, the molecular blueprint that dictates its structure and function. These changes can occur due to various factors, such as errors during DNA replication, or exposure to radiation or certain chemicals.

Let's take our *Cordyceps* as an example. One day, a random mutation occurs in the DNA of an individual fungus, altering its genetic code. This mutation might be detrimental, beneficial or neutral. If the mutation proves advantageous – for instance, making the fungus more resistant to heat – it will enhance its chances of survival and reproduction. Natural selection, often described as the 'survival of the fittest', is how advantageous traits become more common in a population over generations. In our example, the mutated fungus, with its newfound heat resistance, now thrives in a warmer environment. It reproduces more successfully than its non-resistant peers, passing the advantageous mutation to its offspring. Over time, the fungi population shifts, with the resistant strain becoming predominant. This process is not limited to fungi. It occurs in all living organisms, driving the evolution of species. In the wild, a mutation that gives a predator better camouflage might help it avoid detection by prey, increasing its hunting success. Similarly, a mutation in a plant that improves drought tolerance can ensure its survival in arid conditions, allowing it to propagate and outcompete less hardy plants.

Natural selection can also act on more subtle changes. For instance, there's a classic example of slight variations in beak shape among finches in the Galápagos Islands, allowing different species to specialise in eating different types of food. This reduces competition and promotes coexistence. These variations, first noted by Charles Darwin, are classic examples of how mutation and natural selection fuel the diversity of life. While mutations occur randomly, natural selection is anything but random. It systematically favours traits that enhance an organism's fitness – its ability to survive and reproduce in a given environment. Over countless generations, this relentless process sculpts species, refining their adaptations to the niches they occupy.

Mutation and natural selection are the twin engines of evolution, continuously shaping the natural world. However, the rate at which mutations and evolution occur can vary significantly between organisms. Just compare fast-reproducing organisms like fungi to longer-lived organisms like us humans. The *Cordyceps* fungus, for instance, has a relatively short generation time, often completing its lifecycle within days. This rapid turnover, high reproductive rates and the ability to produce numerous spores facilitate a high mutation rate and quick adaptation to environmental changes. This swift adaptability is evident in their ability to develop resistance to fungicides and other pressures in a relatively short time.

In contrast, more 'complex' organisms such as humans and other mammals have much longer generation times (the average period between an individual's birth and the birth of its offspring), typically spanning several years. For humans, the average generation time is about 20–30 years. Here's an excerpt from my book *Invisible Friends*:

> Greenland sharks, meanwhile, don't reach sexual maturity for 150 years – that's twice the average life expectancy of humans. Five successive generations of humans could live to the current average global life expectancy of 73 years before a Greenland shark even reaches sexual maturity. And just think how many microbial generations could roam the Earth in 150 years.[1]

While mutations do occur in these larger organisms, the rate of observable evolutionary change is slower due to the longer lifespans and fewer offspring per generation. As a result, significant evolutionary adaptations in mammals often take thousands to millions of years to become apparent. This disparity shows how adaptable microbes are. It also highlights the power of microbes to inflict damage by evolving new weapons and evading the weapons of others.

So, in *The Last of Us*, this hypothetical adaptation allows the *Cordyceps* fungus to survive and proliferate within the human body, which, at around 37 °C, typically has a higher temperature than that of insects, which can have an active body temperature as low as 10 °C. This evolution is pivotal to the series' premise. It bridges the gap between the natural infection of insects and the fictional infection of humans.

Zombie-Causing Fungi in Nature

The parasitic *Cordyceps* fungus lurks in the depths of the Amazon rainforest. Fungi typically spread by releasing spores into the air, up to 30,000 per second.[2] *Cordyceps* are no different in this respect. They release plumes of tiny spores that float in the humid air. Just one spore can potentially 'take root' and colonise a large patch of land, a larger organism or, indeed, a population of larger organisms. We often view fungi as 'feeding' on dead organic matter. However, some *Cordyceps* have a slightly different strategy and prefer to settle on unsuspecting living things, and in the following infamous example, the 'living thing' is an ant.

Despite the ubiquity of mutualistic symbioses (or 'friendly' interactions), we can also view the rainforest as a battleground for survival where organisms fight for their place. And fungi are no different.

One day, a single *Cordyceps* spore might land on a carpenter ant's body. The spore adheres to the ant's exoskeleton. It's a microscopic invasion unnoticed by the ant's colony. As the spore germinates, it penetrates the exoskeleton using mechanical pressure and enzymatic breakdown of the cuticle; it breaches the ant's defences, growing mycelium that infiltrates her body. This is just the beginning of a grim trajectory. The fungus then exerts a powerful biochemical influence over its host. A few days later, the ant starts to behave strangely (see the forthcoming sections). A mycelium network has infiltrated her muscles, and the fungus floods the ant's brain with chemicals. The neuroscience element is particularly fascinating.

One study found that the zombie ant's brain is actively preserved despite significant biochemical changes during the manipulation event.[3] The thing that preserves the ant's brain is a fungal-derived compound called ergothioneine, which is highly elevated in the zombie ant's central nervous system during infection. In other words, the fungus, which doesn't invade the central nervous system itself, secretes compounds that invade and preserve the brain. Presumably, this happens so that the fungus can manipulate the ant's behaviour for as long as needed. The success of the fungi often depends on their ability to consume the host while keeping the insect intact. This allows the insect to remain active enough to move towards favourable conditions for spore production and release.

The fungus manipulates the ant's behaviour in three steps:

1. Wandering
2. Summiting
3. Substrate adherence

Wandering Behaviour

Cordyceps-infected ants display a heightened, almost random wandering behaviour that hampers effective foraging efforts. This wandering causes the ant to stray from its colony before its family members notice the infection. The behaviour is no mere happenstance. It's all part of the fungus's manipulation strategy. If healthy ants in the nest notice an infected individual, they will attack and neutralise them as part of their 'social immunity'. It's much like our immune cells rushing to destroy an incoming pathogen. Therefore, the fungus must operate under the radar. Its survival depends on the infected ant stealthily but aimlessly wandering away from its nest. There are no ant biochemical equivalents of "See you later, guys", as communication is also hampered.

What are the mechanisms of this behavioural manipulation? Researchers are still unravelling the mechanisms, but the fungus likely releases chemical compounds called secondary metabolites. These compounds disrupt the ant's central nervous system, either mimicking or blocking the neurotransmitters that flick the switch on their behavioural pathways. In *Cordyceps*-infected ants, the fungus essentially rewires their genetic circuitry, dialling up or down the genes tied to the ant's hunger response. The fungus finds the control panel. It then tinkers with the settings to manipulate its host's actions to suit its own needs.

The fungal pathogen's depletion of the host's nutritional reserves can trigger a starvation state in the host. This induced starvation compels the host to alter its behaviour to seek out and replenish its nutrient stores, thereby influencing its wandering behaviour.

Moreover, fungal overgrowth interferes with the antennae of *Cordyceps*-infected ants, locking them in a bent L-shaped position.[4] Antennal movements are important for ant communication and navigation. They use their antennae to perceive the world around them. Therefore, fungal obstruction of the antennae likely inhibits these behaviours and contributes to the wandering behaviour that encourages the ant to move away from the nest.

Summiting Behaviour

So, the ant is now a pawn of the *Cordyceps* and is compelled to leave the colony's safety. The next manipulation step is 'summiting'.[5] During this phase, the ant is driven by an irresistible urge, implanted by the fungus, to climb vegetation. This behaviour ensures that the ant reaches an optimal height for the fungus to continue its lifecycle, positioning it in a prime location to eventually release its spores and infect new hosts. Researchers think that summiting increases transmission through more effective wind dispersal of spores.

Light-seeking behaviour accompanies the ant's upward movement. This behaviour points us to a potential mechanism. Given the consistent circadian timing of the manipulated ant's behaviours, it's likely that fungal compounds target the host's circadian clock. Light, the great stimulator of life, influences this biological ticker. Light signals synchronise the clock with the 24-hour day-night cycle in many organisms, including ants. Specialised light-sensitive cells in the eyes detect changes in light, sending signals to the brain. These signals help to reset and synchronise the internal clock with external environmental cues. This signalling ensures that bodily processes and behaviours, such as sleep-wake cycles and feeding, align with the appropriate times of the day.

The circadian clock in insects, including ants, is in the brain. Specifically, the hypothalamus and a cluster of neurons control the primary circadian pacemaker. In many insects, specialised neurons within the brain called 'clock neurons' partially control the circadian clock. These

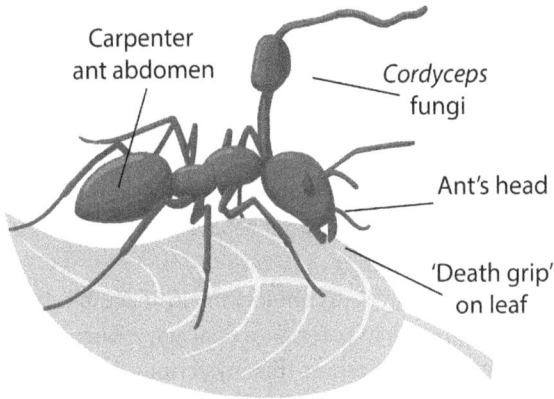

Carpenter ant abdomen

Cordyceps fungi

Ant's head

'Death grip' on leaf

Cordiceps-*infected ant*

neurons generate and regulate rhythmic signals, influencing behaviour and physiological processes such as feeding, activity and rest.

Some genes controlling the circadian clock appear faulty in zombie ants. Therefore, the fungus may secrete compounds that alter gene expression and, thus, light-dependent behaviours. But this doesn't fully explain how the ants seek light or height; it just means that it might play a role. To further unravel the mechanisms, we should consider phototaxis and gravitaxis. 'Photo' comes from the Greek for 'light', and 'gravi' comes from the Latin for 'heavy' or 'gravity'. 'Taxis' comes from the Greek for 'movement' or 'orientation'. Therefore, 'phototaxis' refers to movement in response to light, and 'gravitaxis' relates to movement in response to gravity.

Phototaxis and gravitaxis are important ways that organisms navigate using light and gravity. Phototaxis involves detecting light with their eyes and moving towards or away from it based on what they sense. Gravitaxis is about sensing gravity to help move up or down. These natural responses help organisms find the best spots for food, shelter or survival.

Cordyceps fungi may manipulate these systems using chemical compounds to alter ant behaviour.[6] In phototaxis, fungal compounds could mess with an ant's light-sensing signals, causing it to move in unusual ways – towards or away from light – to end up in a spot that's perfect for spreading the fungus's spores. For gravitaxis, the fungus might tinker with the ant's gravity-sensing signals or brain processing, pushing the ant to climb upwards. This summiting behaviour is vital for the fungus's lifecycle; again, it's all about helping it reach a height that maximises the spread of its spores.

Substrate Adherence Behaviour

Once the ant reaches a certain height during summiting, the third behavioural change kicks in: a biting behaviour known in the scientific community as the 'death grip'.[7]

This behaviour involves the ant locking its mandibles onto a leaf or twig, ensuring it remains in place for a final act. A blend of mechanical and chemical forces likely causes this mysterious death grip. Scientists exploring this phenomenon have discovered that the overgrowth of fungal tissue in the legs and mandibles plays a crucial role. The fungus weaves around the mandibular muscle cells, taking up about 40% of the space and causing muscle fibres to spread apart. It also secretes compounds that directly interact with the muscles. This hints at a chemical influence on the lethal grip.

As the ant's life fades, the *Cordyceps* fungus prepares for its dramatic exit. A stalk erupts from the ant's head, growing upwards and outwards, ready to release a new generation of spores into the wind. Now equipped to infect new victims, these spores drift down to the forest floor, continuing the cycle of life and death. Calling it a grizzly process would be an understatement. The *Cordyceps* fungus has fine-tuned its lifecycle to exploit its host, ensuring its spores reach optimal dispersal points.

Other types of fungal manipulation exist in other insects, too, such as thermotaxis (movement in response to heat) and increasing sexual behaviour to enhance fungal transmission via direct contact, to name a couple.

This brain invasion and manipulation are all fascinating and gruesome, but what's the likelihood of a similar fungus evolving to harm humans like in *The Last of Us*?

We can say that the likelihood of the 'zombie ant' fungus, *Ophiocordyceps*, evolving to infect humans is low, given the considerable biological barriers. The fungi are very selective, having evolved mechanisms tailored to their insect hosts' unique biology and behaviour over millions of years. This specialisation includes penetrating the ant's exoskeleton (its armour), manipulating insect-specific cellular processes and producing chemicals that interact with the insect's physiology to control behaviour. These adaptations are deeply rooted in the evolutionary history shared between the fungus and its insect hosts, making a sudden leap to infect humans highly improbable.

The fungus is also restricted by the host's body temperature. As mentioned, a marvellous array of biochemical processes keeps our bodies at around 37 °C – a temperature at which most fungi simply cannot grow. You may say, "Okay, but what about climate change? Won't a warming planet cause the fungi to adapt to warmer conditions, like our bodies?" Climate change is unlikely to cause a mutation that allows the fungus to thrive in or even tolerate such a dramatic temperature change. However, some fungi can and have been adapting to be slightly more tolerant of warmer conditions – *Candida auris* is one example.

The physiological differences between insects and humans are substantial. Insects possess an exoskeleton, open circulatory systems and immune responses that differ fundamentally from the human body's defences. The complex interactions between *Cordyceps* and its insect hosts involve developing specialised fungal structures and producing specific chemicals to manipulate host behaviour. These mechanisms might not function in the vastly different environment of the human body.

Human tissues, immune responses and cellular structures likely present insurmountable obstacles for the fungus, which is fine-tuned to exploit insect vulnerabilities. Even if mutations occur, the fungus will still face the challenge of overcoming the human immune system and finding ways to propagate within human tissues, which are not conducive to its current infection strategies. But never say never, right?!

Environmental factors also play a crucial role in the improbability of *Cordyceps* infecting humans. The fungus thrives in specific tropical and subtropical habitats with abundant insect hosts. These environments provide the precise conditions necessary for fungal growth, spore dispersal and host infection cycles. Urbanisation will engulf most of the world's population in the next few decades. Towns and cities are novel ecosystems. They're characterised by different climates, hygiene practices, living conditions and land cover, and may not support the lifecycle requirements of *Cordyceps*.

Therefore, while the idea of a 'zombie ant' fungus evolving to infect humans is intriguing and popular in sci-fi, the scientific reality presents numerous, most likely insurmountable barriers, at least for the foreseeable future.

However, incorporating climate change into the storyline, *The Last of Us* has a valuable message about human diseases. Although this specific case of zombie cannibalism is far-fetched, the story harbours a truth in that climate change causes far-reaching and often unpredictable consequences. The narrative suggests that pathogens previously restricted to specific hosts or environments might evolve to exploit new opportunities as the planet warms, and in this way, emerging pathogens could pose novel threats to human health.

Indeed, fungi have already evolved to infect humans, and climate change might exacerbate the spread and virulence of these pathogens.

Existing Human Fungal Diseases

We don't need to venture into the realms of sci-fi to encounter a grizzly, gory and life-threatening fungal frenzy. Global deaths from fungal diseases now far exceed those from malaria – historically, humanity's biggest killer. But they rarely feature in the news headlines.

So, who are the key players in this mouldy affair?

Several groups pose a major problem (including the increasingly problematic 'black fungus'). Let's talk about a couple of them: *Aspergillus* and *Candida*.

First, we have two species of *Aspergillus* fungi: *A. fumigatus* and *A. flavus*. You've probably noticed by now that I'm quite fond of etymology. Here's the lowdown for this fungus: the genus name *Aspergillus* comes from the Latin word *aspergillum*, and it's so named because when viewed under a microscope, the fungus resembles an aspergillum – an implement used to sprinkle holy water. The species name *fumigatus* derives from the Latin *fumigare*, meaning 'to smoke', referring to the smoky-grey colour of the conidia (spores) this fungus produces. The species name *flavus* comes from the Latin word for 'yellow', indicating the yellow-green colour of the conidia produced by this species.

Imagine walking through an ICU ward. A patient is battling their existing condition. But they're also fighting an unseen enemy that infiltrates their lungs – a fungal infection often overlooked until it's too late. Doctors who are familiar with the usual suspects might not immediately recognise the presence of *Aspergillus*. It's a mould that typically thrives in soil and decaying vegetation, waiting for an opportunity. In people with weakened immune systems, like organ transplant recipients or those undergoing chemotherapy, the delay in diagnosis is often deadly. Current tests only catch a third of these infections, leaving the fungus to spread stealthily to other organs. Each

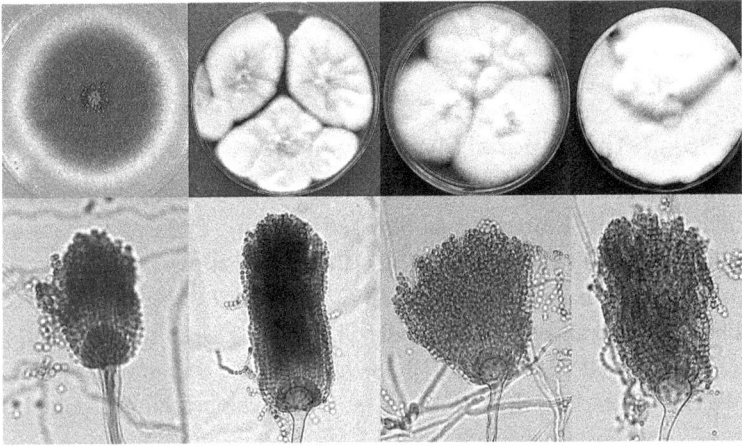

Aspergillus fumigatus *fungi in Petri dishes (top row) and under the microscope (bottom row) (SA Pathology / University of Adelaide)*

year, over 2.1 million people contract invasive aspergillosis, with a staggering 85% of these cases ending in death.[8]

Now, let's go to a different hospital ward. This time, a patient is dealing with *Candida*. This fungus has a different strategy. Often a harmless resident of the gut microbiome, this yeast can turn treacherous – from friend to foe – when the body's defences falter. *C. albicans*, typically a benign tenant on mucosal surfaces, can invade the bloodstream, leading to sepsis.

And then there's *C. auris*, a newcomer first identified in 2009.[9] This fungus is notorious for resisting multiple drugs and clinging stubbornly to hospital surfaces. It reminds me of the 'sit-and-wait' ambush parasites we covered in Chapter 4; only this time, the parasite is invisible.

It's not just the infections that are dangerous; it's their resilience and ability to exploit the gaps in healthcare, especially in settings where resources are scarce. With around 700,000 deaths annually and current blood tests detecting only 40% of these infections, there's a desperate need for improved detection and treatments.[10]

But are these pandemics?

Such fungal infections are creeping across continents, affecting countless lives. Their chronic, widespread nature and growing resistance to antifungal treatments give them the menace of a pandemic. They're a bit like nefarious shadows that follow other diseases. They thrive in those who are already weakened by illness – a storyline that feels all too familiar. So, in a way, they're pandemics of secondary infections. Some might say they're pandemics in their own right.

The National Mycology Reference Centre

"I must talk to a fungal disease expert," I said to myself. I searched the internet for the nearest fungal disease laboratory, thinking I'd have to travel to Melbourne or Sydney (725–1,400 kilometres away). However, as luck would have it, one of Australia's main labs is in Adelaide, where I live, and I arranged to meet medical mycologist Sarah Kidd.

At the National Mycology Reference Centre, Sarah gave me a tour. We entered the lab, where workers tested their colourful fungal colonies to see what species they are.

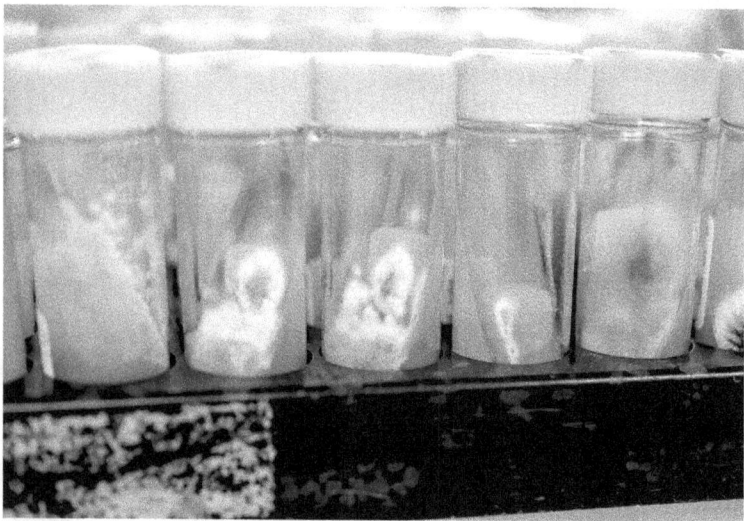

Tubes with fungal pathogens from sputum samples
(SA Pathology / University of Adelaide)

Sarah processed a sample 'under the hood' – a large cabinet that protects samples from contamination and protects the lab workers from the fungi. She used a special 'sticky tape' flag to collect the fungus from a tube. It looked like a see-through square on the end of a cocktail stick. She then placed it on a microscope slide, added a staining and preservative agent, and looked at it through the lens of a microscope.

"Ooh, I wasn't expecting that," she said.

This fungus needed further identification using mycology reference books, potentially followed by molecular methods like the PCR test scientists used during the COVID-19 pandemic. She gestured me to the workbench, and I looked through the eyepieces and saw a vivid purple image with a scattering of bean-shaped cells. The cells seemed so simple and benign, but they were probably the insidious agents of a disease deep in the caverns of someone's lungs.

Sarah then took me to a large incubator used to speed up the growth of the fungal samples. She said, "Most fungi don't survive at thirty-five degrees Celsius or above. This is good for us because they won't survive in our bodies. However, many can survive and cause infection in slightly cooler areas such as the skin and the eye's cornea."

I took photos of the samples and equipment, and we chatted in Sarah's office. "Can we talk about antifungal resistance?" I asked. Like antibiotic resistance, antifungal resistance is a growing threat. Spraying crops with fungicides to prevent plant diseases dramatically increases the resistance of fungal pathogens to a group of antifungal drugs called 'azoles'. Many agricultural fungicides are closely related to human antifungal drugs. In Europe, farmers use azole-based fungicides prolifically.

Sarah said, "*Aspergillus*, a common mould that thrives in soils, is becoming increasingly resistant to azole drugs, which are critical for treating human infections. This growing rate of resistance isn't just happening by chance; the widespread use of agricultural fungicides is driving it. These chemicals, often azole-based themselves, create a 'cross-resistance' effect, where the moulds exposed to fungicides in the environment also become resistant to clinical azoles." This overlap means that the very fungicides intended to protect crops are inadvertently making it harder to treat *Aspergillus* infections in people.

Sarah tells me that people simply walking in the countryside can now acquire drug-resistant fungal strains. *Aspergillus*-induced mortality rates are usually around 50%, but if it's an azole-resistant strain, it's more like 90%.

Another big problem, Sarah tells me, relates to the new drug Olorofim. The FDA hasn't approved it yet, but this will likely happen next year. Olorofim is a 'breakthrough' treatment for *Aspergillus* in humans, even the azole-resistant strains. So, it could save thousands of lives each year. There's a problem, though, at least in Australia. Authorities have approved a similar fungicide on strawberry farms. Evidence suggests that *Aspergillus* could become more resistant to Olorofim if exposed to the fungicide. This issue was eloquently covered in a recent news article titled 'The deadly price of a perfect strawberry'.[11]

Many fungi-related deaths are preventable. Indeed, any disease driven by structural social conditions (e.g., poverty) is preventable. Despite the fact that millions of people die from fungal diseases, they often receive far less attention and funding (less than 1.5% of all infectious disease research funding) than their bacterial or viral counterparts.[12]

But why?

Firstly, fungi are eukaryotic organisms, like humans. This makes it challenging to develop treatments that target the fungus without harming human cells. "Fungal and human cells have similar cell membranes and organelles in their cells," Sarah said. As a result, antifungal drugs must be highly selective to minimise toxicity to human cells. This adds a layer of complexity, which can put off funders and researchers, although I know plenty of scientists who love a challenge.

People may also perceive fungal infections as less threatening than bacterial and viral infections. Think about it. How many times have I said something like, "This disease mostly affects people with weakened immune systems" in this book? This fact may influence people's perception, making fungal diseases seem less relevant to the general population. Historically, medical research has prioritised bacterial and viral diseases. Have these microbes featured more prominently in pandemics and outbreaks? Yes. Think of the plague, Ebola and COVID-19. As a result, funding agencies and researchers have focused more on these areas.

"Do you think there's less public awareness and recognition of fungal diseases?" I asked Sarah, feeling pretty confident that the answer would be yes.

She said, "Many people are unfamiliar with the symptoms and impacts of fungal infections. This can lead to a lower perceived need for research and funding. But it's not just the general public who find it challenging to spot symptoms – doctors often misdiagnose these diseases." This is a great point. Fungal infections can be challenging to diagnose due to nonspecific symptoms and the need for specialised laboratory tests. This difficulty in diagnosing fungal diseases means they often go unnoticed, leading to fewer reported cases and people underestimating how widespread and severe they really are.

Infectious disease specialist Dr Justin Beardsley recently said, "Fungi are the 'forgotten' infectious disease. They cause devastating illnesses but have been neglected so long that we barely understand the size of the problem."[13]

Despite the challenges, people like Sarah are making waves. In addition to the lab's vital role in diagnosing fungal diseases, they work hard to produce educational resources, including books and workshops, to raise awareness. They even have a smartphone app called Medical Fungi to help clinicians make treatment decisions.

Here's hoping this chapter contributes, even in a small way, to raising awareness about the escalating threats posed by fungal diseases. It's not *The Last of Us* yet. However, remember the key message is that climate change and unsustainable human activities can drive microbial mutations. These mutations can turn minor problems into major crises.

CHAPTER 9

Protozoan Perils: Should We Eradicate Mosquitoes?

If you think you are too small to make a difference,
try spending the night with a mosquito.
—African proverb

I t's six o'clock in the evening, and you're setting the table in the garden, ready for a relaxing barbecue. The sun has its hat on, the air is filled with the songs of birds – the scene is idyllic.

But then… "Eeeeeeeee."

You violently spin around to escape the noise, flapping your arms and accidentally slapping the side of your face with the spatula you're holding. The noise continues. The flapping, spinning and slapping continue.

The distinctive high-frequency noise you're so desperately trying to escape is produced by the rapid beating of tiny wings, which can flutter up to 600 times per second. And who do the wings belong to? The dreaded mosquito, of course!

Though often seen as mere nuisances that ruin barbeques, mosquitoes play complex roles in ecosystems. As pollinators, they contribute to the reproduction of various plants, including some we rely on for food.[1] Their larvae serve as a crucial food source for aquatic organisms like fish and amphibians, while adult mosquitoes are prey for birds, bats and other insects. Thus, mosquitoes are interwoven into the fabric of many food webs, influencing the health of ecosystems.

However, mosquitoes are also infamous for their role in disease transmission. Species like *Aedes aegypti* and *Anopheles gambiae* are vectors for deadly diseases like dengue fever, and some we've already spoken about, like Zika virus, malaria and West Nile virus.[2] These diseases pose significant threats to human health, causing countless deaths and widespread suffering annually. The impact on public health and economies in affected regions is profound, driving extensive research into mosquito control and disease prevention strategies.

This brings us to the ethical debate. Should we eradicate mosquitoes? On one hand, eliminating mosquitoes could save countless lives and prevent suffering from mosquito-borne diseases. Advances in genetic engineering, such as releasing genetically modified mosquitoes that reduce populations, make this a tangible possibility. On the other hand, might eradicating mosquitoes disrupt ecosystems, leading to unforeseen consequences? It's feasible that losing mosquitoes might deprive other species of food sources, potentially causing cascading effects throughout food webs.

It's a complex ethical dilemma. We must balance the immediate benefits to human health against the potential long-term ecological impacts. We'll dig into the details soon, but first, let's talk about protozoa.

Protozoa

Let's pretend you have a pond in your garden and a microscope in your house. If you actually have a pond in your garden and a microscope in your house, great! You can nip outside and do the following for real.

You've collected a small amount of pond water with a pipette or spoon, and now you're looking at a single drop under a microscope. You might see an amoeba shapeshifting through the water, extending its pseudopodia (an arm-like projection) to engulf food particles. Nearby, a paramecium might glide gracefully, propelled by the rhythmic beating of its cilia, sweeping bacteria and other nutrients into its gullet. Meanwhile, a flagellate whips its tail-like flagellum to propel itself forward. It navigates the microscopic landscape with precision. These gliding and spinning creatures are protozoa.

Protozoa are single-celled organisms from the kingdom Protista. Their name, derived from Greek – *protos* for 'first' and *zoon* for 'animal' – paints them as early, almost prototype versions of animal life. If you have a microscope, you can spot them just about everywhere, drifting through ponds, oceans and soils or even nestled within the bodies of other creatures. Equipped with cilia, flagella or pseudopodia, protozoa are like microscopic acrobats. They glide and wriggle their way through their environments. They're hunters, too. They engulf their prey – bacteria, algae or other tiny organisms – by phagocytosis, which literally means 'cell eating'. They play dual roles as both predator and prey in microbial food webs.

Beyond their life as microscopic hunters, protozoa are crucial recyclers. They break down complex organic matter into simpler nutrients that plants and other organisms absorb. In this way, protozoa help keep the wheels of life turning. They enrich soils and help sustain aquatic ecosystems. Some protozoa are symbiotic, living in close association with other organisms. For example, certain protozoa reside in the guts of termites and ruminants, aiding in the digestion of cellulose and other complex carbohydrates. This 'friendly' relationship is crucial for the nutrition of these animals.

However, not all protozoa are benign. Several species are pathogenic and can cause diseases in humans and other animals. Notably, the protozoan *Plasmodium*, transmitted by *Anopheles* mosquitoes, causes malaria, a major health concern in many tropical regions. Other pathogenic protozoa include *Giardia lamblia*, which causes giardiasis (an intestinal infection leading to symptoms like diarrhoea, abdominal cramps and nausea), and *Entamoeba histolytica*, responsible for amoebic dysentery.

Plasmodium

The evolution of *Plasmodium*, the protozoan parasite responsible for malaria, stretches back millions of years. It's deeply entwined with the evolutionary history of its mosquito vectors and vertebrate hosts. We think the origins of *Plasmodium* date back at least 100 million years to the Cretaceous period, around the same time when early mammals and birds were diversifying.[3] Genetic studies suggest that *Plasmodium* evolved from a common ancestor shared with other parasites known for their complex lifecycles and ability to infect a wide range of hosts.

The split that led to modern *Plasmodium* species likely occurred when their ancestral parasites began exploiting blood-feeding insects as vectors, allowing them to move between vertebrate hosts more efficiently. Over time, *Plasmodium* species adapted to specific hosts (including *Homo sapiens*) and vectors, leading to the diversity of malaria parasites seen today.

Among the various *Plasmodium* species, *P. falciparum* and *P. vivax* are the most significant in human malaria. *P. falciparum*, responsible for the most severe and deadly form of malaria, likely jumped to humans from gorillas (*Gorilla* spp.) more than 50,000 years ago, coinciding with when humans first migrated out of Africa.[4] This close association with human hosts and habitats aided the parasite's spread and persistence. It was a rare spillover event, but one that would lead to millions upon millions of deaths.

Genetic evidence suggests *P. vivax* may have infected early apes and hominins in Africa before spreading to Asia and other parts of the world.[5] Scientists previously thought that *P. vivax* emerged in Asia.

There was a paradox because mutations conferring resistance to *P. vivax* occurred at a high level in the region where this parasite was absent (in Africa). However, recent evidence suggests a plausible scenario where an ancestral *P. vivax* strain was capable of infecting humans, gorillas and chimpanzees (*Pan troglodytes*) simultaneously in Africa (rather than jumping between many species like *P. falciparum*) until around 30,000 years ago, when a mutation emerged and eliminated *P. vivax* from humans in the region.[6] Human migration and the adaptation of mosquito vectors to various climates and environments aided the spread of mosquitoes to Asia.

Plasmodia have a complex lifecycle that could rival most movie thriller plots. The journey begins in an infected female *Anopheles* mosquito. When she takes a blood meal from a human, she injects *Plasmodium* sporozoites – the infectious form of the parasite – into the bloodstream. These sporozoites rapidly travel to the liver, where they invade liver cells (hepatocytes). There, they undergo asexual replication and mature into merozoites. After several days (or weeks, depending on the species), the liver cells rupture, releasing merozoites into the bloodstream. These then invade red blood cells, initiating a destructive cycle of infection, replication and red cell rupture that causes the symptoms of malaria. In some species, such as *P. vivax* and *P. ovale*, a dormant form called a hypnozoite can form, leading to relapses months or even years later. Some merozoites differentiate into sexual forms called gametocytes, which circulate in the blood. If another mosquito bites the infected person, it ingests these gametocytes, allowing the cycle to continue in the insect's gut.

What makes *Plasmodia* particularly menacing is their ability to evade the immune system. They avoid detection and destruction by constantly altering the expression of proteins on infected cells. This cunning strategy makes the parasite a master of disguise. It also complicates the body's immune response and presents significant challenges for vaccine development.

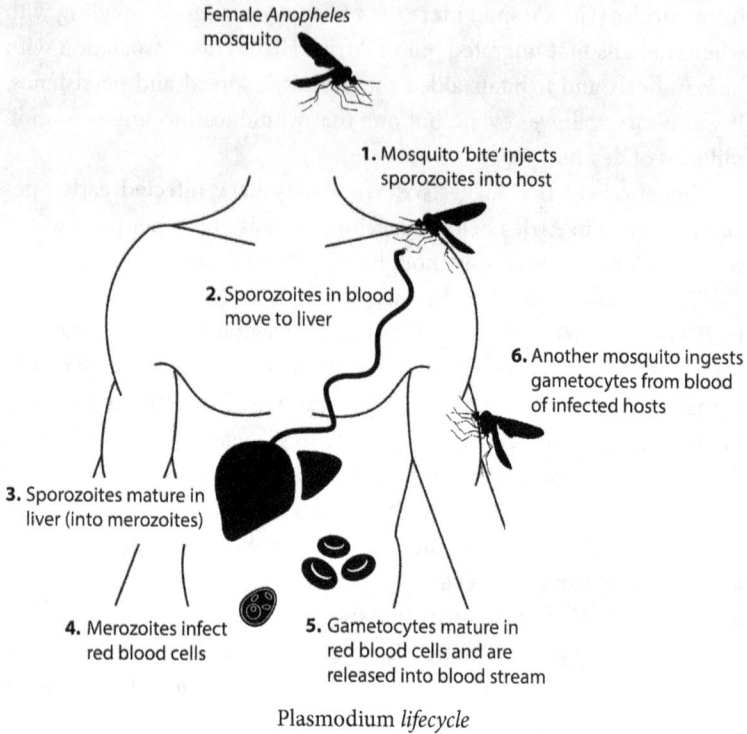

Female *Anopheles* mosquito

1. Mosquito 'bite' injects sporozoites into host

2. Sporozoites in blood move to liver

6. Another mosquito ingests gametocytes from blood of infected hosts

3. Sporozoites mature in liver (into merozoites)

4. Merozoites infect red blood cells

5. Gametocytes mature in red blood cells and are released into blood stream

Plasmodium *lifecycle*

Malaria

It all begins with a 'bite'. A mosquito's proboscis penetrates the skin, and from this tiny puncture, so much suffering can follow. A single mosquito bite can inject thousands of *Plasmodium* sporozoites into the body. That 'eeeeeeeee' is an annoyance at your barbecue in the malaria-free UK. But now imagine that you were dealt different cards in life. Certain demographic factors turn the annoyance into a matter of life and death.

For instance, in a different life, you may live in Sub-Saharan Africa, where 94% of global malaria deaths occur.[7] You might also be a child under five. Children under five are the most vulnerable, accounting for around 70% of all malaria deaths worldwide.[8] You may also live in a low-income community. Poverty-stricken areas often lack adequate healthcare facilities, access to preventative measures (like bed nets) and effective treatment. According to the World Health Organization, there were an estimated 249 million malaria cases worldwide in 2022, resulting in approximately 631,000 deaths.[9]

The term 'malaria' has its roots in the Italian language, derived from the medieval Italian words *mala aria*, which means 'bad air'. This no-menclature dates back to when people believed the noxious vapours emanating from swamps and marshes caused the disease.

Symptoms typically appear 10–15 days after the infectious bite. The infected person experiences episodes of high fever with intense shivering and sweating, and then a subsequent return to normal temperature. The episodes often cycle every 48–72 hours, depending on the specific *Plasmodium* species involved.

Alongside these classic symptoms, patients commonly experience severe headaches, muscle aches, fatigue, nausea, vomiting and abdominal pain. As the disease progresses, untreated malaria can lead to anaemia due to the destruction of red blood cells and jaundice from liver involvement. In severe cases, complications can include cerebral malaria, manifesting as seizures and confusion, acute respiratory distress and organ failure.

Malaria has plagued humanity for thousands of years. Indeed, the earliest known references date back to 2700 BCE in ancient Chinese writings.[10] In ancient Egypt, malaria's impact is evident in the writings and medical papyri from around 1550 BCE, such as the Ebers Papyrus, which outlines treatments for fever, a common symptom of malaria.[11] Furthermore, feverish conditions, possibly linked to the disease, were depicted in ancient Egyptian art. The ancient Greeks, including the famous physician Hippocrates, documented malaria in their medical treatises around 400 BCE.[12] Hippocrates meticulously described the in-termittent fevers and the characteristic timing of the fever cycles. He distinguished between different types of fevers that likely corresponded to various *Plasmodium* species.

Archaeological findings provide additional evidence of malaria's ancient presence. DNA analysis of skeletal remains found in regions of the Roman Empire, particularly in sites such as the Roman harbour town of Lugnano, revealed traces of *P. falciparum*, the deadliest malaria parasite. These findings suggest that malaria was a significant health issue in ancient Rome, contributing to the decline of populations and possibly influencing the empire's eventual fall. Indeed, several articles on the internet pose the question: 'Did malaria bring Rome to its knees?'[13]

Today, malaria costs Africa an estimated £9 billion annually in lost productivity, healthcare costs and decreased economic growth.[14] And, of course, it kills many thousands of people each year. However, it's an unlikely candidate for the 'next pandemic'. This is because malaria is transmitted through the bite of infected *Anopheles* mosquitoes rather than through person-to-person contact. This transmission mode limits rapid and widespread dissemination – characteristics of pandemics, which often involve airborne or highly contagious pathogens that can easily spread among people. Additionally, malaria is largely confined to specific geographic regions where these mosquitoes are prevalent.

A Day in The Life of an *Anopheles* Mosquito

As dawn breaks over the humid marshlands, a female *Anopheles* mosquito stirs. Driven by an insatiable urge to nourish her developing eggs, she takes flight and the 'eeeeeeeee' begins. Guided by the warmth, carbon dioxide and other compounds emitted or exhaled by her targets, she navigates towards a cluster of human dwellings at the edge of the marsh.

Her first bite of the day pierces the skin of an unsuspecting sleeper, but the sleeper stirs, swatting her away. The interruption has left her hungry and determined; she needs a full meal to ensure the survival of her offspring. As the day progresses, she rests in cool, shaded spots, avoiding the midday heat. Her body is built for twilight and dawn, when the air is cooler, and humans are more likely to be still and unaware of her presence. As dusk approaches, she resumes her quest, flitting from one target to another.

To make it easier for her to feed, as she bites, she secretes saliva that contains a mix of anticoagulants, vasodilators and anaesthetics. The anticoagulants prevent the blood from clotting, so she can draw blood without obstruction, and the vasodilators dilate blood vessels, increasing blood flow. The anaesthetics help numb the area, making her target less likely to feel her bite immediately.

Finally satiated, she flies back to the marshlands, her abdomen swollen with blood. She will digest this meal over the next few days, converting it into nutrients to develop her eggs. Soon, she will lay her eggs in the marsh's stagnant waters, continuing the lifecycle.

In optimal conditions, the average lifespan of a female *Anopheles* mosquito is around three weeks.[15] Even with a shorter lifespan of around two weeks, a female mosquito can still infect multiple people. After the mosquito has an infectious blood meal, it takes about 8–14 days (the incubation period) for the malaria parasite to develop within the mosquito before it can be transmitted to another person. A female mosquito will feed on human blood until she's full and then nourish her eggs. A few days after her blood meal, she can lay 100–200 eggs.

So, we know the mosquito transmits the parasite that causes malaria, but what do we do about it? Should we try to eradicate mosquitoes?[16]

It's important to point out that there are approximately 3,500 species of mosquitoes worldwide. Most of them don't bother humans. Many are vital pollinators that help sustain the biodiversity that we rely upon. It's also important to say that of the 3,500 known mosquito species, fewer than 6% are known to regularly feed on humans to help develop their eggs and provide the nutrients required to flourish. Moreover, of the small proportion of mosquito species that bite humans, only about half are known vectors of pathogens that cause human diseases.[17]

Therefore, trying to eradicate *all* mosquito species would be:

a.   Utterly unethical
b.   Utterly impractical
c.   Utterly foolish from an ecological perspective

However, that leaves us with the question: 'Should we eradicate the species that transmit the malaria-causing *Plasmodia*?'

Views on this issue typically fall into the 'yes' and 'no' camps. As always, there are probably some 'in-betweeners' too.

The 'Yes' Camp

In the battle against malaria, efforts to educate communities on using treated nets and employing various tactics to avoid mosquito bites are continuous. These measures are crucial, yet they require relentless vigilance and resources. But what if there was a more straightforward, more definitive solution? What if we could eradicate an entire species of disease-carrying mosquitoes?

Around half of the world's population is at risk of mosquito-borne diseases. That's a huge proportion. Now, imagine a world where children no longer suffer from feverish nights, where parents are free from the constant fear of losing their child to malaria and where entire regions are liberated from the economic burden of this disease. If you can imagine this, you can probably see how eradicating a disease-carrying species becomes palatable to some. Instead of ongoing education and prevention efforts, wouldn't it be more effective to eliminate the threat at its source by making disease-carrying mosquitoes extinct? At the very least, the potential to save millions of lives and transform the futures of countless communities makes it a discussion worth having.

Eradicating an entire species is known as 'specicide'. In a *New York Times* interview, biologist Olivia Judson said, "We should consider the ultimate swatting."[18] Olivia has advocated for the specicide of 30 mosquito species, arguing that this drastic measure could save one million human lives while only reducing the genetic diversity of the mosquito family by a mere 1%.

In the early twentieth century, Fred L. Soper spearheaded a campaign to eradicate the *A. aegypti* mosquito, a vector for yellow fever and malaria.[19] Soper was an ardent advocate for vector eradication over disease control, initially working with the Rockefeller Foundation in Brazil and later as director of the Pan American Health Organization (PAHO), his ambitious efforts aimed not just at reducing mosquito populations but eliminating them. Vector eradication was a strategy that marked a significant shift in public health tactics.

Soper's approach to mosquito eradication included several notable campaigns. In the late 1930s and mid-1940s, he attempted to eradicate the highly efficient malaria vector *A. gambiae* from Brazil and Upper Egypt. Utilising pre-DDT insecticides and anti-larval methods, Soper succeeded in halting malaria epidemics in these regions. However, the results were ambiguous, as the *gambiae* mosquito had only recently arrived in Brazil and Egypt and was not fully embedded into the local ecosystem.

From 1946 to 1950, Soper led a campaign to eradicate the indigenous malaria vector *Anopheles labranchiae* on the island of Sardinia using the newly discovered DDT. While malaria was eliminated, the mosquito population persisted, revealing the challenges of eradicating well-established species. Moreover, many people are now aware of the negative impacts of DDT on ecosystems (refer to *Silent Spring*, mentioned in Chapter 2). Soper's most ambitious project aimed to eradicate *A. aegypti* from South and Central America. Despite initial success, with many countries declared mosquito-free, the campaign ultimately faltered due to reinfestation and logistical challenges.

If nothing else, these past efforts highlight the immense complexity of attempting to eradicate a species. But have technological advances and our understanding of mosquito ecology changed the game?

In a modern-day effort to combat mosquito-borne diseases, scientists at Oxford University and the biotech firm Oxitec have genetically modified males of the *A. aegypti* species.[19] These genetically modified males carry a gene that prevents their offspring from developing properly. This causes them to die before reaching maturity, reducing the disease-carrying mosquito population. Between 2009 and 2010, approximately three million modified mosquitoes were released in the Cayman Islands. The results were striking. Oxitec reported a 96% reduction in the mosquito population compared to nearby areas. Following this success, a similar trial in Brazil is underway and has already achieved a 92% reduction in mosquito numbers. These trials offer a glimpse of a future where we could significantly curtail or even eradicate mosquito-borne illnesses.

In 2018, researchers developed a CRISPR-based gene drive targeting the doublesex gene in *A. gambiae* mosquitoes, the primary vector for the *P. falciparum* parasite responsible for malaria.[20] By de-activating this gene in female mosquitoes, researchers caused them to develop both male and female organs. This rendered them infertile and their proboscis was unable to pierce human skin. While male mosquitoes solely feed on flower nectar and are unable to transmit *Plasmodium*, female *Anopheles* require blood to lay eggs. When the modified doublesex gene was introduced into a caged population of *Anopheles* mosquitoes, the population collapsed within seven to eleven generations. This showcased the technique's potential to eradicate the species and malaria. Some see it as a promising method. However, it's not without controversy.

Jo Lines, a leading expert in malaria control and vector biology at the London School of Hygiene and Tropical Medicine, emphasised the potential for mosquito extinction as a solution to disease transmission in a *Guardian* interview: "There are extinction options. It wouldn't be easy, but we shouldn't forget about it." Lines believes that eradicating *A. aegypti* is particularly urgent. "There is no visible end to this except a war against *A. aegypti*," he warns. "Otherwise, this will go on for a thousand years."[21]

The challenge is clear to the 'yes' camp. Without decisive action, the threat posed by mosquitoes will persist indefinitely, and a comprehensive and aggressive approach is needed to eradicate this resilient vector once and for all.

Wise and Borry reviewed the moral status of mosquitoes and the ethics behind eradication.[22] They explain that while we often ascribe a low or no moral status to mosquitoes, we face an ethical quandary regarding their eradication. This quandary is complicated by the distinction between killing individual mosquitoes and purpose-fully eradicating an entire species. They point to philosopher David DeGrazia's models, which highlight this nuance. His two-tier model grants full moral status to humans and lower but existent status to other sentient beings. Conversely, the sliding-scale model ranks organisms based on cognitive, affective and social complexity, placing humans at the top and non-sentient beings at the bottom.

World-renowned ethicist Peter Singer supports this hierarchical approach.[23] It fully considers sentient and self-aware organisms while reducing the ethical importance of supposed insentient ones. Biologist Michael Soulé and philosopher Tom Regan, however, argue for the intrinsic value of all species, citing their evolutionary heritage and potential.[24] Yet, it's debated whether insects, including mosquitoes, have enough sentience to warrant protection. Singer sees no immediate need for insect rights advocacy.

When considering biodiversity, the impact of eradicating some mosquitoes appears minimal. Only about 30–40 of the 3,500 mosquito species transmit malaria.[25] Thus, targeting these species would reduce mosquito biodiversity by around 1%. Specifically, eradicating the malaria-transmitting *A. gambiae* species used in gene-editing trials would potentially have an even smaller impact.

The 'No' Camp

The bold idea of mosquito eradication isn't without controversy. We must carefully consider the ecological consequences of removing a species from the environment and the ethical implications of wielding such power over nature. To those in the 'no' camp, the precedent set by eradicating one species raises major concerns. For instance, it could justify the eradication of other species.

Based on our past environmental behaviour, I can envisage this evolving into a shifting ethical baseline where we might begin to eradicate mere 'nuisance' organisms. It could be non-disease-causing but barbecue-ruining mosquitoes. It could be badgers because they dig holes in your lawn. It could be the hooting owl that lands on the lamppost outside your house each night. It could be the bats you don't like roosting in your roof space. It could be the raccoons rummaging through your rubbish bins or the deer nibbling on your garden plants. This moves us deep into the realm of 'designer' ecosystems, which admittedly we're already creating to a certain extent – think built environments, urban parks and arable fields, but this would be a whole new level. The ethical considerations are immense. Who decides which species are expendable? How do we balance human benefits against ecological integrity? While some argue that sacrificing mosquitoes to save human lives is morally permissible, we must consider the broader ecological implications.

The potential cascading impact of such species eradication and the unforeseen consequences of crossbreeding have raised significant concerns. Critics argue that such drastic interventions could disrupt ecosystems or cause the unintended spread of modified genes to other species. To date, scientists have only tested the technique in controlled environments. However, the Bill and Melinda Gates Foundation has invested heavily in the gene-drive solution, and Bill Gates predicts the first gene-drive mosquitoes will be ready for release by 2026.[26]

Gene-drive technology manipulates the genetic makeup of organisms to ensure that specific genes are passed on to nearly all offspring. This goes far beyond the usual 50% inheritance rate seen in traditional genetics. By leveraging the precision of CRISPR-Cas9, scientists can insert, delete or modify genes within the DNA of targeted species, ensuring that these changes spread rapidly through populations. The practical applications of gene drive are profound.

A primary concern is the unintended extinction of other organisms that depend on the targeted species for food or other ecological functions. There's also the risk of such genes spreading to unintended populations or even other species through horizontal gene transfer. If this happened, it could have unforeseen and possibly irreversible consequences. The release of genetically modified organisms into the wild also raises questions about consent, especially for communities that rely on these ecosystems for their livelihoods. Scientists and policymakers must navigate these complex ethical landscapes to ensure that the benefits of gene drive outweigh the potential drawbacks.

Some argue that other insects would quickly fill the ecological roles of mosquito species as food and pollinators. However, this replacement could bring its own set of challenges. Phil Lounibos, an expert in mosquito ecology, cautions that the insects filling the ecological void left by mosquitoes could be even more problematic. "Mosquitoes could be replaced by an insect equally, or more, undesirable from a public health viewpoint," he warns in a BBC article.[27] These replacements might spread diseases further and faster than mosquitoes do today. This could exacerbate the very issues we seek to resolve. Could this potentially escalate endemic insect-borne diseases into a pandemic or give rise to novel infectious diseases? These are questions that don't currently have confident answers.

Perhaps one of the most intriguing perspectives is that of science writer David Quammen. He wrote an article for *Outside* magazine called 'Sympathy for the Devil', suggesting that mosquitoes have historically limited human encroachment on nature.[28] In the article, Quammen says: "The chief point of blame, with mosquitoes, happens also to be the chief point of merit: they make tropical rainforests, for humans, virtually uninhabitable." These rainforests, which are home to a significant portion of the world's biodiversity, are under constant threat from human activities. Yet, Quammen asserts: "Nothing has done more to delay this catastrophe over the past 10,000 years than the mosquito."

The question of eradicating a species transcends scientific and environmental considerations; it's deeply philosophical. Some argue that it's morally indefensible to deliberately wipe out a species, regardless of the danger it poses to humans. This viewpoint is grounded in the intrinsic value perspective, which holds that all living beings have inherent worth independent of their utility to humans. People in this ethical camp consider every species as having a right to exist simply by virtue of being part of the web of life on Earth.

This takes me back to my environmental ethics lectures at university. Famous philosophers in this camp include Aldo Leopold and Arne Næss – founder of the deep ecology movement. The moral argument emphasises the inherent worth of each species, asserting that their existence should not be contingent on their utility or the threat they pose, but on their intrinsic value as living entities. This view also acknowledges how different species within an ecosystem rely on each other in many ways.

Each species plays a unique role in maintaining the integrity of its ecosystem. Eradicating one species could have unforeseen and potentially detrimental ripple effects throughout the entire system. Furthermore, the intrinsic value perspective calls for respecting the right of species to exist and thrive in their natural habitats, viewing the deliberate eradication as an ethical injustice and a violation of this right. Therefore, the idea of eradicating mosquitoes poses some major moral dilemmas.

Social Equity

While the media attention focuses on whether we should eradicate one or many species of mosquitoes, there's the elephant in the room: social equity. We know that most people die of malaria because of poverty. Therefore, addressing social equity could provide the ultimate solution to the problem of malaria rather than resorting to eradicating mosquito species. Improving healthcare infrastructure in malaria-affected areas ensures that people can access effective diagnosis, treatment and prevention measures such as bed nets and antimalarial drugs. Investing in education campaigns can increase awareness about malaria prevention and control. This can empower communities to take proactive measures against the disease. Enhancing economic opportunities can reduce poverty, which is closely linked to higher malaria transmission rates, as wealthier communities can afford better housing and protective measures, reducing their vulnerability.

Additionally, improving housing, sanitation and water management can reduce mosquito breeding grounds and lower the risk of malaria transmission.

People living without high accumulated stress will probably have more robust immune systems to fend off diseases. The same goes for those with access to diverse and highly nutritious food. Supporting research into vaccines, better treatments and innovative mosquito-control methods can offer sustainable solutions to malaria without the ethical and ecological concerns of species eradication. This is all about 'systems thinking' and a holistic approach to addressing complex issues. Unfortunately, this is something we struggle to do, especially when it requires some countries to collaborate for the benefit of other countries.

From an ethical standpoint, ensuring that social equity aligns with human rights principles will allow everyone to live a healthy life free from preventable diseases. This approach also respects the intrinsic value of all living beings, avoiding the ethical dilemmas associated with eradicating a species. *Theoretically*, by focusing on social equity, we can address the root causes of malaria sustainably. But in *reality*, are we willing to do this?

Why can't We Eradicate *Plasmodium* Rather than Mosquitoes?

This is also being studied. The complex life cycle of the parasite makes eradicating it a less realistic option. But what about reducing or preventing the parasite transmission from mosquitoes to humans? Investors are interested in this approach. Scientists at Imperial College London have genetically engineered mosquitoes to produce compounds that slow the growth of malaria-causing parasites in their gut, preventing transmission to humans.[29] This innovation has significantly reduced the possibility of malaria spread in lab settings. If proven safe and effective in real-world conditions, it could become a powerful tool against malaria. The modification can be combined with gene-drive technology to ensure it's widely inherited among mosquito populations.

The team is gearing up for field trials. Their models suggest that these genetic modifications could significantly lower the number of malaria cases in African regions. Partnering with experts in Tanzania, they'll test the modifications under real-world conditions. By combining this genetic approach with existing malaria control efforts, they hope to substantially impact the fight against the disease.

Another innovative strategy involves making mosquitoes resistant to the *Plasmodium* parasite using naturally occurring organisms. The Eliminate Dengue programme in Australia, which uses bacteria to stop mosquitoes from spreading dengue fever, could potentially provide a blueprint for tackling other mosquito-borne diseases.

We're certainly playing an evolutionary game with *Plasmodium* and their mosquito hosts. As the world stands on the brink of potentially eradicating one of its deadliest diseases, we must carefully consider the ethical implications of releasing genetically modified mosquitoes into the 'wild'. The promise of a malaria-free future is enticing. Still, it demands a thorough examination of the potential risks and benefits to ensure that the technology doesn't lead to unforeseen ecological consequences.

CHAPTER 10

Ring-a-Ring o' Roses and Bats with White Noses

Ring-a-ring o' roses / A pocket full of posies /
A-tishoo! A-tishoo! / We all fall down!
—Nursery rhyme

It was the winter of 2006. In a cave in Albany, NY, the light of a caver's head torch reflected off the damp walls and revealed a group of bats huddling together. It was a pretty typical sighting, but one thing did stand out: the bats had a fluffy white frosting on their noses. The cavers didn't understand the significance of this sighting.

A year later, Alan Hicks, a wildlife biologist with the New York State Department of Environmental Conservation, raised an alarm.[1] After hearing reports of dead bats in caves near Albany, Alan and a bunch of researchers found around 10,000 bats of the *Myotis* genus, including little brown bats, *M. lucifugus*, and Indiana bats, *M. sodalis*, dead and dying in a few caves in New York. Hicks collected several ailing bats and took them to Melissa Behr, an animal disease specialist at the New York State Department of Health.

Behr couldn't figure out why they were dying. She and her team noticed some white fluff on the bats' noses and wings, but it disappeared after a few hours of triage. The white fluff was fragile and vanished at the slightest touch. Behr decided to get closer to the bats, inside two abandoned mines, to study them in situ. Her team worked so close to the bats that she could reach out and gently collect a bat, stabilise its head and collect a sample of the white stuff. Back in her lab, she put the sample on a microscope slide, and her lab's mission suddenly became: What is this white stuff?! She suspected it was a fungus, but at the time, no one in the New York State Department of Health had ever seen one like it.

Samples of the 'white stuff' were sent to microbiologist David Blehert and his colleagues, who retrieved sections of its genetic code in 2008. They found that the DNA resembled that of cold-loving fungi, for instance, those living in Antarctic soils. The fungus from the bats flourishes at temperatures between five and ten degrees Celsius, which just so happens to be the temperature range at which bats go into hibernation.

I spoke with David to find out more. Early in the disease investigations, he suspected the fungus didn't kill otherwise healthy bats all on its own; instead, the infection likely caused the bats to wake up too often during hibernation. This premature waking burned up the bats' fat reserves too quickly, leading to winter mortality.

News of the dying bats and the likely culprit – the fluffy white fungus – spread. Hearing the news reminded the cavers of their sighting in 2006, and the significance of the strange white frosting on the bats' noses became apparent. They scoured through the photos they'd taken that day and found one of the bats. Knowing this was important evidence, they shared the photo with local wildlife biologists and researchers. The cavers' photo is still North America's earliest evidence of white-nose syndrome.

In 2009, the fungus was officially named *Geomyces destructans*. After further genetic analysis, it was later renamed *Pseudogymnoascus destructans*. The species name '*destructans*', meaning 'destroying', refers to the fungus's devastating impact on bat populations.[2]

Once scientists cracked the fungus's DNA code, they could test for it elsewhere. By the winter of 2008, the fungus had spread to at least 33 caves within a 210-kilometre radius of the original discovery site at Howes Cave. Alan Hicks, the wildlife biologist, realised how serious this epidemic was. Bat populations were getting completely wiped out in some caves. In 2008, he remarked, "I'll be surprised if some of the sites we visited last year aren't at zero, or very near zero, this winter."[3] He was right. In New York and Vermont, the number of bats in hibernation caves has declined by more than 95%.[4]

The authors of a white-nose syndrome study said:

> WNS combines some of the worst possible epidemiological characteristics, including a highly virulent pathogen with density- and frequency-dependent transmission, an environmental reservoir, long-term persistence in hibernacula and susceptibility of multiple hosts.[5]

Cave-dwelling bat with white-nose syndrome in 2008 (USGS: public domain)

The perfect storm had arrived in North America.

In the immediate aftermath of the discovery of white-nose syndrome, researchers gathered in a whirlwind to piece together the epidemiological puzzle of the fungal pathogen.

Where did it come from?

Had it always been in North America, just waiting for a flawless convergence of environmental conditions before striking vulnerable hibernating bats?

Was the fungus a recently evolved species or strain?

The answer to the last two questions was… no.

Across the Pond

Over in Europe, researchers soon began collecting samples to see if the fungus was either a newbie or native. One study demonstrated the presence of the fungus in eight countries, spanning over two thousand kilometres from west to east.[6] The researchers analysed data from a bat hibernation site over two years and compared it with data from various European locations. They found that the fungus first appeared around the chilly month of February. It reached its highest levels in March and could still be detected in some bats in May or June. This was a consistent pattern across Europe.

Importantly, this study and others found the fungus over large areas of the European continent without associated mass mortalities in bats. Scientists have also found infected bats in Asia – again, without mass mortalities like in North America. The fungus did infect the bats in Europe and Asia. However, the symptoms were mild. This phenomenon is like when one of the viruses that cause a common cold infects healthy humans. It's inconvenient. Yet so long as we don't have a compromised immune system, we'll survive relatively unscathed.

The evidence pointing to a European (or Palearctic) origin was growing. Bats in Europe and Asia were infected by the same pathogen as bats in North America, but this didn't affect their population numbers.

It's worth describing the difference between *resistance* and *tolerance* here. You may say, "Hold on, aren't they essentially the same thing?" The short answer: no. The longer answer: the host's resistance reduces the amount of a pathogen (also known as 'pathogen load'), while tolerance limits the damage caused by a pathogen. These two defence mechanisms affect the spread of infectious diseases and how hosts and pathogens evolve together. Resistance helps the host by attacking the pathogen, thereby reducing its presence in the population. Tolerance, however, protects the host without necessarily harming the pathogen (it can even have a positive effect), which means the pathogen can still be prevalent.

European bats don't have a strong resistance to white-nose syndrome, much like their North American counterparts. However, they can tolerate the fungus well, as their populations remain stable or grow despite high infection rates. This tolerance suggests that the fungus is probably native to Europe and that European bats have co-evolved with it over a long period. They've evolved mechanisms to survive the infection without severe harm.

In 2017, scientists swabbed 138 nineteenth- and twentieth-century bat specimens housed at the National Museum of Natural History, Washington, DC.[7] The bat specimens were from North America, Europe and East Asia. They wanted to see if the swabs would pick up any DNA from the white-nose syndrome fungus, proving that the fungus wasn't a recent hitchhiker from Europe. They sampled dry skins and intact bodies stored in 70% ethanol, and swabbed bat rostra and wings.

They ran their samples through a PCR machine and analysed the results. The negative results went on and on. It was a long shot: the bats would have had to encounter and be infected by the fungus, and the fungal DNA would have had to remain unravaged by the hands of time.

But then... jackpot! One of the specimens' target DNA sequences matched that of the white-nose syndrome fungus. It was from the skin and skull of a male Bechstein's bat (*Myotis bechsteinii*) collected on the 9th of May 1918 – a nearly 100-year-old specimen. But where was the bat collected from? A big clue is that Bechstein's bat is native to Europe and Southwest Asia. Museum collectors found this bat in the Forêt Domaniale de Russy, Centre-Val de Loire, France, about 200 kilometres southwest of Paris.

Bechstein's bats inhabit forested areas, and they're one of the UK's rarest mammals (probably related to the massive loss of forest across the country!). I had the great pleasure of recording Bechstein's bats on surveys when I lived in England. They mostly rely on trees for hibernation through the winter, seeking out sheltered locations such as abandoned woodpecker holes to tuck themselves away safely.

However, on rare occasions, they're known to hibernate in caves, which seems to be the fungus's favourite haunt. Forêt Domaniale de Russy means 'Russy National Forest' in French. This means our Bechstein's bat was collected from a forested area, as we would expect. This got me thinking: does the fungus infect bats that hibernate in trees?

I dug into the literature and failed to find any studies mentioning this – they mostly focused on how the fungus prefers the cold and humid conditions of caves and underground mine environments. So, the hibernation habits of Bechstein's bats probably reduce their exposure to the primary environments where the fungus thrives, but they're not entirely free from risk. The widespread nature of the fungus and its ability to persist in different environments (albeit less commonly outside of caves) means there's still a chance they could pick up the fungus. The fungal pathogen is a generalist. This means it can potentially infect any bat species hibernating under the right microclimatic conditions. Still, I was surprised to learn the bat in our museum collection was a Bechstein's.

I suppose there's a chance that the fungal DNA results represent a 'false positive'. In other words, the DNA could have come from another source and been inadvertently transferred to our Bechstein's bat whilst it was being handled by the museum staff. However, the authors of the study state:

> This sequence is unlikely to represent
> *P. destructans* from a recently collected infected

bat from North America because recently
collected specimens have been purposefully
stored with care in a separate room within the
USNM mammal department, away from the
historical bat collection. Furthermore, none of
the historical samples of bats collected in North
America, which are more likely to be cross-
contaminated with potentially infected specimens
than European specimens, tested positive for
P. destructans.[7]

The French bat is the earliest known evidence of the white-nose syndrome fungus occurring in Europe. Moreover, the researchers found no evidence of the fungus in bats collected in eastern North America between 1861 and 1971. Both pieces of information bolster the current narrative that the fungus is a European native recently transported to North America.

Researchers have now studied the genes of the fungus in detail. Typically, a newly introduced pathogen shows limited genetic diversity, indicating that it recently arrived and went through a 'demographic bottleneck'. This means only a few individuals founded the new population. In contrast, a pathogen with high genetic diversity suggests it has been present in the area for a long time, supporting the idea that it's native to that region. Researchers found a shallow genetic diversity among North American fungal samples.[8] They also found that genetic diversity among European fungal samples was substantially higher than in North American samples. Moreover, the fungal genomes from these geographic regions are distinct from those isolated in China.[9] This molecular finding provides another layer of evidence that someone or something from Europe introduced the white-nose syndrome pathogen to North America.

Intriguingly, the white-nose syndrome fungus might be the descendant of a plant-associated fungus. Indeed, the mechanisms by which the fungus invades bats are very similar to plant fungal-disease mechanisms. Researchers conducted a genetic analysis and estimated an approximate 90% probability that the ancestors of the white-nose syndrome fungus were plant-associated fungi.[10]

This means we may need to survey plants in Europe and Asia to find the fungus's true origin.

Back to North America

So, we know the white-nose syndrome fungus infects both North American and European bats, and the latter have more tolerance. But how exactly does the fungus kill the North American bats?

The fungus predominantly affects bats while they're hibernating. And when they hibernate, their immune systems are less active. During hibernation, bats enter a state called torpor. They significantly lower their metabolic rate to conserve energy. It's like putting your smartphone into energy-saving mode to prolong battery life during periods without charge.

We know the cold temperatures in caves and mines are ideal for the growth of the fungus. In these conditions, the fungus invades and colonises the skin tissues of bats, particularly on their wings, muzzle and the membrane connecting their tail and hind legs. But it's not just a passive companion. It causes significant damage to the skin, resulting in lesions. These lesions are often more than superficial; they can penetrate deep into the tissues, leading to severe irritation and physiological stress. As the fungus progresses, it disrupts the normal functions of the skin, the first line of defence. The bats' skin is crucial for maintaining water balance and regulating body temperature.

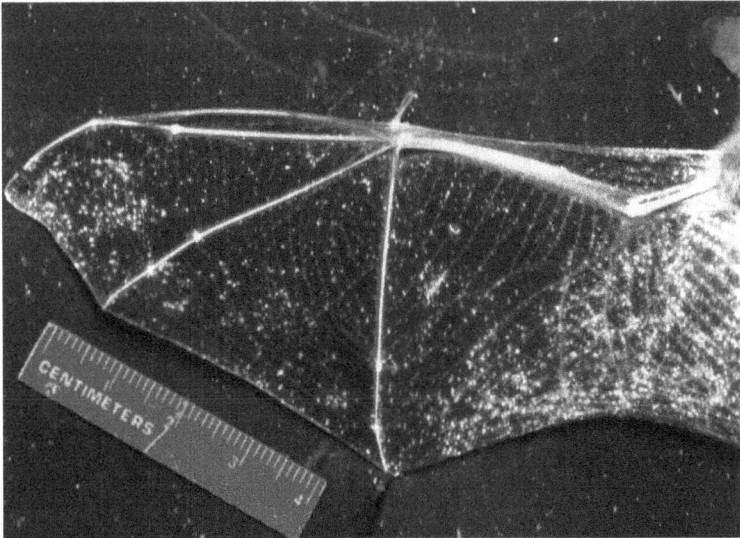

*Long-wave UV illuminating lesions associated with white-nose syndrome
(USGS: public domain)*

As alluded to earlier, the most detrimental effect of the fungus is its impact on hibernation. The infection causes bats to wake up more frequently from their torpid state. This arousal increases their metabolic rate and depletes their fat reserves much faster than usual. Hibernating bats rely on these fat reserves to survive the winter, when food (insects) is scarce. It reminds me again of working with hedgehogs during my early days in research. To survive hibernation, hedgehogs require a healthy layer of brown fat (which is also known as brown adipose tissue, or BAT). Sadly, if they don't pile on enough pounds during autumn, they might not have enough energy to wake up in spring. (Incidentally, the brown fat is rich in mitochondria – the tiny organelles that provide cellular energy. The mitochondria contain iron, giving the tissue its brown colour.)

A similar thing is going on with the bats. During hibernation, bats rely on stored energy reserves. Brown fat, being metabolically active, helps the bats efficiently burn these reserves to produce the necessary heat without relying on muscular activity, which is minimised during hibernation. However, frequent arousals from torpor increase energy expenditure. And guess what? The fungus causes frequent arousals. Since bats cannot find sufficient food during winter to replenish their energy, they eventually starve. This starvation, combined with the physical damage from the fungus, leads to high mortality rates.

Is it a Wildlife Pandemic?

In 2011, ecologists estimated that up to 6.7 million bats in North America had died of white-nose syndrome.[11] But is it a wildlife pandemic, or 'panzootic'?

Given that the fungal disease infects bats right across Europe and Asia (albeit causing relatively mild symptoms) and is now ravaging populations across North America, we can arguably call it a wildlife pandemic.

But what do I mean when I say "across North America"? How far has the fungus spread since its discovery in the dark and damp New York cave in 2006?

By 2008, scientists had found the disease in 16 locations in Connecticut, New York, Vermont and Massachusetts. By 2009, over 40 locations had white-nose syndrome, reaching as far south as the border of Virginia and Tennessee, over 1,200 kilometres away. By 2010, the disease was firmly in Tennessee and had spread north into Canada.

It was also suspected to have spread west to Missouri and Iowa, 1,500 kilometres from the epicentre. Between 2011 and 2015, it spread rapidly across the eastern and northern states in the USA and Canada.

Then, in 2016, scientists first documented white-nose syndrome way out towards the western coast of the USA, in King County, Washington.[12] This was a shocking leap across the country. The disease can pass from one bat to another, but it's also suspected to spread when people carry it on their clothing and equipment. This dramatic jump to the West Coast was almost certainly due to human-caused transmission, possibly by cavers.

The following year, the disease reached Texas in the south and spread across the Midwest, with suspected cases in California and Mexico. Today, white-nose syndrome has been confirmed in at least forty out of fifty US states and nine out of ten Canadian provinces.

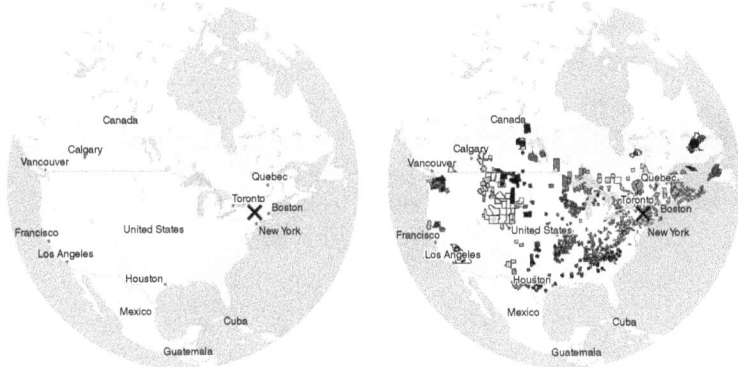

The spread of white-nose syndrome in North America from the epicentre in 2006 (left) to 2025 (right)[13]

There are some concerns that the fungus could spread to the colder regions of South America. However, David Blehert told me:

> Certain caveats must be met for the fungus to establish in South America. I hypothesise that any species of non-migratory bat that lives in a temperate climate and uses long-term torpor to survive winter conditions may be susceptible to white-nose syndrome. However, I also note that different species of hibernating bats that occupy a

habitat that supports the fungus exhibit differential susceptibility to the fungus. Additionally, for the fungus to reach South America, there would either need to be sufficient connectivity among bat populations to facilitate long-distance bat-to-bat spread or some other event (e.g., human-mediated translocation) that resulted in introducing the fungus to a region of South America with a climate conducive to fungal persistence.[14]

David also said,

> While the disease continues to negatively impact bat populations, there are signs that bats are developing natural resistance. Moreover, public awareness of the impacts of this disease has resulted in an increased interest in bats. This has been beneficial for their conservation.

This increased interest in bat conservation is very welcome. Moreover, David's team and others are working on a promising bat vaccine, as I'll discuss in Chapter 19. Here's hoping we can breathe life back into bat populations and protect them from further harm.

Sylvatic Plague

I'd like to end this chapter with a very different disease. I'll be brief because we've already discussed this one (albeit in humans) in the Introduction. I am, of course, referring to the plague caused by *Yersinia pestis*. However, this version is known as *sylvatic* plague – from the Latin *sylvaticus* or 'of the forest'. It essentially means a plague that affects and circulates in wild animals.

The black-footed ferret (*Mustela nigripes*) – have you ever heard of it?

It's a sleek and slender mustelid (the same family as badgers, pine martens and stoats) native to the grasslands of central North America. Historically, its range extended across the Great Plains, from southern Canada through to the western United States, including states such as Wyoming, Montana and South Dakota, and expanding into northern Mexico. It's primarily encountered in prairie ecosystems, characterised by open, grassy landscapes that provide its main prey, prairie dogs (*Cynomys* spp.).

Agricultural development, urban expansion and other land-use changes have drastically reduced the prairie ecosystem. This habitat destruction has directly impacted the suitable living environments for ferrets. They also rely heavily on prairie dogs for food and burrows for shelter, yet the widespread poisoning and extermination of prairie dogs, considered pests by many landowners, have led to a significant decline in prairie dog populations. This has the knock-on effect of threatening the ferrets' survival.

However, sylvatic plague also threatens ferrets. The disease affects both ferrets and prairie dogs – it can decimate the latter, reducing the ferret population. The bacterium *Y. pestis* was not originally native to the continent. Scientists believe it was introduced in the late nineteenth and early twentieth centuries.

Yersinia pestis

In the early twentieth century, black-footed ferrets were widespread across the Great Plains, but extensive habitat destruction and prairie dog eradication programmes led to significant population reductions.[15] By the 1950s, many thought the ferrets were extinct. Yet in 1981, a small population was discovered in Meeteetse, Wyoming. However, sylvatic plague and canine distemper (a highly contagious viral disease) decimated this population. By 1987, only 18 individuals were left in the wild, prompting emergency captive-breeding efforts. As the black-footed ferret population dwindled to near extinction, the remaining population also suffered from low genetic diversity. This 'bottlenecking' made them more susceptible to diseases and reduced their ability to adapt to changing environments.

Today, around 300–400 black-footed ferrets live in the wild.[16] Their numbers fluctuate largely due to sylvatic plague. Back in the noughties, while writing a university assignment on black-footed ferrets and sylvatic plague, I contacted Travis Livieri, director of Prairie Wildlife Research, who has dedicated nearly 30 years to researching and recovering the ferret. He told me how black-footed ferrets are highly susceptible to sylvatic plague, which can be transmitted to them in several ways.

The primary mode of transmission is through the bites of infected fleas. Fleas feeding on infected prairie dogs can carry and transmit the bacteria to ferrets. If you remember from the Introduction, infected fleas harbour *Y. pestis* in their digestive tracts, and when they bite a host, they regurgitate the bacterium into the wound, thereby infecting the host. Another avenue for plague transmission to black-footed ferrets is through direct contact. Ferrets can contract the disease by interacting with infected prairie dogs or their carcasses. This is a highly likely scenario given that prairie dogs are the first choice on the ferrets' menu! In addition, in rare cases, ferrets might inhale the bacteria if they are close to an infected animal.

The impact of sylvatic plague on black-footed ferrets is severe and often fatal. Once infected, ferrets typically show symptoms like lethargy, fever and swollen lymph nodes within a few days. The disease progresses rapidly, leading to septicaemia (blood infection) and death. Due to their high susceptibility, untreated ferrets often die within a few days of infection.

Conservationists have tried several approaches to combat sylvatic plague, including vaccination, flea control and regular monitoring and quarantining. Oral and injectable vaccines help protect black-footed ferrets and prairie dogs from plague (more on this in Chapter 19). Dusting prairie dog burrows with insecticides also helps control flea populations, reducing the risk of plague transmission. Moreover, Travis told me that regularly monitoring ferret and prairie dog populations for signs of plague and quarantining infected areas helps manage outbreaks.

Sylvatic plague affects various wildlife populations, particularly rodents and their predators, across multiple continents, including North America, Africa and Asia. However, its classification as a pandemic or panzootic can be nuanced. Due to its wide geographic spread and substantial impact on wildlife populations, sylvatic plague exhibits some pandemic-like features. Yet, it's probably more accurately described as an *endemic* disease within specific ecological contexts. The disease remains stable in certain areas, causing regular but localised outbreaks rather than the widespread, uncontrolled spread that's typical of pandemics.

These two devastating diseases – white-nose syndrome and sylvatic plague – affect different species in distinct ecosystems. Yet, both reveal a common thread. Human activities exacerbate the natural world's vulnerability to hidden threats. White-nose syndrome has devastated bat populations far from the region where the pathogen originally evolved. Similarly, sylvatic plague has devastated rodent communities and the predators that depend on them.

A single pathogen, an infinitesimal being, perhaps more than a thousand times smaller than a poppy seed, can ravage entire ecosystems.

CHAPTER 11

The Devil's Work

*The Devil whispered in my ear: You are not strong
enough to withstand the storm. Today, I whispered
back: I'm the storm.*

—Adharanand Finn, *The Rise of the Ultra Runners*

W hat is a species? This seemingly simple question has puzzled
scientists for centuries. It's a question that, on the face of it,
seems so straightforward. Yet if this question could swagger
into a buzzing Midwest saloon (bear with me), it would sap the life out
of the party, silence the pianist and draw stares of confusion from every
corner. Depending on the day and the saloon, it might even spark a
brawl, with chairs flying and bottles shattering against the walls.

But we must have a 'widely accepted view'; otherwise, we'd be stuck
in a quagmire with chaos and endless debate. So here goes: a species is a
group of organisms that can interbreed and produce fertile offspring under
natural conditions, and share similar characteristics and genetic makeup.
Each species plays a unique role in its ecosystem; they fill millions of
distinct niches across the planet. Some species might have a few thousand
individuals in natural conditions, like snow leopards. In others, there may
be hundreds of millions, like the common frog (*Rana temporaria*); a billion,
like straw-coloured fruit bats (*Eidolon helvum*); hundreds of billions, like
the bristlemouth fish (*Cyclothone* spp.); and trillions, like bacteria.

Excluding microbial species, which likely number in the hundreds
of billions or more, there are an estimated 8.7 million species on the
planet.[1] Amphibians – those cold-blooded vertebrates that typically
live both in water and on land, including frogs, toads, salamanders and
caecilians – comprise about 8,000 of these species.[2]

Now, imagine 501 amphibian species with distinct characteristics
and ecological roles facing a malady that has driven many to extinction.

That's a dramatic decline in one in every sixteen species. In the olden days, a blight so powerful would be considered the devil's work. But in reality, it's a microscopic fungus that has brought entire populations to their knees. It has altered ecosystems and challenged the survival of amphibians worldwide.

Chytrid Fungus

What is this fungal disease? It's known as a tongue-twister of a word, chytridiomycosis, and it's caused by two fungal species, *Batrachochytrium dendrobatidis* and *B. salamandrivorans* (*Bsal*).[3] The former fungus primarily affects frogs and toads and is found worldwide, including in North and South America, Europe, Africa, Asia and Australia. The latter fungus primarily affects salamanders and newts. It was initially detected in Europe, particularly in the Netherlands, Belgium and Germany; however, scientists think it originated in Asia, with the pet trade being a significant vector for its spread.

And what are the symptoms? The fungi infect the keratinised (hardened) outer layer of the amphibian's skin. This is especially harmful since amphibians rely on their skin for critical functions such as respiration and maintaining electrolyte balance. They also thicken the skin. This thickening impairs the amphibian's ability to absorb water and electrolytes. This disruption can lead to a condition known as hyperkeratosis, where the skin becomes excessively thickened.

I mentioned electrolytes. Well, the fungus causes significant electrolyte imbalances, particularly in reducing sodium and potassium levels in the blood. These imbalances can lead to cardiac arrest. The disease can also lead to respiratory distress. In addition, infected amphibians may show signs of lethargy, loss of appetite and abnormal behaviours such as excessive skin shedding and sitting out of water for extended periods. If the infection is severe, it can lead to death. Mortality rates can be extremely high in susceptible species, leading to rapid population declines.

The Amazing (But Fragile) Skin of Amphibians

Take a deep lungful of air. Each time you do, oxygen from the air passes through the alveoli into the bloodstream. Carbon dioxide from the blood is transferred to the alveoli to be exhaled. We take some 20,000 breaths daily, about 7.5 million each year.[4] That's around fourteen thousand litres of air a day, over five million litres a year. Maintaining this continuous contraction of our diaphragm, intercostal muscles and lungs takes a lot of energy.

Let's also consider drinking. Adults are advised to drink about 2.7 to 3.7 litres of water daily.[5] That's approximately 985 to 1,350 litres per year, or over 100,000 litres or 400,000 cups of tea in an average lifetime. That's quite a substantial amount of drinking.

Now imagine a different reality, one in which we didn't have to do either of these laborious tasks. We still need air in this new reality, but we don't have a huge pair of lungs and the never-ending muscular contraction that comes with them. But if we still need air to function, how does it get from the surrounding environment and into our body's cells? Well, we just breathe through our skin, of course! After all, it's the largest organ in the body, constantly in contact with the air. We also don't need to put the kettle on 400,000 times because we absorb water differently, too, through our magical skin. All makes sense, right?

Okay, let's jump back to our normal reality. Breathing through our skin is insufficient for larger organisms because the skin's surface area can't provide enough oxygen to meet our metabolic needs. I mentioned alveoli earlier. Well, these probably evolved for this reason. We have a vast network of alveoli that provides a much larger surface area for gas exchange, allowing efficient oxygen uptake and carbon dioxide removal to support higher energy demands. Moreover, having such permeable skin would mean our bodies were more susceptible to breaches by pathogens and other harmful agents unless we evolved specialised features to protect ourselves.

Amphibians, our slimy, warty friends and the protagonists of the first part of this chapter, actually do breathe through their skin. Amphibian skin is a true wonder of the natural world. It embodies a range of adaptations critical to these creatures' survival. This skin is thin and highly permeable to gases, allowing for the efficient exchange of oxygen and carbon dioxide with the environment. This ability is particularly crucial when amphibians are submerged in water. It enables them to continue breathing even when their lungs are not in use. Some salamanders are completely lungless. They use a blood vessel absent in other animals; they shunt their blood supply from the heart to the skin. When salamanders are still, you can see pulsing in their throat as they pump air in and out of their mouths.[6]

Moreover, amphibian skin is essential for maintaining hydration. Amphibians can absorb water directly through their skin, much like we did in our imagined example above. This process eliminates the need for drinking in the traditional sense. Indeed, many frogs and toads have a special drink patch on the underside of their body. In toads, even though the little skin patch constitutes only about 10% of the total skin area, it's responsible for more than 70% of their total water uptake.[7] Specialised cells manage the uptake and balance of water and electrolytes, facilitating this ability to drink through their skin. This adaptation is vital for their survival, particularly in fluctuating environments where water availability can vary.

Amphibian skin has tools to protect against pathogens. It secretes diverse antimicrobial peptides that protect against bacterial, fungal and viral infections. These natural antibiotics are a key component of the amphibian immune system. They provide a first line of defence in the often damp and pathogen-rich environments they inhabit. Additionally, many amphibians produce toxic chemicals through their skin glands. These toxins can range from mild irritants to potent neurotoxins, deterring various predators. The bright, vivid colours and striking patterns seen in species such as poison dart frogs are aposematic signals, warning potential predators of their toxic nature. In addition to these functions, amphibian skin plays a role in thermoregulation, helping these ectothermic (cold-blooded) animals manage their body temperature. The skin's permeability allows for evaporative cooling. Its colouration can help absorb or reflect sunlight. However, amphibian skin is also sensitive to environmental changes. Amphibians are often called 'canaries in the coal mine' because their permeable skin makes them particularly susceptible to pollutants, novel pathogens (those it hasn't co-evolved with) and changes in environmental conditions – making them an important indicator of ecosystem health.

Now, let's return to our made-up 'reality' where we breathed and drank through our skin. Life's going great. We're just lazing around, 'breathing' and 'drinking', and our skin's doing all the heavy lifting. But one day, we find a fungus spreading across our skin, smothering the very organ that allows us to function. The fungus messes with our electrolytes. It thickens our skin – the hyperkeratosis mentioned earlier. The electrolyte-interrupting and skin-thickening hamper our 'breathing' and 'drinking'. The pathogen is essentially choking us. We eventually succumb to the disease.

This is scary, right? It's what's happening to our slimy, croaky, warty friends across the world.

Personal Anecdote

Back when I earned my keep conducting ecological surveys across the damp corners of the UK, one species dominated the spring survey season: the great crested newt (*Triturus cristatus*). It's the biggest newt in the UK. It's almost jet black, with spotted flanks and a striking, mottled orange belly, which, for ecologists, acts as a fingerprint as each orange pattern is unique to the individual. It has warty skin, and males have long, wavy crests along their bodies and tails with flashy silver stripes during the breeding season. The great crested newt is highly protected by law in the UK. This is due to enormous population declines in the last century across its range in Europe (it's actually doing okay in the UK). Habitat destruction caused by human activities, such as pond drainage and hedgerow removal, primarily drives these declines.

In 2015, reports suggested that the newt and salamander version of the chytrid fungus (*Bsal*) contributed to a 95% decline in the salamander population in Belgium and the Netherlands.[8] Ecologists found that the fungus had infected and killed four salamanders imported to the UK. This is quite a scary situation for the great crested newt. If it escapes into the wild, the fungus could decimate the great crested newt populations and other newt species in the UK, including the smooth newt (*Lissotriton vulgaris*) and alpine newt (*Ichthyosaura alpestris*).

Great crested newts and their uniquely patterned underbellies

In 2019, researchers collected swab samples from 2,409 wild newts in ponds across the UK, all of which tested negative for chytrid fungus.[9] Additionally, monitoring of newt mortality incidents between 2013 and 2017 found no presence of the fungus. However, the pathogen is prevalent in British captive amphibian collections. Consequently, researchers emphasise the critical need to heighten awareness of effective biosecurity measures, especially among those keeping captive amphibians, to prevent the spread of the fungus to wild populations.

During the breeding season, great crested newts live in a cluster of ponds called a meta-population. Individuals can travel between different ponds. This behaviour peaks when seeking new mates. The movement between ponds helps maintain genetic diversity and population stability by allowing newts to colonise new ponds, find mates and escape from unfavourable conditions. However, it also means that if individuals become infected, they could transport the fungus to various ponds in the cluster. This could make the disease spread like cracks in thin ice.

Another way it could spread across meta-populations is via people who walk in and around ponds. And who might do this? Ecologists surveying great crested newts!

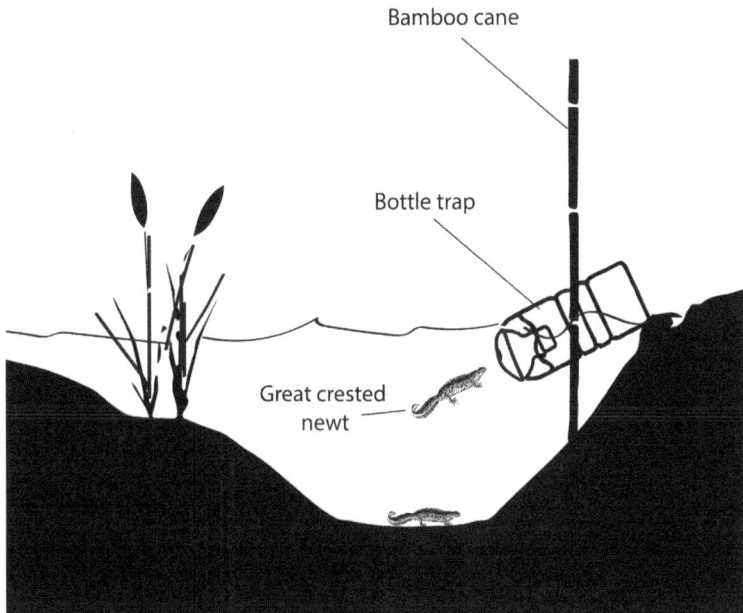

Great crested newt bottle trap – a potential disease vector

This is why ecologists must decontaminate their footwear and equipment between surveys. Traditionally, ecologists use 'bottle traps' as a surveying method. These are two-litre plastic bottles with the top cut off and placed in the reverse direction into the bottle, with a long bamboo cane pushed through to hold it in the pond. The newts swim into the large funnel opening to the bottle and typically fail to swim back out. A large air bubble is left to ensure the newts survive unscathed. The trapping allows ecologists to check the bottles and record the number of newts, providing an estimate of the population size in the pond. However, the bottles and bamboo canes are placed into the water and reused in different sites across the country. So, you can probably see how the equipment could become a disease vector. Ecologists travel up and down the country, placing these bottle traps into ponds. The best practice is to thoroughly disinfect the bottles between surveys, but I've encountered some very slack practices in my time. Complacency is the breeding ground for disease. Where vigilance falters, contagions thrive.

Chytrid Down Under

It's a sunny day. I'm sitting on a beanbag writing this chapter and looking over Belair National Park in Adelaide, South Australia. From my research, I know that the chytrid fungus arrived here in Adelaide way before me. Since its arrival, the disease has caused dramatic population declines in numerous species. It has even driven seven species to the brink of extinction in Australia, including the unique gastric-brooding frog (*Rheobatrachus silus*) and the southern day frog (*Taudactylus diurnus*).[10] Indeed, the amphibian chytrid fungus was first recorded in Australia in the 1990s.[11] However, chytrid fungus has spread across parts of Australia and the Americas since the 1970s. We just didn't realise it. Unfortunately, this meant much of the damage had already been done before its discovery in the 1990s.

There's been considerable discussion about whether a pathogen alone could cause such widespread destruction. However, researchers recently said, "Watching uninfected frog populations crash with the disease's arrival has convinced most sceptics."[12] While amphibians have been significantly impacted by habitat loss, the enigmatic and rapid disappearance of frogs in protected and remote mountainous regions over the last 30 years is primarily attributed to the additional threat posed by the fungus. As always, it's likely a combination of these factors, culminating in a perfect storm scenario.

Origin and Evolution

It's tricky to work out the origin of this disease because in some regions, such as southern Africa and Asia, the fungus exhibits characteristics of an endemic disease. In other words, it's been present in these areas for a long time without causing widespread epidemics, suggesting a stable relationship with the local amphibian populations.

For example, the fungus has been found in museum specimens from these regions dating back to the early 1900s. These samples indicate it's been part of the local ecosystem for at least a century. This long-term presence without large-scale die-offs implies that local amphibians may have developed some level of resistance or tolerance to the fungus. Either that or environmental conditions in these regions mitigate the impact of the disease. It could also mean that the fungus has since developed more virulent strains, which triggered the wildlife pandemic.

A study by Erica Rosenblum at the University of California involved researchers worldwide submitting fungal cultures to her team.[13] They used DNA sequencing methods on dozens of fungal samples and found that most fell within the 'Global Pandemic Lineage' cluster – the one that has spread widely and contributed to the mass amphibian declines. The samples revealed an unexpectedly high genetic diversity, indicating that this fungus lineage is probably thousands of years old. We can be confident in the fungus's age because genetic variations accumulate over extended periods. As a lineage ages, mutations increase its diversity. In other words, ample time is needed for these genetic variations to develop and persist.

This finding challenges the notion that the recent emergence of the frog chytrid is due to the evolution of a new, virulent strain of the fungus. Instead, its spread is likely a consequence of globalisation. Plane, train, ship and car hopping have aided the fungus's spread through trade to areas with naïve frog populations. Additionally, there are localised lineages of the fungus that appear to be endemic to specific regions. However, these lineages may have also been inadvertently transported, further complicating the situation. More research is needed to compare these lineages to determine their behavioural differences, particularly their virulence. The study concludes that the results are complex and do not pinpoint any single region as the disease's origin.

How Humans Have Spread Chytrid

It's no coincidence that, much like us, chytrid has been a relentless wanderer. Humans have undoubtedly facilitated the global spread of the disease through the amphibian trade, the movement of infected wildlife and the inadvertent movement of the fungus itself. Again, think of plane, train, ship and car hopping. It's what we've excelled at in the Anthropocene.

Researchers have examined museum amphibian specimens and have uncovered an early case of chytridiomycosis in the African clawed frog (*Xenopus laevis*). The specimens date back to 1938.[14] They originated from South Africa, though the African clawed frog is widespread across Sub-Saharan Africa. The frog has played a crucial role in molecular and developmental biology research. Most notably, it was first used in pregnancy testing in the mid-1930s. A pregnant woman's urine would induce egg production in a female African clawed frog. This discovery led to the widespread export of the frogs from Africa to nearly every continent worldwide. Tragically, African clawed frogs can carry the chytrid fungus without showing any signs of infection. As a result, exporting this seemingly healthy amphibian inadvertently introduced the fungus to new regions, infecting local amphibian populations. It took over half a century for humans to realise the catastrophic consequences. By then, the disease had evolved into the global wildlife pandemic (panzootic) that scientists and conservationists are battling today.

The fungus spreads through direct contact among amphibians or through infected materials in their environment. It's easily spread by human activity. Boots, clothes and equipment can harbour fungal spores, so ideally, we should clean these items before visiting amphibian habitats. Indeed, a biologist at the US Fish and Wildlife Service once said, "You always have to assume you could be carrying the fungus."[15]

Signs of Improvements?

A study on Panamanian frogs has revealed that some species appear to be evolving resistance to the disease.[16] It's challenging for a host species to evolve resistance because fungal pathogens can evolve more rapidly to overwhelm their defences. Pathogens have short lifecycles. They can quickly respond to natural selection. However, the wild frogs in Panama may be evolving better defences against the pathogen. Jodi Rowley of the Australian Museum once told *Wired*:

> Maybe we should be spending more time and
> effort making sure that other factors are not

hindering the frogs—so the streams are protected from deforestation and pollution and things like that, so they're able to perhaps take on the disease themselves.[17]

I could not agree more.

Danielle Wallace at the University of Melbourne studies chytrid fungus in Australian frogs.[18] The whistling tree frog (*Litoria verreauxii*), with its emerald-green cloak, has been hit hard by the fungal disease. The frog has faced a devastating decline, losing over 80% of its former range. The chytrid fungus has proven to be exceptionally lethal, killing nearly all mature frogs after their breeding season. This pathogen is so effective at its deadly work that few adult frogs survive beyond their reproductive phase, leading to a critical drop in population numbers. However, the frog may have recently evolved a hidden ace up its sleeve. Now, in the throes of infection, the frogs crank up their reproductive engines. Research from Danielle's lab showed that infected males boost their sperm counts and females produce a bounty of eggs. It's as if the frogs say, "If we're going down, we're going down swinging."

Evolution is tweaking the frog's reproductive dials. It also seems to be dabbling in the aesthetics of frog courtship. Danielle has found that infected males exhibit a surprising transformation in controlled lab environments and in the wild. They become more colourful and vibrant. The duller greens morph into a palette more akin to a tropical cocktail. This is about survival; it's about sex appeal. Through a natural selection response, the chytrid fungus inadvertently enhances the males' breeding displays, making them more attractive to potential mates. Whether or not this change in frog behaviour protects their populations in the long term is yet to be determined, but here's hoping.

A Devilish Dilemma Down Under

It would be imprudent to title a chapter 'The Devil's Work' without including a somewhat obvious candidate – the Tasmanian devil (*Sarcophilus harrisii*), and the recent evolution of a mega-disease in this species.

My initial understanding of the Tasmanian devil was far from ideal. It was shaped by the *Taz-Mania* animated television series that aired between 1991 and 1995. The show was a spinoff of the classic *Looney Tunes* character, the Tasmanian Devil, affectionately known as Taz. Set in the fictional land of Tazmania, the series follows the adventures and misadventures of Taz and his family. Taz is depicted as a hyperactive, tornado-spinning character with an insatiable appetite and a limited vocabulary, often expressing himself through grunts, growls and gibberish.

The real Tasmanian devil is a carnivorous marsupial that's native to Tasmania and renowned for its ferocious temperament and distinctive vocalisations, which include spine-chilling screeches, growls and snarls, so the cartoon wasn't too far from the truth on this facet. With its stocky build, black fur and prominent white markings on its chest, the Tasmanian devil resembles a cross between a small, sturdy dog and a wombat. Despite its modest size (although it's now Tasmania's largest native carnivore), the devil has a voracious appetite, often consuming entire carcasses, bones and all. It's a fluffy ball of tenacity.

A healthy-looking Tasmanian devil

With this tenacity comes a bite. The devils tend to bite each other primarily as a form of social interaction and to establish dominance. These interactions are mainly observed during feeding and mating. When devils congregate around a carcass, competition for food can intensify, leading to aggressive behaviours and biting as they jostle for the best position. During the mating season, males may also bite each other and potential mates to assert dominance and secure mating opportunities. Additionally, biting can occur as a defensive mechanism when devils feel threatened or challenged. This aggressive behaviour is part of their natural instinct to ensure survival and reproduction within their competitive environment. However, this natural tendency to nip each other has led to a crisis of enormous proportions.

In the mid-1990s, a disease began ravaging the Tasmanian devil population. Known as devil facial tumour disease, it was first documented in 1996 near Mount William in northeastern Tasmania.[19] It's an aggressive cancer that manifests as grotesque tumours on the face and neck. It hinders the devil's ability to eat and ultimately leads to death by starvation or secondary infections.

And there's one thing that's unique about this cancer. Unlike most cancers, devil facial tumour disease is transmissible. It spreads through the Tasmanian devils' natural biting behaviour. As devils nip and bite each other during feeding or mating, they often break the skin, creating open wounds. The cancer cells from an infected devil can be transmitted through these bites, allowing the tumour cells to graft onto the healthy tissue of the bitten devil. This direct transmission of living cancer cells is unique and devastating, as it bypasses the immune system's usual defences.

Interestingly, the cancer can be transmitted from the biter to the bitee (new word?) and from the bitee to the biter. Those most likely to bite and compete aggressively for resources and mates are the ones we would normally consider the most dominant or the 'fittest' individuals. However, research has shown that the disease disproportionately afflicts the fittest devils.[20] Usually, diseases predominantly affect the weakest members of a population, like those with weakened immune systems or the elderly during flu season. But this disease primarily affects animals that are otherwise very 'fit' in an evolutionary sense. The likely reason is that socially dominant devils, which initiate most aggressive and mating encounters, are the ones contracting the disease, as it spreads through biting. Quite the twist in the 'survival of the fittest' narrative.

How did the Cancer Evolve?

The cancer first evolved from a single rogue cell in a female Tasmanian devil. This renegade cell underwent mutations, becoming cancerous and gaining the ability to transfer between individuals. It's believed to have originated in the Schwann cells, which are part of the peripheral nervous system.[21] They play a crucial role in the maintenance and function of nerve cells.

Schwann cells produce the myelin sheath, a protective covering that wraps around nerve fibres and facilitates the rapid transmission of electrical signals. These cells also help repair our nerves by providing support and guidance to regrow damaged nerve fibres.

The ability of Schwann cells to promote nerve repair and their essential role in nervous system health make them vital for proper neural function. However, their ability to proliferate rapidly may also be their most dangerous facet. These cells, once cancerous, became capable of living independently and spreading through the bites that devils naturally inflict on one another during social interactions. This original tumour clone has since propagated through the population, having a devastating impact on the species.

In a bizarre turn of events, the Tasmanian devil faces not one but two types of facial tumour cancer. The first and most prevalent type was discovered in 1996. However, when researchers were developing an understanding of it, a second type emerged in 2014. This new variant behaves differently. It mutates three times faster and is genetically distinct.[22] The exact origin of this second cancer type is unknown, but it's thought to be a recently evolved cancer.

Facial tumour diseases might be a natural aspect of Tasmanian devil ecology. These cancers may have occurred historically and could emerge again in the future. Many incipient facial tumours might die out before detection. This makes it challenging to fully understand their origin and prevalence throughout time.

However, human disturbance and damaged ecosystems may have led to compromised immune systems in Tasmanian devils, facilitating the evolution of facial tumour disease. When an animal's immune system is compromised, it's less able to identify and destroy abnormal cells. This can allow the cells to proliferate unchecked. In the case of Tasmanian devils, environmental stressors such as habitat degradation and pollution could contribute to their compromised immune systems. This weakened state may have influenced the initial development and mutation of cancer cells, leading to the emergence of transmissible tumours.

The conversion of extensive dry forests within the Tasmanian devil's core range into a mosaic of grazing land and forest remnants has likely increased grazing prey species like wombats, macropods and brushtail possums.[23] This shift, beginning with pastoral development in the early 1800s, may have led to unnaturally high devil population densities. Such conditions are ideal for the spread of diseases, including the current infectious cancer epidemic. This human-induced environmental change may have inadvertently set the stage for the spread of the facial tumour disease.

You've probably noticed I wrote "may" quite a few times in the previous paragraphs. This is just a reflection of the complexity of the disease and the research that still needs to be done. Nonetheless, as researchers recently said, "These diseases' devastating impact on their host species is exacerbated by anthropogenic threats including loss of habitat and roadkill."[24]

Impact of the Disease

The impact has been catastrophic. Since its discovery, the disease has wiped out nearly 80% of the wild devil population, pushing the species to the brink of extinction. The rapid decline prompted immediate scientific and conservation efforts to understand and combat the disease.

In response to the disease, Tasmanian devils have altered their breeding habits. Affected individuals now begin breeding at younger ages. Due to their reduced life expectancy, they often participate in only one breeding cycle. Previously, females bred annually at age two and continued for about three years. Now, breeding begins at age one, with many dying shortly after due to the disease. It's probably the first known case of an infectious disease leading to increased early reproduction in a mammal.[25]

Devil facial tumour disease is the odd one out in this section of the book. This is because it's unlikely to give rise to a wildlife pandemic. However, it remains a critical issue for Tasmanian devils. Our actions, both now and in the future, will determine the devil's future. We can learn from wildlife pandemics to help conserve this iconic species.

The good news is that many people have banded together to protect the devil.[26] In August 2023, the Devil Ark at Barrington celebrated a significant milestone: the birth of its 500th Tasmanian devil since the project's inception in 2011. This achievement marks a vital contribution to the conservation efforts aimed at protecting and increasing the population of this endangered species. Researchers are also developing better diagnostic tests and oral bait vaccines, which we'll discuss later in the book.[27]

CHAPTER 12

It's Just the Flu, Right?

We live in a dancing matrix of viruses; they dart,
rather like bees, from organism to organism, from
plant to insect to mammal to me and back again
—Lewis Thomas, *The Lives of a Cell*

D iana Bell is a conservation biologist at the University of East Anglia, UK, who studies emerging infectious diseases. There's one question that people often ask her: "When do you think the next pandemic will be?" Diana often says, "We are in the midst of one – it's just afflicting a great many species more than ours."[1]

She's referring to the highly pathogenic strain of avian influenza H5N1, otherwise known as 'bird flu'. The virus has killed many millions of birds and unknown numbers of mammals, particularly during the past three years. The H5N1 flu virus subtype emerged in a Chinese poultry farm in 1997. It quickly leapt across the species barrier and into humans in Southeast Asia, with a mortality rate of up to 50%.[2]

In 2005, Diana's research group made a startling discovery at Cuc Phuong National Park in Vietnam, when H5N1 claimed the life of an endangered Owston's palm civet (*Chrotogale owstoni*) – a small to medium-sized nocturnal mammal with an elongated body, short legs and a bushy tail. This incident occurred within the confines of a captive breeding programme to preserve this rare species. The exact mechanism by which the palm civets contracted bird flu remains a mystery. Unlike the captive tigers (*Panthera tigris*) in the region, which were infected by consuming diseased poultry, the Owston's palm civets primarily consume earthworms. This dietary difference suggests that the transmission route for the civets differed from the more straightforward pathway observed in tigers.

As we learned in Chapter 6, there are two broad classifications of the avian influenza virus: low pathogenic avian influenza (LPAI) and highly pathogenic avian influenza (HPAI). Influenza A viruses are categorised based on the types of two proteins on their surface: haemagglutinin and neuraminidase – the proteins or 'keys' that unlock different parts of the immune system.

There are subtypes and subtypes of subtypes, and if you're like me, the squishy ball of tofu in your head is now working overtime – but stay with me for a second. Within the H5N1 subtype of influenza A viruses, further classification occurs based on genetic sequences, dividing the viruses into *clades* and *subclades*. It's not imperative to retain this information, but just think of these as branches and sub-branches on a family tree. For instance, H5 2.3.4.4b is a specific subclade within the H5N1 subtype. This strain is notorious for causing large-scale fatalities among various species of wild birds. It can also affect other non-avian wildlife (such as palm civets), posing significant threats to biodiversity. In contrast, some other highly pathogenic avian influenza strains are less likely to cause disease in wild birds and mammals.

Diana's team's discovery of the H5N1 virus in endangered Owston's palm civets spurred them into further action. They recognised the potential implications of this finding and went on a hunt to compile all confirmed instances of fatal bird flu infections across various species. They aimed to gauge the extent of the threat that H5N1 poses to global wildlife.

Was this the beginning of a devastating trajectory?

By the end of 2005, confirmed H5N1 bird flu infections were limited to a few zoos and rescue centres in Thailand and Cambodia. However, Diana's team's 2006 analysis revealed a broader threat: nearly half (48%) of bird orders had at least one species with a reported fatal bird flu infection.[3] This amounted to around 84% of all bird species, many already globally threatened. The team deduced that H5N1 strains circulating at the time were likely highly pathogenic across all bird groups. Their findings also highlighted that critical habitats for these species, including Vietnam's Mekong Delta, were near the reported poultry outbreaks.

The Mekong Delta is a vast network of rivers, swamps and islands covering approximately 40,500 kilometres in southern Vietnam. It's a crucial area for bird conservation, providing a stopover for migratory birds and hosting resident and migratory bird populations. Notable birds include the Sarus crane (*Grus antigone*) and the lesser adjutant stork (*Leptoptilos javanicus*). Both species rely on the delta's habitats for nesting and feeding. The delta is also a sanctuary for several endemic and endangered non-avian species, including the Irrawaddy dolphin (*Orcaella brevirostris*), the Siamese crocodile (*Crocodylus siamensis*) and various rare fish. As you can imagine, a major H5N1 outbreak near this biodiversity hotspot was a recipe for disaster.

As the palm civet case suggested, birds weren't the only ones at risk. Mammals known to be susceptible to bird flu during the early 2000s included primates, rodents, pigs, rabbits, domestic cats (*Felis catus*) and large carnivores like Bengal tigers (*Panthera tigris tigris*) and clouded leopards (*Neofelis nebulosa*). Diana's 2006 paper illustrated the alarming ease with which the H5N1 virus leapt across species barriers. The paper's findings were not merely theoretical but highlighted a largely unheeded reality: the virus might one day produce a pandemic-scale threat to global biodiversity.

Unfortunately, Diana and her team were correct.

Bird Flu's Unrelenting March across the Globe

Bird flu continues its deadly sweep two decades later. It ravages species from the high Arctic to mainland Antarctica. In recent years, the virus has surged across Europe and infiltrated North and South America, leaving a trail of devastation. Bird flu's resurgence has been swift and brutal. Millions upon millions of poultry have succumbed to the virus. This has illuminated the severe economic and food security impacts of our dependence on intensive agriculture – an atrocity of modern humanity. But the carnage doesn't stop at domestic birds. A multitude of wild birds and mammals have been affected.

Influenza Virus

A recent study revealed the staggering breadth of the virus's impact. Since 2020, deaths from bird flu have been reported in at least 50 mammal species in 26 countries.[4] The landscape of avian influenza changed dramatically in 2020. Outbreaks in poultry and wild birds surged, marking the onset of a global crisis. By 2021, mass mortality events were reported across Europe, signalling a rapid global spread.

The world found itself in the throes of a panzootic. This particularly virulent clade of the H5N1 virus leapt the Atlantic, reaching North America around October 2021. Remarkably, just a few months later, the virus made another jump to North America, this time crossing the Pacific. This jump highlights the virus's ability to cross species and continental barriers. This rapid spread has sparked serious concerns about the virus mutating further and jumping more easily between different species.

The study found that some new mutations could help this avian pathogen replicate in mammals. It appears to be changing and adapting fast. The South American strain of the virus was a 'reassortant', meaning it had a mixed genetic heritage. Half of its genome was related to the European H5N1 virus that had spread via wild birds from Europe to North America in late 2021. The other half came from less pathogenic viruses. These circulated among wild birds in the Americas, likely picked up as H5N1 moved across the USA before reaching Peru. While you might expect that incorporating genes from a less pathogenic virus would reduce the virus's lethality, this was not the case. Instead, the genome has spawned new evolutionary pathways for H5N1, potentially enhancing its ability to adapt to mammals.

Into the Ocean

It would be forgivable to assume that oceanic animals are somewhat protected from the virus. However, since 2020, bird flu has cast its net over marine life, causing deaths in 13 species of aquatic mammals. Among the hardest hit are South American sea lions (*Otaria flavescens*), porpoises and dolphins, with mass die-offs reported in the thousands. According to an article published in May 2024, after arriving in South America, the virus travelled over six thousand kilometres in only three months.[5] Its route appears to have been along the Pacific coast of South America to the southernmost tip of Tierra del Fuego. Over 650,000 seabirds and 30,000 sea lions died in Peru and Chile alone, and 18,000 southern elephant seal pups died in Argentina. An astounding 40% of all Peruvian pelicans (*Pelecanus thagus*) also died. The pelicans exhibited various neurological symptoms before succumbing to the disease. These included disorientation, ataxia (lack of coordination), circling, nystagmus (uncontrolled movements of the eyes) and torticollis (twisted neck). To those watching on in Peru, it was shocking to see thousands of sea lions washing up on the beaches, many already dead. Others exhibited neurological symptoms similar to those seen in the pelicans.

Switching hosts from birds to mammals involves a series of critical mutations that enable the virus to latch onto different cell receptors. Once it does this, it essentially hijacks the host's cellular machinery for replication. In Peru, research has shown that several H5N1 viruses isolated from sea lions exhibited two key mutations (I won't add the names of the mutations because they're just another jumbling of numbers and letters!).[6] These mutations enhance the virus's ability to replicate in mammalian cells.

Scientists have discovered similar double-mutant H5N1 strains in Chile's human and sea lion populations.[7] They're puzzled by the infection routes of H5N1 in sea lions and other marine mammals in South America. Possible pathways include contaminated beaches or seawater and scavenging of deceased birds. It's easy to imagine that a lot of bird guano is spread across cliffs, beaches and water. It accumulates in large quantities in areas where seabirds roost or nest, such as coastal cliffs and island beaches. The bird flu virus can be present in guano. Infected birds shed the virus through their droppings, which can contaminate the environment, including water sources where other animals swim, play, mate and feed. This contamination potentially spreads the virus to other birds and mammals that encounter the infected guano.

Yet, a critical question remains: has the virus spread from mammal to mammal? A recent article discussed how, in theory, we could detect mammal-to-mammal transmission, where clustering patterns of H5N1 DNA sequences from mammals might indicate direct transmission.[8] However, large numbers of background DNA sequences from wild birds are necessary to confirm these patterns and avoid errors. This process ensures that mammalian clusters aren't simply due to random sampling.

The authors say that during the COVID-19 pandemic, many countries, including those in the global South, ramped up genomic sequencing capacity for SARS-CoV-2. However, this increase did not extend to other respiratory viruses, including H5N1. Consequently, there has been limited large-scale sequencing for H5N1 despite its impact on wildlife. Between 2021 and 2023, fewer than 7,000 H5N1 sequences were available globally from wild birds, poultry and mammals (including humans) combined. This contrasts sharply with the high number of SARS-CoV-2 sequences from a single state like Delaware, USA, with a population of just one million.

A recent study from Uruguay has revealed a startling new development in the spread of bird flu.[9] Sea lions were dying here before mass bird deaths occurred. This suggests that mammal-to-mammal transmission might be driving outbreaks in coastal South America. This discovery indicates that the virus has adapted to spread among mammals more efficiently than previously thought.

Huge Scare for the California Condor

The California condor (*Gymnogyps californianus*) has a fascinating lifestyle and a history dating back around 40,000 years, to the Pleistocene epoch.[10] These magnificent birds have impressive wingspans of over three metres. This makes them the largest birds in the Western Hemisphere.[11] An ancient South American relative of the condor, *Argentavis magnificens*, may have been the largest flying bird to have ever lived. These gargantuan birds had a wingspan of seven metres – that's 3.68 times the length of me.

California condors soared alongside – and indeed probably ate – mammoths and sabre-toothed cats. They scavenged for large animal carcasses across North America. However, by the twentieth century, their population had drastically declined due to a barrage of human-induced threats: habitat loss, lead poisoning from ingested spent ammunition, shooting, egg collection and environmental contaminants like DDT, which weakened their eggshells.

By 1982, only 22 California condors remained in the wild.[12] They teetered on the brink of extinction.

This dire situation prompted a controversial but necessary captive breeding programme. Between 1982 and 1987, the last wild condors were captured and brought into the programme led by organisations including the US Fish and Wildlife Service, the San Diego Wild Animal Park and the Los Angeles Zoo. The goal was to breed condors in captivity and reintroduce them into the wild. The captive breeding programme was a success. By the early 1990s, condors were being reintroduced into their natural habitats in California and later in Arizona and Baja California, Mexico.

Despite the success, the reintroduction faced challenges, including the persistent threat of lead poisoning. This threat led to efforts to ban lead-based ammunition in condor habitats. Conservationists also had to teach captive-bred condors to avoid modern hazards like power lines. Young condors were raised with minimal human interaction, using condor-like puppets to feed them. The conservationists also used GPS tracking devices to monitor their movements.

Thanks to these efforts, the California condor population has slowly but steadily increased. By 2020, there were over 300 condors in the wild and another 200 in captivity.[13] However, the birds face ongoing threats, including lead poisoning and habitat destruction.

Then, in 2023, along came H5N1.

In early March of 2023, a California condor settled on a cliff face in Arizona and stared into the distance for days. Tim Hauck, the condor programme director with the Peregrine Fund, thought the bird must have lead poisoning.[14] The condors feed on the dead bodies of coyotes (*Canis latrans*) and deer, which are often killed by hunters firing lead bullets. Hauck's team managed to trap the bird to do a closer health inspection. They noticed something not usually seen in a lead-poisoned condor. The bird had cloudy eyes.

The team immediately phoned a local vet, who confirmed that corneal oedema – the inflammation causing the cloudiness – was a bird flu symptom. They rushed the condor to a wildlife centre for emergency tests, treatment and care. While they awaited the results, news came that someone had spotted a dead condor at the bottom of the cliffs. The team sent the body for a fast-tracked necropsy. They knew this was something to fear. The condors were still critically endangered, and highly pathogenic avian influenza threatened their existence. The necropsy lab confirmed their fears: the condor had fallen victim to H5N1.

By April 2023, at least 21 California condors had died of bird flu.[15] This might not sound like a huge number, but it was a major blow to the condor. Indeed, it represented a 7% loss of the wild population. Just think about that in human terms. Earth has nearly eight billion people. Now, imagine the combined human population of Brazil and the USA wiped out in one season!

It's not hard to imagine the wave of panic that set in, especially as condors died almost daily for the subsequent few weeks. It was still quite chilly, and the condors nested in damp caves. These were ideal conditions for the flu virus to thrive. Yet, as the hot and sunny days returned, the virus seemingly vanished, and condors stopped dying.

In the wild, avian flu has traditionally extinguished itself by rapidly killing the infected, leaving behind survivors armed with antibodies that might provide some protection against future infections. But H5N1 is breaking the mould. Unlike its predecessors, this strain kills birds slightly more slowly, allowing them to spread the virus over greater distances. This is one reason it's so dangerous. Before the bird succumbs to the disease, chances are, it has inadvertently spread it to several other victims.

Ashleigh Blackford, the California condor coordinator for the US Fish and Wildlife Service, said the disease has an outsized impact on the condors because they're so social – they live in extended family groups. This behaviour makes the disease more communicable. She said they may have inadvertently cultivated it in the cool and damp caves where they nest.[16]

If only they practised social distancing.

Despite a relatively large chunk of the population succumbing to the virus in the spring of 2023, it could have been so much worse. Highly pathogenic avian influenza (such as H5N1) can kill up to 90–100% of domestic poultry it infects within 48 hours.[17] Recognising this, the race was on to convince the US Department of Agriculture to authorise an avian flu vaccine.

I don't know about you, but this made me think, "Hold on. Do we already have a suitable vaccine for H5N1 in birds?" The answer is yes. However, for political and practical reasons, it has never been authorised for wild birds or poultry in the USA. One reason is that if scientists inoculated the vaccine in poultry, there would be no sure way of knowing which birds were infected and which had been vaccinated. This ambiguity could lead to more difficulty in predicting and managing outbreaks. However, after many hours of discussion and deliberation, the US government decided to authorise the vaccine. They recognised the threat the virus posed to the condors' existence.

Having the vaccine provides hope, but it's not a cure-all. There's uncertainty about its full effectiveness, and it won't eliminate the virus from the wild. The hope is that it will provide the condors with a boost in immunity.

Before the vaccine could be used, it needed to be tested to ensure it was safe for the condors. The vaccine was developed with birds in mind, but not this species, so there's potential for adverse reactions. In May of 2023, federal biologists embarked on a mission at the Carolina Raptor Center in Huntersville, North Carolina, where they captured 28 wild black vultures (*Coragyps atratus*) to test the vaccine. They divided the birds into three groups: eight served as the control group with no vaccine, ten received a half-dose followed by another half-dose twenty-one days later, and the remaining ten were given a single shot.

The results of the test were promising. None of the black vultures exhibited adverse reactions. They didn't even show signs of mild bruising or swelling like humans might after a COVID-19 or tetanus shot. The vaccine's effectiveness was equally encouraging, with 90% of the vultures that received two doses developing some level of immunity, compared to 70% of those who received only one shot.[18]

Buoyed by these positive outcomes, researchers have now begun vaccinating condors in captivity. So far, 20 condors have received their first shots as part of the trial. Veterinarians meticulously monitor their health and collect blood samples to determine if they're responding to the vaccine as well as the black vultures did. They're also evaluating the effectiveness of combining the primary shot and the booster into a single injection, as opposed to the traditional two-step approach. If this is successful, it will minimise the number of times the condors need to be captured, reducing the stress on the birds... and the conservationists!

Bird Flu in Australia

In May 2024, a different strain of influenza A (H7N3) hit a poultry farm near Meredith in Victoria, Australia. H7 outbreaks are not a recent phenomenon in Australia. The earliest recorded H7 outbreak in the country was an H7N7 outbreak in Melbourne, Victoria, in 1976. The most recent occurrences were in 2020, affecting free-range farms in Lethbridge, Victoria.

Notice a pattern in the previous paragraph? I mentioned Victoria three times. It seems like Victoria is a hotspot for bird flu outbreaks in Australia. But why might this be?

Victoria has a land area of around 227,000 square kilometres, and it has around 250 large poultry farms.[19] This is the second-highest number of farms amongst Australian states. The state with the most poultry farms is New South Wales, with 360 in its land area of around 801,000 square kilometres. So, Victoria has 2.45 times more poultry farms per square kilometre than New South Wales. This high concentration is likely contributing to Victoria being a bird flu hotspot, because the proximity and high density of birds in poultry farms create ideal conditions for the virus to spread rapidly, as I'll discuss later in the chapter.

Victoria also lies along key migratory paths for various bird species. Migratory birds can carry avian influenza viruses long distances, introducing them to local poultry populations. Moreover, the climate in Victoria, which can be cool and wet, may facilitate the survival and transmission of avian influenza viruses. These conditions can help the virus persist in the environment, increasing the risk of outbreaks.

The H7N3 strain currently affecting birds in Australia is highly pathogenic, causing severe illness in poultry and wild birds. Scientists think that wild birds are the primary source of the virus, and they inadvertently transmit it to farmed or domestic poultry. Additionally, the virus can infect other animals, such as pigs (*Sus domesticus*) and horses (*Equus caballus*).

Unfortunately, the May 2024 outbreak on the farm near Meredith spread to other farms. As of June 2024, seven farms across Victoria had highly pathogenic bird flu strains. One farm infected by the disease housed between 150,000 and 200,000 egg-laying chickens (*Gallus gallus domesticus*). One million birds were culled to help stop the spread. Around 40,000 farmed ducks (*Anas* spp.) were also culled on another farm near Meredith after bird flu was detected.

How Intensive Agriculture Creates the Perfect Storm

Once upon a time, humans farmed animals and plants in a low-intensity and sustainable way. At some point in history, things changed. The era of intensive agriculture was born. When and how did this happen? And why does this influence the emergence, transmission and amplification of pathogens?

The early twentieth century saw the beginnings of more intensive farming practices with advancements in agricultural science, mechanisation, and synthetic fertilisers and pesticides. The introduction of tractors and other machinery increased the scale of farming operations. Essentially, it became easier to grow more and rear more at scale.

The period after World War II marked a significant acceleration in the development of intensive agriculture. The ironically named Green Revolution, which began in the 1940s and 1950s, introduced high-yield crop varieties, chemical fertilisers, pesticides and irrigation techniques. The name is ironic today because intensive farming practices are rather damaging to the environment. They're about as green as a plastic factory.

Innovations during this period led to a dramatic increase in agricultural productivity. During the 1950s and 1960s, agriculture in high-income countries, particularly North America and countries in Europe, became increasingly industrialised. Synthetic inputs (fertilisers and pesticides), mechanisation and selective breeding of crops and livestock expanded. This period also saw the consolidation of smaller farms into larger business operations. In the 1970s and 1980s, intensive livestock farming, also known as factory farming, became more prevalent. This involved raising large numbers of animals, such as chickens, pigs, and cattle (*Bos taurus*), in confined spaces to maximise production efficiency. In the following decades, the globalisation of the food supply chain and advances in biotechnology, including genetic modification of crops, have further intensified agricultural practices. So-called 'precision farming' techniques, which use data and technology to optimise farming practices, have also become more common.

So, that's the history in a nutshell. In evolutionary terms, intensive agriculture is a very recent phenomenon. This means we've created novel environmental conditions and selection pressures that the world has never seen before. But how do these conditions represent a perfect storm for viral diseases like influenza?

Firstly, intensive agriculture involves keeping large numbers of animals in close quarters. This high density facilitates rapid transmission of viruses among the population, as infected individuals can easily encounter others. It's a physical distance issue. Recall we were doing the opposite during the early days of COVID-19 – we called it 'social distancing'. Simply put, reducing physical proximity makes it harder for the viral particles (virions) to colonise a given host.

The high density of animals also creates an environment where viruses can quickly evolve and adapt, potentially leading to more virulent or transmissible strains. Think about it: when animals are packed closely together, a virus can spread rapidly from one to another. This constant exposure gives the virus more opportunities to mutate and become stronger. Each new host provides the virus with a slightly new environment to adapt to, increasing the likelihood of mutations that enhance its survival and spread. In such conditions, it's not just the speed of transmission that's a concern but also the chance for the virus to develop into something even more dangerous. Indeed, viruses, especially RNA viruses like influenza, have high mutation rates. Intensive farming environments, with their dense populations of susceptible animals, create ideal conditions for viruses to mutate and mingle with other strains – sometimes giving rise to more virulent and easily spread forms.

The conditions in intensive farming settings, such as limited space, high population density and artificial environments, can also lead to stressed animals. Stressed animals have weakened immune systems. This makes them more susceptible to infections. Stress is a silent saboteur. It's another reason these environments are hotbeds for pathogens.

There's also the issue of genetic uniformity. Commercial farming often involves raising animals with similar genetic backgrounds to ensure consistent production. This genetic uniformity means that if a virus can infect one animal, it can likely infect many others in the same population with similar efficiency. Without genetic diversity, the entire population is more vulnerable to being wiped out by a single disease outbreak. This is because there are fewer chances for some individuals to have natural resistance to the virus.

What could add to this cauldron of doom? Well, intensive farms can also attract wild animals. These can be natural reservoirs for many viruses. This increases the risk of viruses jumping from wild populations to domestic animals. Moreover, the high density and movement of animals in intensive farming operations mean that once a virus is introduced, it can spread quickly through the entire population and potentially to other farms through trade and transport. I'll talk more about trade later in the book, but intensive agriculture is often part of a global network involving the trade of live animals, animal products and equipment. This global movement can carry viruses across great distances, fuelling the emergence of outbreaks in wildlife – and potentially sparking pandemics in humans.

I find it baffling that we continue this type of farming. We know it's inordinately unethical from an animal welfare perspective. We know it contributes to climate change and ecosystem degradation. We know it increases the risk of wildlife and human pandemics. Yet, we continue to buy into it.

While wildlife plays a role in the emergence of new diseases, such as those caused by novel influenza strains, the conditions created by industrial agriculture facilitate the spread and mutation of viruses, increasing the likelihood of disease outbreaks. We must tackle the root causes of this persistent drive for high productivity and low costs.

And what are the root causes of this drive? Economic pressures, consumer demand for cheap food and the pursuit of profit by large agribusinesses all contribute. Due to economic constraints, producers and consumers often prioritise cost over other considerations. Where there's political will, there's a way to fix this. But who's going to do it first?

Bird Migration and Immunity

As we touched on in Chapter 1, migratory birds can fly many thousands of kilometres to reach their seasonal locations. One species holds the record for the longest migration in the animal kingdom – the Arctic tern (*Sterna paradisaea*). This medium-sized bird travels up to 90,000 kilometres from pole to pole each year, journeying from Greenland in the north to the Weddell Sea in the south. Arctic terns can live up to 30 years, meaning in their lifetime they travel a vast distance, the equivalent of going to the moon and back more than three times.[20]

Understandably, migratory birds often need to make pit stops (unless the bird is the bar-tailed godwit!) to feed and recharge their energy levels, ready to fly these ultra-ultramarathon distances. I remember watching the Brent geese (*Branta bernicla*) landing in the Solent in the UK each winter after travelling around 5,000 kilometres from Siberia. They start arriving in October, and their numbers will reach around 30,000 by January. It was a sight (and sound) to behold. They often pitstop in Scandinavia before they arrive.

Migratory birds don't just take pit stops to recharge their energy levels. They also do it to recover and boost their immune systems to prevent them from succumbing to diseases. Some birds can sustain physical activity at a rate 20 times higher than their base metabolic rate. Most mammals would not survive such exertion due to a breakdown in essential physiological functions. To achieve such physical feats, the birds essentially downregulate their investment in their immune system.

Migration exposes birds to new pathogenic pressures, like when humans pick up unfamiliar bugs when they go on holiday. However, because the immense journey ravages the birds' immune systems, their susceptibility to novel pathogens increases. During stopovers, birds often gather in dense groups – again, the opposite of social distancing! The birds usually share these pit stop locations with other species, too. This can give the pathogens a higher chance of jumping the species barrier. Nonetheless, there's growing evidence that these pit stops, however risky, are also vital for recharging the birds' immune systems.

In a recent study, researchers investigated migrating birds, focusing on common redstarts (*Phoenicurus phoenicurus*) journeying from Europe across the Sahara Desert.[21] They also looked at dunnocks (*Prunella modularis*) and chaffinches (*Fringilla coelebs*) at the same stopover site. The study explored two key aspects of bird immunity: innate and adaptive. We often call innate immunity 'non-specific immunity'. It's part of the vertebrate (including humans) immune system that attacks

Dark-bellied Brent geese in the Solent

and immobilises anything that tries to invade the body. Ordinarily, its regulators prevent it from attacking innocuous substances like dust and pollen. However, if something goes awry with these regulators, allergies can happen. And when something goes majorly awry, the innate immune system can attack the body's own cells, manifesting as autoimmune diseases. Nonetheless, it's a defender that continuously prevents us, as vertebrates, from getting sick. In the bird study, innate immunity – the genetic resistance present from birth – was tested by introducing *Escherichia coli* bacteria into blood samples to observe their spread – a vivid indicator of immune strength. Essentially, if the bird's immune system is poor, the *E. coli* spreads through the Petri dish like ink in water.[22]

Meanwhile, adaptive immunity is built over a lifetime of combating pathogens. The body's adaptive immune system must be exposed to many different types of microbes from an early age to build up tiny armies of immune 'memory' cells. These cells 'remember' the pathogens they are exposed to and can mount a more efficient immune response the next time they try to invade the body. The bird study assessed adaptive immunity by introducing rabbit (*Oryctolagus cuniculus*) blood cells to gauge the birds' ability to recognise and respond to foreign agents.[23]

The results were striking: birds that lingered at the stopover site exhibited enhanced immunity across both measures compared to recent arrivals. This finding supports the hypothesis that the migratory pit stops bolster bird immunity.

There's an urgent need to safeguard these crucial habitats, as disruptions could compromise birds' immune readiness. Indeed, there's a concern that removing or damaging such habitats leaves the birds more vulnerable to diseases like highly pathogenic avian influenza. And the barrage of stressors from climate change and pollution doesn't help.

A recent paper reviewed the effects on birds of some of the common human-made pollutants: light, noise, polluted air, heavy metals, radioactive compounds, pesticides, pharmaceuticals, oil and plastic pollution.[24] The authors found that pollutants released in the environment worldwide affect bird fitness, reproduction and survival. This interconnectedness is why we need to 'think in systems'. Our actions in one realm could cascade through the ecosystem, increasing the chances of wildlife and human pandemics. Each thread in the web of life is connected; tug on one, and the whole web may unravel.

CHAPTER 13

Spongy Brains and Zombie Deer – a Prion Predicament

*Everything is made of folds – the earth, our DNA,
illness and health in the folding and unfolding of
protein molecules.*
—Sharon Small, *Clean Language*

A deer retreats from her herd, isolating herself. She stands listlessly by the water's edge, forgetting the rituals of drinking, her body thinning as the urge to feed wanes. Once bright and aware, her eyes now gaze vacantly, mirroring the slow erosion of her once spritely temperament. She's disorientated. Her movements are erratic, as if she's lost. But she's in a familiar landscape. She's wasting away. Her legs are folding under her – a grim consequence of the misfolded proteins in her brain.

For this story, we could be in the United States, Canada, Norway, Finland, Sweden or South Korea, but let's go where it first began: the Midwest USA. It was 1967. Some major cultural events were happening – it was the Summer of Love, the Outer Space Treaty was signed, and racial tensions led to the Detroit Riots. While this happened, scientists used captive mule deer (*Odocoileus hemionus*) for nutrition research in a Fort Collins, Colorado facility.

The deer exhibited concerning symptoms. They appeared lethargic, displaying signs of despondency such as drooping heads and ears. They lost their appetites, shed weight rapidly and eventually succumbed to emaciation, pneumonia or related complications. Researchers initially assumed the deer were reacting badly to captivity – they were probably stressed, had nutritional deficiencies or picked up toxins, they thought.

Following a regular flow of cases, scientists labelled it 'chronic wasting disease' or 'CWD'. A decade later, chronic wasting disease was recognised as a type of neurodegenerative disorder known as spongiform encephalopathies. It was the cousin of the infamous bovine spongiform encephalopathy or 'BSE', commonly referred to as *mad cow disease*.

I say 'infamous' as if it doesn't require an explanation. It's probably a good idea to talk about bovine spongiform encephalopathy here for four reasons:

1. It's a similar disease to chronic wasting disease, caused by the same type of disease agent.
2. Chronic wasting disease was still largely ignored when bovine spongiform encephalopathy emerged.
3. There are some lessons to learn from bovine spongiform encephalopathy that are very relevant to chronic wasting disease.
4. I have a living memory of the hysteria that surrounded the bovine spongiform encephalopathy outbreak late last century.

Both chronic wasting disease and bovine spongiform encephalopathy are caused by a misfolded protein known as a *prion*. These misfolded proteins turn the bovine and cervine brain into a twisted, sponge-like mass. Unlike bacteria or viruses, prions almost defy conventional biology; they replicate by converting normal proteins into their malformed, deadly counterparts, spreading the disease stealthily through herds. Because it's so unconventional, this is a truly fascinating disease.

Folding

Folding, we've all done it. Whether tucking a piece of paper into an envelope or wrangling a shirt (mine are often *misfolded*, to put it mildly) into a drawer, folding is a familiar task in our daily lives. But it usually has a trivial or cosmetic function. If your shirt drawer is a chaotic mess, it's unlikely that your life will end abruptly or the universe will implode. However, in the realm of biology and biochemistry, folding is a fundamental process that shapes the functioning of proteins, the workhorses of life itself.

We can think of a protein as a long string of beads, each bead representing an amino acid – a small molecule with unique properties.

When the body's biochemical factory synthesises a protein in the cell, it starts as a linear chain of these beads. Folding is the intricate and precise process by which this linear chain of amino acids twists and contorts into its functional three-dimensional shape. This shape is crucial because it determines how the protein interacts with other molecules in the cell and carries out its specific tasks, whether it's catalysing chemical reactions, providing structure to cells or regulating gene expression. All these things are vital to life.

The folding process is not haphazard but guided by the sequence of amino acids, which dictates how the chain folds into its final structure. Proteins fold into complex shapes driven by interactions between amino acids – hydrogen bonds, van der Waals forces (weak, short-range forces of attraction between atoms or molecules) and other invisible interactions. These forces pull the chain into coils, sheets and loops, forming the intricate folds that define a protein's structure.

A prion (misfolded protein)

However, folding isn't always straightforward. Proteins can misfold, leading to dysfunction and diseases known as protein misfolding disorders, which can be infectious. This is where the term 'prion' comes from: **pr**otein and infect**ion**.[1] Researchers have linked diseases like Alzheimer's, Parkinson's and prion diseases to misfolded proteins that clump together, disrupting normal cellular function.[2]

Unlike bacteria, viruses or fungi, prions do not contain genetic material such as DNA or RNA. Prions are strange and dangerous because they're *just* misfolded proteins. Think about a protein in your body that suddenly folds incorrectly, becoming infectious and causing chaos. This misfolded protein is a troublemaker, convincing other normal proteins to fold incorrectly. The process can happen independently, but speeds up when already misfolded proteins are around. As these harmful proteins pile up in the brain, they mess with how cells work. They lead to nerve damage and, eventually, symptoms of severe brain diseases.

Bovine Spongiform Encephalopathy

Before we go deeper into chronic wasting disease, I'd like to talk about its cousin, bovine spongiform encephalopathy. In the 1980s and 90s, British countryside cows unwittingly harboured a menace that took many people by surprise; the hysteria was palpable.

As a child in the 1990s, I recall hearing stories of how eating beef burgers could turn you into a zombie. In 1997, UK councils announced a ban on beef in some 2,000 schools.[3] This was a somewhat tardy decision by the government, especially given that we'd known about the disease for several years. At the time, scientists had suspected a link between bovine spongiform encephalopathy and its human manifestation, Creutzfeldt-Jakob disease (CJD), for at least two to four years. I won't talk too much about the human form of the disease (after all, this is the wildlife pandemic section of the book), but it has played a role in our understanding of wildlife diseases, so I mention it for context.

Few food-related diseases have caused as much fear and upheaval as bovine spongiform encephalopathy. It seemingly emerged in the 1980s, but this story goes back several centuries. The most popular theory for how bovine spongiform encephalopathy began is that it came from another prion disease, scrapie. In 1732, our ancestors first recorded scrapie in sheep (*Ovis aries*).[4] The understanding of its cause and nature was limited. In fact, knowledge of most diseases back then was rudimentary. Farmers and shepherds observed its symptoms, which included itching, rubbing against objects and behavioural changes. However, the disease remained poorly understood and was often confused with other sheep ailments. The term 'scrapie' likely originated from a British dialect, referring to the sheep's tendency to scrape or rub against objects due to intense itching. Farmers noticed that affected sheep would gradually waste away, losing weight and eventually dying, which distinguished scrapie from more acute diseases.

A passage from agricultural literature from 1759 reads:

> Some sheep also suffer from scrapie, which can
> be identified by the fact that affected animals lie
> down, bite at their feet and legs, rub their backs
> against posts, fail to thrive, stop feeding and
> finally become lame. They drag themselves along,
> gradually become emaciated and die. Scrapie is
> incurable. The best solution, therefore, is for a
> shepherd who notices that one of his animals is
> suffering from scrapie to dispose of it quickly and
> slaughter it away from the manorial lands, for
> consumption by the servants of the nobleman.
> A shepherd must isolate such an animal from
> healthy stock immediately because it is infectious
> and can cause serious harm to the flock.[5]

Later, in 1883, French vets reported the first case of scrapie in a cow, and by the 1920s, rendering, where slaughterhouse remains are processed into animal feed, became widespread among farmers. It was around this time that the first cases of 'classical' Creutzfeldt-Jakob disease were reported – a form of the brain disease that can be transmitted through genetic mutations that predispose people to prion diseases.

In 1957, kuru, a transmissible spongiform encephalopathy (TSE), was found in a tribe of New Guinea cannibals.[6] The disease is character-ised by symptoms that include tremors, loss of coordination and neuro-logical decline. It's also known as laughing sickness due to pathological bursts of laughter – another disease symptom.

Kuru was first identified by Australian physician Vincent Zigas. American physician Daniel Carleton Gajdusek and Baruch Blumberg won the Nobel Prize in Physiology or Medicine in 1976 for their work on the disease. Kuru was prevalent among the Fore people, an Indigenous group in Papua New Guinea. Their funeral practices involved ritualistic cannibalism, particularly the consumption of the brain tissue of deceased relatives. The early symptoms of kuru include headaches, joint pain and slight shaking of the limbs. As the disease progresses, affected individuals experience loss of muscle coordination, difficulty walking (ataxia), tremors and emotional instability. In the advanced stages, individuals may suffer from severe tremors and difficulty swallowing and eventually become bedridden, leading to death within a year or two of the onset of symptoms. A prion causes kuru. Like other prion diseases, it causes spongiform changes in the brain tissue, characterised by sponge-like holes, leading to brain deterioration.

At its peak, kuru had a significant impact on the Fore people, leading to high mortality rates, especially among women and children, who were more involved in ritualistic cannibalism. The incidence of kuru declined after the practice of cannibalism was discouraged. However, the disease lingered due to a remarkably long incubation period of ten to over fifty years. Today, there have been no confirmed deaths since the 2000s.

The study of kuru contributed significantly to understanding prion diseases, including other disorders such as bovine spongiform encephalopathy. By the 1970s and 80s, scientists thought scrapie had reappeared in cattle as bovine spongiform encephalopathy, a jump in the species barrier and evolution of a distinct disease.

In February 1985, the first signs of bovine spongiform encephalopathy were confirmed.[7] A cow known as 'Cow 133' exhibited symptoms including weight loss, head tremors and lack of coordination, eventually succumbing to the disease. A clinical report subsequently identified the symptoms of a novel, progressive spongiform encephalopathy in cattle. At first, its effects appeared insidious – subtle changes in cow behaviour, unexplained stumbling and a health decline – much like the chronic wasting disease in the Coloradan deer. As the disease progressed, affected cows exhibited serious neurological symptoms, including tremors, loss of coordination and an eerie vacant stare. The discovery sent shockwaves through the agricultural world. Cows, seen as a product of agricultural abundance, became potential vectors of an outbreak.

The following year, in 1986, the disease was officially recognised. Government ministers were informed about the novel disease, and the consumption of meat and bone meal was identified as "the only viable hypothesis for the cause of BSE".[8] It was kuru all over again, but this time in non-human animals. Researchers linked the disease to the practice of feeding cattle with meat and bone meal that contained the rendered remains of other cattle – essentially forced cannibalism – which led to the spread of prions. In June 1988, the Bovine Spongiform Encephalopathy Order was swiftly passed, banning the use of certain types of meals for cattle.

In May 1990, the then agriculture minister, John Gummer, claimed that beef was "completely safe" and even appeared on a TV show encouraging his four-year-old daughter, Cordelia, to bite into a beef burger. However, between 1993 and 1995, four cases of Creutzfeldt-Jakob disease were reported in dairy farmers whose cow herd was affected by bovine spongiform encephalopathy, and in 1995, the first known victim of variant Creutzfeldt-Jakob disease, 19-year-old Stephen Churchill, died, followed by three others.[9]

In 1996, the health secretary, Stephen Dorrell, officially announced that there was a "probable link" between the cattle disease and variant Creutzfeldt-Jakob disease, and so, contrary to Gummer's claim back in 1990, eating beef might not have been so safe after all. The EU imposed a global ban on all British beef exports.

Variant Creutzfeldt-Jakob disease cases peaked in the UK in 2000, and the last known case in the UK was detected in 2016. However, Richard Knight, a senior neurologist based in Edinburgh, said there were still people "silently infected".[10] The incubation period could be 30 years or longer.[11] A report by a committee of MPs in 2014 said there could be tens of thousands of people carrying the prions, and blood transfusions were likely a key source of transmission.[12] There's currently no test to discover if someone is infected with the prion disease, and there is no cure.

Back to Chronic Wasting Disease

As bovine spongiform encephalopathy unfurled across the pond, chronic wasting disease in deer and other hooved wild animals in the US received limited attention. In the early 2000s, concerns emerged about the potential spread of prions from wild deer to cattle and humans. This concern was heightened by the country's vast amount of meat consumption. Indeed, in the US, the annual per capita meat consumption is one hundred and forty-four kilograms, compared to, for instance, just four kilograms in Bangladesh.[13]

Chronic wasting disease began to appear in farmed elk (*Cervus canadensis*), and in the early 2000s, the disease breached the Rockies. It began to take animals on the mountains' western slopes by storm, and then it unexpectedly appeared in Wisconsin, the 'dairy state', at least a thousand kilometres to the east. Finally, the government began to take notice. By 2003, chronic wasting disease had been found in 12 US states, either in wildlife or farmed animals or both. By 2024, it had been reported in 32 states, covering all four regions – the West, Midwest, South and Northeast.[14]

Incidentally, hunters and their families unknowingly eat up to 15,000 infected animals annually, a number that increases each year as the disease spreads.[15]

Should people be concerned? I would be. Why? See the previous section on bovine spongiform encephalopathy!

Chronic wasting disease has a prolonged incubation period, typically 18 to 24 months, during which infected animals exhibit no obvious symptoms and appear normal. This makes it difficult to spot in the early stages. The most noticeable sign of the disease is gradual and significant weight loss. Then the animal's once alert ears droop, their coat losing its healthy sheen. As the disease progresses, more sinister symptoms emerge. Some animals start to show excessive salivation and difficulty swallowing, their tongues lolling out of their mouths, and they seem to lose the innate ability to graze. In the final stages, the deer stumble and fall, their hind legs weakening, and their heads sway with a lack of control. The neurological unravelling is now painfully evident.

How did Chronic Wasting Disease Evolve?

The exact origins of the disease remain a mystery, and pinpointing the precise moment or manner of its emergence may never be possible. We know it was initially observed as a condition affecting captive mule deer in wildlife research facilities in Colorado during the late 1960s. Whether it first emerged here is unknown. Computer models suggest that the disease could have been present in free-ranging mule deer populations since at least the 1960s. There's also a hypothesis that scrapie in domestic sheep may have been the source, like in bovine spongiform encephalopathy.[16]

Scrapie has been recognised in the United States since 1947, and there's the (unproven) possibility that deer may have contracted the disease from scrapie-infected sheep roaming the pastures, particularly along the front range of the Rocky Mountains, where intensive sheep grazing was common in the early 1900s. Scrapie has been experimentally transmitted to elk and white-tailed deer, showing they are susceptible to the disease, i.e., it can jump the species barrier.[17] Given the severity and distinctiveness of the symptoms, it is plausible that the condition was not widespread before the twentieth century – one might expect that such a striking disease would have been recognised earlier if it had been more prevalent.

Like many other emerging infectious diseases, it's increasing with human-induced habitat destruction and high-intensity agriculture. Approximately 80% of the world's agricultural land is dedicated to raising livestock, such as grazing land or cropland used to grow animal feed.[18] In the United States, around 99% of livestock are estimated to be factory-farmed, including over 98% of chickens, turkeys (*Meleagris gallopavo domesticus*) and pigs.[19]

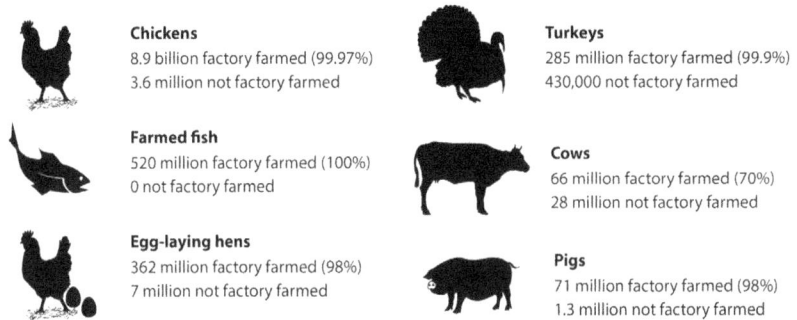

Chickens
8.9 billion factory farmed (99.97%)
3.6 million not factory farmed

Turkeys
285 million factory farmed (99.9%)
430,000 not factory farmed

Farmed fish
520 million factory farmed (100%)
0 not factory farmed

Cows
66 million factory farmed (70%)
28 million not factory farmed

Egg-laying hens
362 million factory farmed (98%)
7 million not factory farmed

Pigs
71 million factory farmed (98%)
1.3 million not factory farmed

United States factory farm statistics from 2017 (Our World in Data)[20]

Nearly 44% of the world's habitable land – about 48 million square kilometres – is dedicated to agriculture, an area roughly five times larger than the United States. Of this, one-third is used for growing crops, while two-thirds serves as grazing land for livestock.[21] This is a lot of land with a high animal density, which opens opportunities for novel pathogen emergence and spillover events.

How is it Transmitted?

Like bovine spongiform encephalopathy, chronic wasting disease is primarily transmitted through direct contact with infected animals or their bodily fluids: saliva, blood, urine and faeces. Infected deer, elk or moose can transmit the disease to other animals during social interactions or communal feeding.

In addition, the prions can persist in the environment for years, and soil, plants and water contaminated with prions from infected bodily fluids can serve as sources of infection for healthy animals. One study analysed the retention of prions and their infectivity in grass roots and leaves incubated with prion-contaminated material.[22] They found that even low concentrations of the prions can bind to the roots and leaves. The authors also found that plants can suck prions up from contaminated soil, transporting them to their leaves. By doing this, the plants can act as carriers of the disease and horizontally transfer it.

So, a deer urinates (or defecates) on some grass. Then, another deer comes along and eats the grass and becomes infected. Alternatively, a deer urinates on a bare patch of soil. Then, a few wind-blown grass seeds land in the soil, grow and uptake the prions. Another deer comes along, eats the grass and becomes infected. These deer then urinate on another patch of grass or soil, and the disease proliferates.

Scientists already knew that prions are good at binding to soil, especially sticky clay-based types. In one experiment, researchers injected soil into animals, and they contracted the prion disease.[23] Several years before, an animal infected with chronic wasting disease had been buried in this same soil. This shows that prions can not only bind to the soil but can persist there for years. Therefore, another scenario is that when infected animals die in the wild – and there may be thousands each year – their prions are released into the soil as they decompose. The plants that grow in the soil transpire and absorb water and nutrients in their roots; they also take up the prions, bringing new life to the ecosystem, but maybe also death.

There's also evidence suggesting that chronic wasting disease can be transmitted from mother to offspring during gestation or through nursing, although this method of transmission is less common.[24]

Has Chronic Wasting Disease Been Recorded Outside of the USA?

At the beginning of this chapter, I mentioned that the story could be set in the United States, Canada, Norway, Finland, Sweden or South Korea. These are all the countries where chronic wasting disease has been recorded to date.

The first case of the disease in Europe was recorded in March 2016, in a free-ranging reindeer (*Rangifer tarandus*) in the snowy landscape of Nordfjella, Southern Norway.[25] A team of researchers had embarked on a mission to study the interactions between reindeer and humans. They aimed to place radio collars on individuals within a herd of about 400 free-ranging reindeer. Using a helicopter for the operation, the crew aimed to temporarily immobilise a reindeer with a dart for collaring. However, their search took an unexpected turn when they discovered a distressed reindeer near where they had initially approached the herd. This female reindeer had separated from the group, and her tracks showed that she was dragging one of her right limbs. The researchers found her lying on the ground, still responsive but in visible distress. She had frothy lips and a feverish body temperature of 41.9 degrees Celsius, and unfortunately, she died shortly after they found her.

Subsequent examination revealed the reindeer was an adult, estimated to be between three and four years old, and in poor body condition, weighing forty-three kilograms with minimal body fat. She displayed hair loss in patches and had suffered from multiple haemorrhages and muscle ruptures, particularly in her hindquarters. Her lungs showed signs of congestion and oedema, and she had damage to the right cranial lobe.

The emergence of chronic wasting disease in Norway raises questions about its origins. While the importation of disease-infected deer is a known transmission route, Norway has stringent laws prohibiting the importation of live cervids, with rare exceptions for moose imported to zoos from Sweden. The red deer farmed in Norway all stem from wild local populations, making it unlikely that chronic wasting disease came from legally imported animals. However, the possibility of illegal imports from North America, where chronic wasting disease is prevalent, cannot be entirely ruled out. One study suggests this is a noteworthy consideration, given that Finland's white-tailed deer population originated from a 1934 import of four does and one buck from North America.[26] However, they underwent extensive testing between 2003 and 2015. All 643 deer tested *negative* for chronic wasting disease.

Another potential source of the disease contamination could be hunting urine baits imported from North America. However, there's currently no information confirming the use of such baits in Norway. For those of you who haven't come across urine baits, they typically consist of urine from a specific species or synthetic compounds that mimic the scent of urine. The strong, recognisable scent of urine is a lure for animals, drawing them to a particular location. Many hunters prefer natural urine due to its authenticity, but some have suggested this form may harbour the prions.[27] The synthetic form will most likely be prion-free. Urine can be sprayed around an area to create a scent trail or to concentrate the scent in a specific spot, slowly released over time to maintain a constant scent. It can also be soaked on a cloth and placed strategically to attract animals.

The exact route of chronic wasting disease introduction in Norway is still a subject of investigation. However, recent research has questioned the import route, because the prion is a novel strain in many of the samples collected from Norwegian reindeer.[28]

Kuru in Reindeer?

Some have suggested that the prions could have jumped the species barrier from sheep (i.e., scrapie), but another intriguing hypothesis exists.

I recently came across a paper titled 'Antler cannibalism in reindeer.'[29] The paper's introduction includes the following line: "Here, we documented the unique and bizarre phenomenon of extensive gnawing of intact antlers among the reindeer of the affected population in Norway." 'Osteophagia' (literally 'bone-eating') is the scientific term for bone and antler consumption. It's a fascinating and well-documented behaviour among ungulates. Typically, these animals gnaw on shed antlers or bones from carcasses, often driven by a need to address mineral deficiencies. This behaviour is not unique to bones and antlers; herbivores also engage in geophagy ('soil-eating'), the ingestion of soil, as well as predation and carnivory, particularly when they lack essential minerals.

I wrote about this phenomenon in humans in my first book, *Invisible Friends*.[30] Indeed, researchers have documented societies where eating soil is fairly common, with children consuming dirt well into their teenage years and in relatively large quantities. Recent studies have revealed that pregnant women frequently buy and eat soil sticks, known locally as 'pemba' and 'kichugu', sold in the markets of Central and East Africa.[31] The locals make these sticks from soil sourced from house walls, termite mounds and the ground. This practice has a surprisingly long history around the world. In some countries, you can even find packets of soil in various flavours, such as black pepper and cardamom, and in some grocery stores, it's marketed as 'pregnancy clay'. Theoretically, consuming soil could offer benefits, including the intake of beneficial microbes and micronutrients, and the same concept applies to gnawing on bones and antlers – they're packed full of nutrients such as calcium and phosphorus.

However, in the antler-gnawing study, the researchers recorded something different: extensive antler cannibalism among reindeer in populations infected with chronic wasting disease occurring even before the deer shed their antlers. In other words, reindeer were nibbling away at their friends' antlers while they were still on their heads!

This behaviour has increased significantly over the past decades. It has coincided with the emergence of chronic wasting disease and involves the ingestion of vascularised antlers (antlers with an active blood supply). As we've discovered in the previous examples, consuming tissues from a member of the same species poses a risk for the emergence of prion diseases. The study's authors say the widespread antler cannibalism among these reindeer raises the intriguing possibility of a 'Kuru-like' origin for chronic wasting disease in the European reindeer population.

A male reindeer with a beautiful set of antlers

What Caused this Behavioural Change in the Reindeer?

Given that the antler cannibalistic behaviour has increased in the last few decades, the question then becomes: what triggered this change? The general consensus is that the reindeer practise osteophagia to replace or supplement essential minerals, particularly calcium and phosphorus. Timing is important, too. Male deer will snack on antlers when they need minerals to grow their own antlers, and females will do it for a calcium fix when pregnant and lactating.[32]

Has something in the environment recently changed to make reindeer need these nutrients so much that they now resort to antler cannibalism? I can't find any information on why this behaviour change occurred, but let's do a thought experiment.

Suppose the reindeer's natural diet begins to lack essential minerals due to changes in vegetation or soil quality; they might be driven to seek alternative sources, including live antlers. Why might this happen? Climate change and harsh weather conditions could affect the availability and distribution of food resources. Human-induced habitat alterations could also be a factor. For instance, logging, mining and urban development can degrade and fragment habitat. This can reduce the availability of food sources and push the reindeer to adapt their feeding behaviour. Moreover, if reindeer are stressed from environmental pressures, they may engage in atypical behaviours like antler cannibalism. At a fundamental level, prolonged stress increases the body's metabolic needs.[33] But is there any evidence for any of this?

Climate change is already affecting reindeer in these regions. For instance, researchers recently wrote:

> The increase in the frequency of the freeze-thaw
> and rain-on-snow events – Goavvi in Sámi or
> 'the bad year caused by the ice and snow on the
> reindeer grazing pastures' – has increased from
> once every 50–100 years to frequencies on decadal
> times. These events have devastating consequences
> with reindeer losses as the herd cannot reach
> their food: the lichen beneath the frozen surface
> contains the essential carbohydrates of life.[34]

Another study suggested that enhanced ultraviolet-B rays (due to climate change) will likely increase the concentration of plant-based phenolics, decreasing forage quality and choice.[35] Phenolics can make the plants less palatable and interfere with the digestion and absorption of nutrients.

Evidence also suggests that the forest landscape mosaic in Scandinavia is becoming less suitable for reindeer herding. Logging and fragmenting old-growth forests serve to reduce the amount of available lichen, the reindeer's favourite food.[36]

I'm unsure if this is a major contributor, but what about antler collection? People love to go out and collect fallen reindeer antlers. Here's a quote from a travel website: "Finding your own reindeer antler is a great experience and the perfect trophy to remember your adventures in the North."[37] Some (non-peer-reviewed) sources claim that female reindeer keep their antlers and drop them in a place where they need them the most (i.e., they gnaw on them for nutrients), as the harsh Arctic environment leaves little room for error for species that live there. Whether or not antler collection contributed to the changing behaviour of the reindeer is an open question. But if people are collecting their antlers, the reindeer can't return to them to gain the nutrients.

What about the prions themselves? Am I looking at the relationship the wrong way around? Could the prions cause the deer to change their behaviour to increase their transmission? Is this why the increase in antler-gnawing behaviour coincided with the increased prevalence of chronic wasting disease?

While no direct evidence is currently available to support the notion that prions can influence behaviour in such a targeted way, it's not entirely outside the realm of possibility, given the known effects of prion diseases on the brain and behaviour. Prion diseases cause significant neurological damage. This can lead to behavioural changes, although these are usually symptoms of disease progression rather than adaptive changes selected by the prion itself. In theory, if the prions were to cause behaviours that increased the likelihood of transmission (such as antler gnawing), this could enhance the spread of the disease. However, it's unlikely that prions have a mechanism for actively selecting or promoting specific behaviours; any such changes would likely be incidental results of brain damage. Indeed, increased antler gnawing among infected reindeer could be an incidental behaviour caused by neurological damage or nutrient deficiencies.

Suppose this behaviour happens to increase prion transmission. In that case, it might seem as if the prion is selecting for this behaviour. Still, viewing it as a coincidental outcome rather than a directed evolutionary strategy would be more accurate. What if chronic wasting disease led to a nutrient deficiency, which triggered the behaviour? The selective advantage for the reindeer would be the easy access to nutrients (in nearby members of the same species), and the advantage of enhanced transmission to the prion would still be incidental.

Human activities driving climate change and ecosystem degradation can change the health and behaviour of the reindeer and other cervids. It's all interconnected. Novel diseases in wildlife emerge as a direct consequence of our impact on ecosystems and climate. Widespread ecosystem degradation has occurred across the planet, including in the United States, where chronic wasting disease was first recorded. Indeed, as I discuss in another book, *Treewilding*,[38] half of all deforestation in history has occurred since 1900, and settlers in the United States wiped out vast areas of ecosystems in such a short timescale. These large-scale changes have global repercussions.

Globalisation is also driving the emergence of prion diseases – infected deer from Canada were the source of chronic wasting disease cases found in South Korea. The disease was first detected in South Korea in 2001, in elk at a privately owned deer farm. Since then, scientists have discovered the disease in various captive deer and elk populations nationwide.

Climate change, ecosystem degradation, travel and trade – they're all contributing to the spread of spongy brains in wildlife across the world. We must recognise that the health of our planet and its inhabitants is inextricably linked to our actions.

CHAPTER 14

Roots, 'Shrooms and Webs

Fungi constitute the most poorly understood and
underappreciated kingdom of life on Earth.
—Michael Pollan, *How to Change Your Mind*

The horizon was once a shimmering emerald spectacle; now it's black and wilted. Beneath the canopy, leaves wither and fall, littering the forest floor. Branches, once sturdy and strong, now snap like brittle bones. Ash dieback, they call it. It's a disease that's taking European ash trees (*Fraxinus excelsior*) by storm.[1]

My childhood memories of ash trees are rich and diverse. I would climb them and perch in their branches with the squirrels (*Sciurus carolinensis*), bluetits (*Cyanistes caeruleus*) and goldfinches (*Carduelis carduelis*). I remember watching their seeds (also known as 'keys') spinning to the ground in the late autumn, and later finding out you can pickle and eat them. I remember thinking about how tall and straight the trees grow, and learning how this, along with their strength and elasticity, is why woodworkers choose them to make snooker cues. I recall being amazed by their resilience. One ash tree in the local nature reserve had been struck by lightning and hollowed out. Much of it was burnt to charcoal. Three decades later, it was still alive, sprouting branches from the patches of unburnt wood.

However, the resilience of ash trees is being overwhelmed by ash dieback, a devastating disease caused by the fungus *Hymenoscyphus fraxineus*. This pathogen, originally native to Asia, likely arrived in Europe through the movement of infected plants and wood products. The fungus spreads through spores released from infected leaves, quickly colonising and killing the tree's vascular system.

The disease dynamics involve the fungus invading the leaf stems, spreading to the branches and ultimately girdling the trunk,

The ash tree from my childhood, hollowed out by a lightning fire, its canopy still in full glory today (recent photograph sent by my mother)

interrupting the flow of nutrients and water, often leading to the tree's death. Ash dieback has already decimated millions of ash trees across the UK. It has transformed once vivid hedgerows and woodlands into shadows of their former existence. The disease's rapid spread and high mortality rate among infected trees threaten the biodiversity and ecology of the affected regions. Indeed, ash trees support a considerable number of invertebrate species. Approximately 89 species (some say more) of invertebrates are associated with them, with 65% found on living trees and the rest dependent on decaying wood or roots. Among these, 34 species use live ash as their sole food source, including various gall mites and insects. Additionally, the tree is crucial for several beetle species that rely on the cramp ball fungus (*Daldinia concentrica*, also known as King Alfred's cakes), which generally prefers to grow on ash or beech trees in Europe. Therefore, a pandemic in trees affects not only the trees but also a whole community of life.

Ash trees have also held a place of importance in human history and culture. In Norse mythology, the ash tree Yggdrasil is considered the World Tree, connecting the heavens, earth and underworld.[2] Therefore, the loss of ash trees due to the ash dieback disease also represents a cultural loss, erasing a living connection to history and tradition.

Despite their devastating impact, pandemics in trees and other plants – let's call them 'phyto-pandemics' (from the Greek *phyto* meaning 'plant') – often fly under the radar compared to human pandemics. They threaten global biodiversity and food security, but often don't garner the same attention or resources.

Plant outbreaks like the blight that caused the notorious Irish potato famine or the modern ash dieback crisis can decimate crops and native species. They reveal a critical gap in our disease control preparedness. It's time to recognise these underappreciated threats and address them with the urgency they deserve.

More on the Disease Dynamics of Ash Dieback

The lifecycle of ash dieback fungus is complex and involves both sexual and asexual reproduction. It produces fruiting bodies known as apothecia on fallen ash leaves, which release airborne spores (ascospores) during the summer. These spores infect healthy ash leaves, leading to the formation of lesions and dieback of branches. Environmental conditions such as temperature and humidity influence the disease cycle, affecting spore release and infection rates. The prolific production of these ascospores aids the spread of ash dieback. The spores can travel long distances in the wind. The pathogen's ability to persist in leaf litter and produce spores year after year ensures its continued presence in infected areas.

Research into the fungus's genetic structure has revealed low genetic diversity among the European population, suggesting it originated from a small number of introduced individuals.[3] Despite this, the fungus exhibits a significant disease-causing ability, likely due to its ability to recombine and adapt quickly.

There's some good news. Genetic studies have identified ash trees that exhibit natural resistance or tolerance to ash dieback.[4] Resistance is a quantitative trait controlled by multiple genes, with over 3,000 DNA markers associated with reduced disease severity. This means that making trees resistant to disease involves picking many genes with small protective effects rather than relying on one big, strong resistance gene.

This game of numbers makes selecting for resistance more challenging. Enhancing resistance relies on both nature's hand and human intervention. In the wild, natural selection slowly tilts the odds in favour of trees that can stand strong against diseases, letting these hardy traits spread over time. Meanwhile, we can step in with breeding programmes to speed things up, scouting for those resilient individuals and giving them a head start.

Recent research has also focused on understanding the inter- action between ash dieback and other pathogens, such as *Armillaria* species – the 'honey fungus', which can further weaken infected trees.[5] Integrating genetic tools and traditional breeding methods offers hope for developing ash populations that can withstand this devastating disease, or phyto-pandemic as I'm calling it.

Plant Doctor

The Zoom screen flickered to life, and plant pathologist Tara Garrard was framed by shelves filled with neatly labelled plant specimens and books. After the usual pleasantries, we jumped into the subject of plant diseases. I had a few questions for Tara, starting with which key pathogens or diseases currently pose the greatest global threats to wild and domesticated plants. Without hesitation, Tara said one word: "Rusts."

Hearing the word 'rust' in this context might seem a little strange for those less familiar with ecology or diseases. The term 'rust' might initially evoke thoughts of the reddish-brown oxidation seen on metals. You wouldn't be too far off in terms of appearance. But this time, we're talking about fungi.

Tara was referring to fungal rusts, menacing adversaries in the plant health realm. They're characterised by distinctive reddish-brown spores, which often resemble rust on metal surfaces – hence the name. And I must admit, I knew little about fungal rusts before writing this book.

They're known for targeting a wide array of wild and cultivated plants. They slip seamlessly into diverse climates – from the steamy depths of rainforests to the swaying blades of grasslands. Using specialised structures known as haustoria, rusts pierce plant tissues, siphoning off nutrients. Leaves yellow, growth is stunted, and when the assault is severe, entire crops can fail, laying waste to fields. Wheat stem rust (*Puccinia graminis*) and soybean rust (*Phakopsora pachyrhizi*) are names synonymous with massive crop losses across the globe.

Tara shared with me a worrying story about myrtle rust in Australia. This fungal disease, caused by *Austropuccinia psidii*, shows just how destructive rust fungi can be to plants. Originally from South America, myrtle rust spread to Australia, Aotearoa (New Zealand) and some Pacific islands. It mainly affects plants in the Myrtaceae family, including well-known species like eucalypts, tea trees and guava (*Psidium guajava*). Myrtle rust stands out because of its bright yellow to orange spores on infected plants, contrasting against their usual greenery. These spores spread quickly through the air, aided by wind and human activities such as the plant trade. Once it takes hold in new areas, myrtle rust can inflict damage in native ecosystems and disrupt farming. It's not just about looks: myrtle rust causes leaf deformities, stunts growth and can even kill plants in severe cases. This disease threatens biodiversity efforts. It also impacts agricultural yields and affects Indigenous communities who depend on these plants for cultural practices and livelihoods.

Tara told me that various other rusts affect domesticated plant species. This puts pressure on the food industry. The rusts have two key facets that make them so menacing:

1. Their immense ability to travel long distances
2. A pigmented coating on their cells that gives them UV protection – pretty useful in Australia!

Farmers are increasing their use of fungicides because the rusts threaten food crops. As I mentioned in Chapter 8, fungicides may control the pathogen in the short term, but they also bring a potent selection pressure into the mix. Since the early 2000s, barley and wheat rusts have evolved to be resistant to fungicides. This resistance concerns food producers as barley and wheat are two major annual crops.

Globally, wheat stripe rust (*Puccinia striiformis*) is one of the top pathogens.[6] The rust travels the world, especially where wheat is grown, which is pretty much on all continents except Antarctica. You can spot wheat stripe rust by its distinctive stripes – tiny, reddish-brown lines that appear on the leaves of infected wheat plants.

How does it spread? The spores of wheat stripe rust float through the air, landing on new wheat plants and starting the cycle of trouble all over again. This matters for those who depend on wheat for their livelihoods and those who just love bread and pasta! Infected wheat plants don't grow as well, and their grain is often suboptimal for making flour. That means in the future there may be less bread, pasta and cereals for all of us.

Scientists are fighting back and trying to develop wheat plants that resist fungal pathogens. Yet, this constant arms race, compounded by the environmental toll of monoculture farming, has led some to advocate for moving away from annual food plants altogether. One proposed solution is an innovative farming method called syntropic agroforestry, which I'll highlight in Chapter 18 and discuss in detail in my book *Treewilding*.[7]

Stubble Spells Trouble

Tara raised another interesting point about farming practices. She said that no-till methods are now a major driver of emerging diseases.

No-till farming is all about keeping things undisturbed. Instead of digging up the earth before planting, farmers leave the soil as it is, with all the leftover bits of last season's crops right on top. This can keep the soil microbiome healthy and help the soil hold on to water better, which is important, especially when it's hot and dry. Around 75% of soils on the planet are currently damaged, a figure that could rise to 90% by 2050 if the current trajectory remains.[8]

Moreover, remember I wrote in Chapter 7 that between 57% and more than 99.9% of all species on Earth live in the soil? This means soil degradation is a major problem for many species, and constant tillage is a key driver. Healthy soil means healthier plants that can grow strong without needing as much water or fertiliser. Plus, all those old crop bits on top of the soil can act like a natural blanket, keeping weeds from popping up too much. By keeping carbon in the ground instead of releasing it into the air, no-till farming may also help to fight climate change. And because farmers don't need to use as much fuel or chemicals, it can save them money, too. Therefore, no-till farming has been promoted in recent years.

However, Tara told me that the pathogens causing the biggest issues are now stubble-borne. Stubble-borne diseases are plant diseases that survive and spread through crop residues or 'stubble' left behind after harvest. These residues can harbour pathogenic fungi or bacteria that infect the next crop when conditions are favourable. These diseases are particularly relevant when crops are grown in succession. For example, wheat stubble can carry pathogens like *Fusarium graminearum*, which causes head blight in wheat (*Triticum* spp.) and barley (*Hordeum vulgare*).[9] This can lead to yield losses and produce toxins that are harmful to humans and livestock. Similarly, stubble-borne diseases in other crops, such as corn (*Zea mays*) or soybeans (*Glycine max*), can include pathogens like *Gibberella zeae*, which causes corn stalk rot and ear rot, or *Phomopsis* spp., which affects soybean stems and pods.[10]

So, we're caught in a bit of a pickle. On one hand, we're all about no-till farming to protect our precious soil – it promotes healthy soils. But then, there's a twist. We've got these emerging plant diseases, and they thrive in the leftover bits of plants on the ground after harvest.

Farmers switched things up more often in bygone times, growing different plants in the same location each season. Nowadays, though, growing the same crop in the same spot year after year is common. However, it turns out this can lead to a problem. The diseases that survive in last year's crop residues can return stronger the next year because they've adapted to the immune defences of the plants growing in the same place. In a way, we're selecting for emerging diseases by creating novel conditions for them to flourish. But then we go ahead and spray the field with fungicides to control the diseases, which serves to select for resistance, which eventually leads to new and untreatable diseases. It's a vicious cycle.

So, what's the solution? Well, 'mixing it up' is vital. Rotating crops – or growing different things each year – could help. It breaks the cycle of diseases getting too comfortable in one type of crop. Plus, we might need to get creative with new farming techniques that bring more variety to our fields. This diversity helps create a healthier ecosystem overall.

Fungi on Fungi and Other Invisible Foes

And what about fungi? We know they cause diseases in animals and plants. But have you ever considered that diseases might sometimes afflict fungi themselves? For example, could a fungal version of a COVID-19-like pandemic strike *Boletus*, the genus that produces the famous porcini mushrooms (also known as 'ceps' or 'penny buns')? Imagine a pathogenic fungus, perhaps an aggressive, mutated strain of *Aspergillus*, infiltrating a population of *Boletus edulis*. This fungal invader could spread rapidly through the soil and decaying leaf litter, attacking the mycelium (the vegetative part of the fungi) and causing widespread damage. Such an outbreak could decimate porcini populations, leading to significant ecological and economic consequences.

Or imagine an outbreak in *Lentinula*, the genus that gives us shiitake mushrooms, caused by bacterial disease or another fungus. Shiitake mushrooms are cultivated globally, often under controlled conditions. However, they're still susceptible to bacterial diseases. A bacterial pathogen could invade a shiitake farm, spread through water or substrate, and cause severe outbreaks. The bacteria could produce toxins or simply outcompete the fungi for nutrients, leading to stunted growth or even death.

A potent pathogen affecting fungi and causing an outbreak on a pandemic scale doesn't seem to have an official name. So, let's call it a 'myco-pandemic' (from the Greek *mykēs* for 'fungus').

Are there any examples of *real* pathogens affecting fungal populations? And what about examples of myco-pandemics?

Firstly, we can look at parasitism in fungi. We can actually look at parasitism in fungi *by* fungi. Mycoparasitic fungi, such as the *Trichoderma* species, are known to attack other fungi.[11] These mycoparasites can latch on to the hosts, penetrate their cell walls and consume their cytoplasm – the gooey stuff inside the cells. This could lead to the collapse of entire fungal communities. We use some strains of *Trichoderma* fungi to our benefit – they help control fungal pathogens on crops. Others are parasitic on beneficial fungi like mycorrhizae and those that produce cultivated mushrooms.

Another notable example of a parasitic fungus is the species *Hypomyces lactifluorum*, which infects and engulfs mushrooms from the genera *Lactarius* and *Russula*, transforming them into the distinct orange, edible 'lobster mushroom'. While this parasitic relationship benefits human foragers by creating a desirable food product, it illustrates how fungi can fall victim to other fungi.

Several bacterial and viral diseases affect fungi that produce food-grade mushrooms, and this has a considerable economic impact. For instance, mushroom production in the US increased from 70.3 thousand tonnes in the 1960s to 416 thousand tonnes in 2017. Demand for organic mushrooms increased from 3.8 thousand tonnes in 2000 to

58 thousand tonnes in 2017. The worldwide net value of mushrooms reached \$19.46 billion in 2016 and has risen since.[12] So, any disease that affects mushrooms on a global scale (a myco-pandemic) could be devastating for the food industry.

One of the most pervasive threats is brown blotch disease caused by the bacterium *Pseudomonas tolaasii*. It was first identified in Europe in 1915 by A.G. Tolaas. It spread rapidly across the world as globalisation and mushroom production increased. The bacterium secretes a toxin that breaks down the mushroom's tissue, causing a yellow and brown blotchy appearance and sometimes pitting the surface.

And what about out in the wild?

Bacteria like *P. tolaasii* probably affect various species that produce the 'mushroom' fruiting bodies we see dotted across the landscape. However, Tara made a good point: historically, little research funding and attention have been given to fungi. Interest in fungi has surged recently, but the mycology field is playing catch-up. She said, "In Australia, very few of our fungal species have been identified, never mind the diseases that afflict them." This is why I think a new term like 'myco-pandemic' is important – it shines a light on the fact that fungi can also be affected by diseases on an intercontinental scale. Scientists estimate that up to 3.8 million fungal species could exist globally, yet only about 120,000 have been officially identified – a mere 3%![13] Therefore, it's possible that fungal species are being wiped out by myco-pandemics before we even get to identify them.

The implications of fungal diseases are significant. In ecosystems, losing key fungal species can disrupt nutrient cycling, as fungi play critical roles in decomposing organic matter and forming symbiotic relationships with plants. Exploring fungal diseases could offer insights far beyond the fungal kingdom itself. The interactions between fungi and their pathogens might reveal universal truths about infection and immunity, shedding light on the mechanics of disease in other organisms. These revelations hold the potential to shape innovative antifungal treatments and agricultural strategies, safeguarding the beneficial fungi that enrich our world while keeping the harmful ones at bay.

What about Microbiomes?

Could a pathogen cause a pandemic-like phenomenon in a microbiome? How might this affect the broader ecosystem? You may have noticed this chapter is getting slightly more radical as we go along. But bear with me – I'm a 'systems thinker', and this is systems thinking!

A microbiome is the entire collection of microbes in a given environment and their theatre of activity. And as we've covered previously, humans have microbiomes in and on their bodies. There's the gut microbiome – trillions of microbes tightly packed into the gastrointestinal tract. There's the skin microbiome, which plays a vital role in the immune system's first line of defence. And there's the airway microbiome, which lines the respiratory tract. The microbes in these communities include bacteria, fungi, viruses and others. They contribute to our overall health and wellbeing by outcompeting pathogens, digesting our food, secreting chemicals that control inflammation and fulfilling many other roles.

The same goes for plants, fungi (including lichens – those beautiful conglomerates of fungi and a photosynthetic partner) and other animals. Plants have distinct microbiomes in their tissues (the 'endosphere' or 'internal world'), on their foliage (the 'phylloplane') and around their roots (the 'rhizosphere'). The microbial communities in and on plants play vital roles similar to those in humans. They protect plants from diseases, break down nutrients into usable forms, secrete chemicals essential for growth and contribute to overall plant health. As mentioned in Chapter 4, we can view animals, plants and fungi as 'holobionts' or a host plus trillions of microbial partners working symbiotically to form a functioning ecological unit.

Could a pathogen cause a disease-like state (or 'dysbiosis') in a microbiome? And could this be transmitted across the world to cause a pandemic-like phenomenon in microbiomes? If this scenario is possible, it's not hard to see how this could cause secondary issues in the hosts because, as I mentioned earlier, the microbiomes are vital to the functioning of the hosts.

There are several candidates for such a situation. First, think of phages – those often-beneficial viruses that keep bacteria in check. Phages are diverse viruses. They come in various shapes and sizes, each with unique ways of attaching to and injecting their genetic material into bacterial cells. Some look like lunar landers with spider-like legs, while others resemble simple rods or spheres. This diversity means that phages can attack a wide range of bacteria, from those living in our intestines to those in soil, plants and oceans. Each phage type has evolved to exploit its bacterial host in specific ways. Yet, despite their important roles in maintaining a biological equilibrium, this diversity of essentially 'predators' means that phages can also 'attack' bacteria that are beneficial to a host.

Evidence suggests that in humans, gut phages may contribute to inflammatory bowel disease in three main ways:

1. Changes in the variety of gut phages can affect overall gut health.
2. Phages can influence the balance of bacterial populations in the gut, potentially harming beneficial bacteria.
3. Phages can trigger inflammation and affect the local immune response, potentially leading to, or worsening, inflammatory bowel disease symptoms.[14]

These pathways highlight the mechanisms by which phage viruses could lead to dysbiosis ('life in distress') in humans, plants and other organisms. We still know relatively little about phage viruses, but they could be candidates for a pandemic-like phenomenon in microbiomes.

We can consider dysbiosis and disease from another perspective too. What if environmental conditions harm the microbiome, allowing a pathogen to invade, proliferate and cause disease?

An example close to home (in our bodies) is when we damage our gut microbiome with poor diets or antibiotics. This creates the conditions for a pathogenic bacterium, *C. difficile*, to invade and secrete toxins that cause gut disease. In a healthy microbiome, the resident microbes can keep such pathogens in check – remember, 'There's no room at the dining table!'

The same goes for plants. An increase in salt levels in the soil, for example, can stress beneficial microbial communities. It can reduce their diversity, abundance and functional roles and weaken plant defences. Pathogens like *Fusarium oxysporum*, which causes wilt, can take advantage of weakened plant defences and proliferate in these conditions.[15] The compromised microbiome and increased pathogen load result in significant crop losses due to wilt and root rot.

Nested Food Webs, Nested Pandemics?

While writing Invisible Friends,[16] I had a lightbulb moment: what if the tiny partners inside us are shaping much bigger ecological realities? These hidden interactions between microbes and their hosts could ripple up through food chains, influencing how whole populations behave and thrive. And if we really embrace the holobiont idea – that microbes and their hosts function as a single unit – it challenges the way we picture ecosystems altogether. Suddenly, those neat layers of the food pyramid start to become rather fuzzy, and microbes take centre stage at every level.

I took this idea to my friend and colleague Martin Breed, and we had a few meetings over Zoom to iron out the details. We concluded that it was better to think about this in terms of *food webs* rather than energy pyramids, which are a tad outdated. And so, the concept of 'nested food webs' was born.

The nested food webs concept explores the layer of interactions occurring at microscopic levels, which may have implications for the broader ecological landscape. Microbial food webs within holobionts are part of bustling ecosystems in their own right. For instance, there are countless microorganisms in a cow's gut, including bacteria and archaea like *Ruminococcus* and *Methanobrevibacter*. They continuously exchange nutrients and chemical signals. However, our spider-like lunar-lander phage viruses can also 'hunt' these microbes, keeping their populations in check. Collectively, these microbial interactions could significantly impact the cow's digestion, health and overall wellbeing. A 'balanced', healthy microbiome, considered to be in a state of 'eubiosis', can enhance the host's immune function, digestion and overall health. Conversely, an 'imbalanced' microbiome, or dysbiosis, can lead to various health issues, making the host more susceptible to diseases and environmental stresses.

We hypothesise that the influence of these microbial interactions extends beyond the individual host to affect macro-level food web dynamics.

Here's a thought experiment to articulate my point. Consider a prey animal, such as a rabbit, with an unhealthy gut microbiome. Its compromised health could make it slower, less vigilant and more prone to fox (*Vulpes vulpes*) predation. Predators might find these weakened rabbits easier targets, potentially altering predator–prey relationships and influencing the population dynamics of both rabbits and foxes.

On the other hand, a healthy microbiome can enhance a rabbit's resilience, making it less vulnerable to predators and influencing predator population dynamics in a different way.

But we don't just consider the dysbiosis–eubiosis axis in nested food webs. For instance, irrespective of this axis, we know that certain resident microbes can influence sexual preference, feeding decisions and the desire to exercise, all of which could affect host fitness and behaviour and, thus, population dynamics.

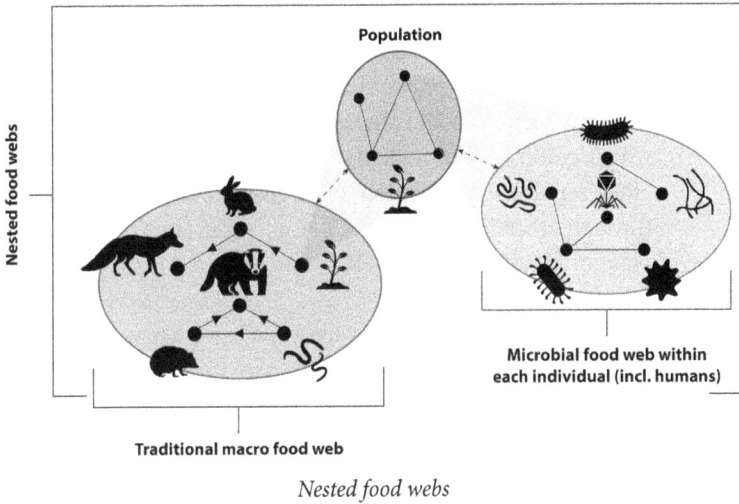

Nested food webs

This nested food web concept highlights the interconnectedness of life across scales – we're talking *food webs in food webs*.[17] The health of an individual organism can hinge on the microbial communities it harbours, which in turn can shape broader ecological interactions and processes.

Another example might go something like this: in a forest ecosystem, the roots of trees, such as oaks (*Quercus* spp.), form symbioses with mycorrhizal fungi like *Rhizophagus irregularis*. These fungi enhance nutrient uptake for the tree, which supports better growth and resistance to pests. Healthy trees can better support herbivores like deer, which are on the menu for predators like wolves (*Canis lupus*). Conversely, trees suffering from unhealthy microbiomes might become more susceptible to infestations, leading to cascading effects throughout the macro food web.

Understanding the nested nature of food webs could lead to a more holistic approach to studying and managing ecosystems. However, there are two reasons I mention it here:

1. Damaged ecosystems can lead to changes in microbial food webs. As mentioned earlier, changing environmental conditions can select for certain microbial species. This change in the microbial community (the combination of bacteria, viruses, fungi, protozoa and so on) could lead to changes in the macro food web – for instance, the health and behaviour of

animal and plant populations. This could influence the population's susceptibility to outbreaks and pandemics across scales.

2. A pandemic-like phenomenon of microbiomes (see the previous section) could change microbial food web dynamics, which could lead to a change in macro food web dynamics. However, the transmission of a given disease could occur through trophic interactions – e.g., when an animal eats a plant or another animal.

So, it's not just individual hosts that could be affected by such a disease in their microbiome, but entire populations and, thus, food webs. This is systems thinking.

I think using terms like 'phyto-pandemic' and 'myco-pandemic' is important to highlight that global-scale diseases also afflict plants and fungi. By framing these issues similarly to human pandemics, we may give them the attention they deserve. Again, it's not all about humans. Yet, by protecting nature and recognising how disease ripples through food webs and ecosystems, we support both human wellbeing and the precious biodiversity we share this planet with.

CHAPTER 15

One Health and Healing Nature

We need acts of restoration, not only for polluted waters and degraded lands, but also for our relationship to the world.

—Robin Wall-Kimmerer, *Braiding Sweetgrass*

The Early Roots of One Health

After days of a lingering cough and constant fatigue, I finally went to the pharmacy. There, I reached for a quick remedy. "I need to get myself back on track," I thought. It's a routine response: treat the symptoms and move on. This is the dominant model of thinking about health in our society. It's the biomedical model. It's the reactive model. But what if we shifted our thinking to a more *proactive* approach instead of just reacting to illness? In this world, we'd consider how the environment affects our health (because we are deeply embedded within it) and how social factors influence our health. We can take it a step further. We can reflect on how our actions shape the environment, influencing the health of wild animals and, ultimately, our own wellbeing. This is also *systems thinking*. 'Reactive' is important (we can't be passive bystanders with these maladies), but so is 'proactive'.

As my friend and GP Gillian Orrow wrote:

> We cannot escape that 'human as machine' and 'healthcare as factory' are implicit metaphors within industrialised healthcare. If people were machines, then a healthcare system designed to 'fix' broken parts would make sense. But if we have inadvertently designed a system based on

> mechanistic thinking to support the health of
> complex, interacting ecosystems, we can begin to
> understand why healthcare, for all its successes,
> is so deeply unsustainable.[1]

We are not machines, and healthcare shouldn't be an assembly line; treating complex beings with mechanistic thinking creates a system that sometimes heals symptoms but fails to nurture true health. We are, in fact, complex walking ecosystems.

To truly thrive, we need a healthcare system that embraces this complexity and nurtures the whole ecosystem – mind, body, community and planet – rather than just addressing individual symptoms. Now imagine doctors, veterinarians, ecologists and social scientists (and the general public) all working together to keep people and the planet healthy. This coalescent force is the essence of One Health. It recognises that the health of people, of non-human animals and of our environment is interconnected.

An old university professor once told me a fascinating story about Karl Friedrich Meyer, a Swiss-born veterinarian and bacteriologist, and his contributions to early One Health thinking. In 1950, *Reader's Digest* invited Paul de Kruif to write a tribute to Meyer, who had become a legendary figure in medical microbiology.[2] De Kruif, a celebrated science writer, described Meyer as "the most versatile microbe hunter since Pasteur" and recounted his adventures in tracking down deadly diseases from his laboratory at the Hooper Foundation for Medical Research in San Francisco.

De Kruif's narrative painted Meyer as a real-life detective, unravelling the mysteries of various diseases. Meyer discovered that botulism was a resilient spore found in soils across the USA and that psittacosis (which always makes me want to say 'pistachios'), or 'parrot fever', was spread by over 50 species of birds. Meyer's work was pioneering, not just for the discoveries but for his approach. He enlisted the help of entomologists, ecologists and climate scientists to solve complex public health problems.

One of the most gripping tales was Meyer's investigation into psittacosis.[3] During the 1930s, psittacosis caused widespread panic, akin to modern fears about avian influenza or SARS. Meyer's meticulous research revealed that this disease was transmitted by parrots and budgerigars, commonly bred in backyard aviaries during the Great Depression. His findings traced the origins of psittacosis back to wild

Scarlet Macaws (Ara macao)*, like other parrots, are at risk of contracting psittacosis*

parakeets imported from Australia. He shed light on how confined conditions in aviaries exacerbated the spread of the disease.

Meyer proposed an innovative solution to control the outbreak. He offered breeders a deal: if they agreed to sacrifice a portion of their stock, he would run tests to ensure the birds were healthy and approve aviaries as free from diseases. This act might seem ordinary and trivial now, but it was highly innovative in those days.

The initiative required Meyer and his team to handle thousands of potentially infected birds, putting their own health at risk. In one particularly tense meeting with breeders in Los Angeles, Meyer highlighted these dangers, stating that they had to "almost put [their] foot in the grave" to solve the problem.[4]

Meyer's story, which Mark Honigsbaum has eloquently covered in more detail, illustrates the application of what we now call the One Health approach, as described earlier.[5] His work helped us understand how diseases can move between wild animals and humans. He showed why it's important to look at the bigger picture when it comes to controlling diseases, considering not just people but other animals and the environment. Reflecting on this story, I realise it was some of the earliest thinking in One Health, long before the term was coined.

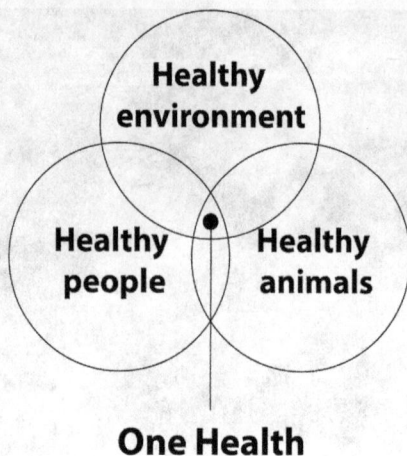

One Health

The One Health triad

One Health is a simple idea with enormous implications.

Recall that way back in Chapter 3, I spoke about wildlife biologist Peggy Eby. Her research was vital to understanding how and why the deadly Hendra virus jumped from bats to horses to humans. Peggy said:

> Our response to COVID-19 has made it clear that vaccines aren't the only answer. While they are very important and containment is crucial, having a better idea of what's causing the spillover in the first place is vital to preventing pandemics.[6]

Eby and colleagues discovered that the loss of winter habitat pushes flying foxes closer to homes and farms, increasing the chances of infection. Human activities create conditions for diseases to thrive.

So, what can we do about it? First, we need to spread awareness of the vital roles flying foxes play in the ecosystem. As mentioned in Chapter 3, they're a keystone species, pollinating trees and dispersing seeds. Without their presence, forests would degrade and become fragmented. This could result in a cascade of impacts on other species. Peggy says:

We need as much support as possible to:

a.   prevent further clearing of the bats' habitat, and

b.   restore their habitat to ensure their populations remain healthy.[7]

When bat populations are healthy, they're less likely to shed viruses due to stress. Healthy bats can only be a good thing. Moreover, when their habitat is in good condition, there's less need for them to migrate into urban and agricultural areas to find food. Again, this is a good thing from a disease spillover perspective.

Peggy said, "Queensland has a terrible record for habitat clearing… but they document it well!"[8] Once every two years, the herbarium assesses the coverage of remnant habitat. It's not a good situation in Queensland, or anywhere else in Australia, for that matter. Australia is one of the world's top deforestation hotspots. Between 2000 and 2017, Australia cleared around 7.7 million hectares of land, one of the highest deforestation rates among high-income nations.[9] The timber industry cleared around 680,000 hectares of forested land in Queensland between 2018 and 2019.[10] Much of this was for agricultural expansion, particularly in grazing and cropping.

As we remove the flying fox habitat, they change their behaviour. They disperse into multiple little roosting groups closer to their food. This means they don't have to travel so far. They shift their diets to introduced plants or green native fruits with low nutritional value. This diet change might increase stress in the bats. It also means the bats aren't fulfilling their natural roles as forest pollinators and seed dispersers. Peggy says, "I think restoration and repair is an idea whose time has really come, and we must deal with this now."[11]

One of Peggy's colleagues, Alison Peel, also works at the forefront of One Health. Alison is an Australian academic specialising in wildlife disease ecology. With a veterinary background, she focuses her research primarily on how environmental changes and human activities influence the transmission of infectious diseases in wildlife. Alison has extensive expertise in studying the genetics of zoonotic viruses, especially within bat populations. She's also done significant work on the Hendra virus in Australian flying foxes.

Alison co-authored a paper in 2024 (led by disease spillover expert Raina Plowright) titled 'Ecological countermeasures to prevent pathogen spillover and subsequent pandemics'.[12] It's all about preventing pandemics by addressing their root causes in nature, focusing on early intervention rather than waiting to respond until after pathogens have infected human populations. The authors argue that while much of the world's pandemic response is geared towards medical solutions, like vaccines and treatments, far less attention is paid to preventing spillovers in the first place. They proposed strategies where spillover risks are high. These include safeguarding wildlife habitats, restoring ecosystems, and carefully managing human activities. This will likely reduce the need for costly medical interventions down the line.

Buffer Zones

The paper by Raina, Alison and colleagues highlights a practical approach: creating 'buffer zones' around bat roosts and ensuring these creatures have ample resources to stay healthy. By reducing stress, we can potentially lower viral shedding rates. This concept of buffer zones seems relatively simple. It could also be a game-changer in preventing disease transmission.

Buffer zones are protected areas surrounding bat habitats where human activities like agriculture, construction and tourism are minimised or restricted. By maintaining these buffer zones, we create safe spaces for bats to thrive without interference. These zones can take different forms depending on the landscape. For instance, in a forested area, buffer zones might include sections of uncut or replanted forest around known bat roosts, creating a barrier between the bats and human settlements or farmland.

In areas where bats live near urban environments or farms, buffer zones could include green belts, wildlife corridors or tree-lined areas that prevent direct interaction between bats and human spaces. For bats that rely on wetland ecosystems, buffer zones could involve protected riparian zones or water bodies where human access is limited to reduce disturbance. In caves or specific roosting sites, buffer zones could mean creating no-entry zones for visitors, with only minimal, managed human access allowed.

To me, this One Health strategy makes complete sense. Keep the bats healthy. Tick. Keep their habitats healthy. Tick. Keep them away from susceptible hosts and built-up areas. Tick.

The authors also highlight the potential downsides of conventional approaches like culling. This might inadvertently increase viral transmission within bat populations and exacerbate spillover risks.

While the benefits of ecological approaches are promising, gaps in our knowledge remain. Most research focuses on medical solutions. This leaves the broader environmental factors that drive pathogen spillover largely unexplored. The team advocates for ambitious studies examining the entire spillover pathway – from environmental disruptions to human infection. By deepening our understanding of how these complex factors interact, we can better target and prioritise prevention strategies that stop pandemics before they start.

Restoring Flying Fox Habitat

Peggy is also at the forefront of flying fox habitat restoration. She recently developed a project called the Habitat Restoration Hub, an online tool that helps track and manage habitat restoration projects nationwide.[13]

Peggy envisaged the hub as a user-friendly platform where anyone – from researchers to volunteers – can consistently share information on restoration projects. Think of it as a communal library for all things restoration. Managers can easily sift through past and present efforts to refine their strategies and tackle big projects more efficiently. The project has a key benefit: understanding how habitat restoration might influence disease patterns.

Restoration Case Studies

In Wodonga, Victoria, efforts are underway to protect the grey-headed flying foxes and their habitats. With temperatures soaring above 40 degrees Celsius during scorching summers, flying foxes need cool, shaded roosting sites near rivers to survive. Incidentally, I walked along the Torrens River in Adelaide in 2024 and was surprised to see hundreds of flying foxes sweeping low over the river's surface. They skimmed the water, wings stretched wide, catching the cool air as they dipped to drink and soak their furry bodies to cool down, barely touching the water before rising back to the treetops. To me, it was a wildlife spectacle. To the bats, the river was a vital reprieve from the heat of the sun. Many bats die if the temperatures are too high for too long. Therefore, a water body close to their roosting sites helps to reduce stress and mortality.

The Wodonga restoration project is underway. It's designed to give these megabats a better chance. The initiative focuses on fencing off key habitats to exclude livestock, allowing native vegetation to regenerate naturally. They aim to restore the river corridor and create diverse roosting sites with layered canopies, which are essential for flying foxes to thrive. This project goes beyond simply protecting the bats – it helps reconnect fragmented habitats and ensure these corridors remain resilient as the climate changes. As the restored areas flourish, they'll provide much-needed shelter for flying foxes and other native species.

Flying foxes hang in trees in the thousands (in Adelaide)

There's another example in Bathurst in New South Wales. The regional council is creatively managing the tension between humans and the vulnerable flying foxes that have made the city's central business district and parks their roosting ground.[14] The council has said 'no' to costly and disruptive measures to drive them out. Instead, they're investing £128,000 to lure the bats back to their natural habitat along the Macquarie/Wambuul River. The project, part of a broader habitat restoration effort, involves planting 3,500 native trees like she-oaks

(*Casuarina* spp.) and ribbon gums (*Eucalyptus viminalis*). The hope is that this new green corridor will entice the flying foxes to move away from populated areas while offering a safer, more suitable environment for survival.

I recently caught up with Wayne Boardman, an academic at the University of Adelaide. Wayne is a wildlife veterinarian and flying fox specialist. I wondered where the bats were feeding around the city and state, and whether they were flocking to restored habitats for food. Wayne said, "Well, first off, I don't think there's much habitat being restored at scale here, which is a shame. However, the bats are keen on feeding on the local exotic plants and non-endemic trees."

This is an interesting point. The flying foxes aren't locally native – they've moved here recently from Eastern Australia. However, many of their favourite food trees were planted here for their beautiful flowers – a decision driven largely by aesthetics. We often talk about 'restoring' an area to a former state. However, we should also pause to consider the implications for today's fauna, which might differ from the fauna of bygone eras. If we remove the non-endemic trees and replace them with species that don't produce the food the bats need, we may increase the stress in the bats. The likelihood of viral shedding then rises. So many things to consider. Perhaps we need to plant swathes of those non-endemic trees (with the bats' favourite food) outside the city to keep the bats healthy but away from dense urban areas?

I've spoken about the benefits of restoring habitats, but are there downsides? One recent study suggested we may see a time lag before the benefits of restoring habitats are realised.[15] We may even see a temporary increase in disease risk as the ecosystem is restored. The authors say it takes time for forest structure and biodiversity to establish and buffer against diseases through the dilution effect (see Chapter 4) and other mechanisms. So, we may need additional strategies to monitor disease risks while restoring the habitat. This all needs further research.

Ecosystem Restoration: A Public Health Intervention?

The key takeaway from this chapter is that ecosystem restoration is not only great for biodiversity, but we can also view it as a public health intervention. Indeed, I co-authored a paper published in the *Lancet Planetary Health* in 2022 that suggested ecosystem restoration is integral to humanity's recovery from the COVID-19 pandemic.[16]

COVID-19 has shown us how fragile the balance between human health and the environment can be. While the pandemic highlighted disparities in healthcare and living conditions, it also underscored how deeply our health is tied to the health of our ecosystems. Our paper argues that restoring degraded ecosystems is more than just an environmental goal – it's a public health necessity. Healthy, resilient ecosystems provide cleaner air, regulate climate and, importantly, can help reduce the risk of diseases spilling over from wild animals to humans.

Projects like reforesting degraded land, restoring wetlands or even creating green spaces in urban areas could be powerful tools for 'building back better'. Do you remember that phrase? It was widely voiced during the pandemic.

However, for 'building back better' to work, policymakers must rethink how they view public health. Vaccines and treatments matter, but so do access to nature, clean air and healthier environments – foundations that support lasting public health. By embracing this holistic approach, we can tackle future pandemics and address long-standing social inequities worsened by COVID-19.

The next decade is crucial. The UN heralded 2021–2030 as the 'Decade on Ecosystem Restoration'.[17] This initiative offers a global roadmap, but for it to succeed, governments, communities, and individuals must commit to making our planet's health as big a motivation as humanity's health.

Humans aren't machines. Neither are wild animals. One Health is the path forward.

CHAPTER 16

Green Prescriptions

I go to nature to be soothed and healed and to have
my senses put in order.
—John Burroughs, *Studies in Nature and Literature*

After months of persistent stress, a patient was prescribed something unconventional: one hour a day in nature. Initially sceptical, she ventured into a nearby park. As she walked, her senses began to attune – the crunch of gravel underfoot, the scent of damp earth, the chorus of birds, the glint of sunlight on leaves. Over time, she noticed her tension easing and her sleep improving.

Her doctor had also encouraged a simple daily ritual. She said: "Jot down five good things you notice while in nature."

This kind of reflective practice sounds so simple. Perhaps too simple? *Simple*, yes, but it engages *complex* brain systems. Such activities likely activate neural networks involved in attention, gratitude and emotional regulation. They can help shift focus to the present, boost mood via the brain's reward system, lower stress hormones like cortisol, and build neural pathways that support resilience and positive thinking. We also know these simple activities can reinforce a positive emotional connection to the natural world.

The patient opened her notebook and wrote:

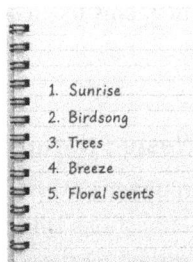

1. Sunrise
2. Birdsong
3. Trees
4. Breeze
5. Floral scents

Days turned into weeks, and her daily visits to the park became a ritual. The vivid colours of wildflowers, the rustle of leaves in the breeze and the ever-changing sky above worked their subtle magic. The patient noticed her mind becoming clearer, and a newfound sense of calm began to settle in.

This wellbeing boost wasn't just a coincidence. 'Green prescriptions', as the doctor called them, are grounded in growing scientific evidence. The natural world, with its complex life, beauty, and diverse stimuli, is capable of restoring both body and mind if we engage meaningfully.

Suppose we view health as an ecological concept – which, from a biological perspective, makes sense, given that we're just another species in the ecosystem – then our understanding of health shifts. Improving human health can be seen as a form of ecosystem restoration (see Chapter 15).

The fundamental premise is that we're all part of 'nature' – each of us made up of atoms just like the rest of the biotic community. We all inhale the exhalations of plants, and our bodies are mini jungles teeming with other living entities. It may sound strange when someone suggests we should 'prescribe nature' to heal our ills. But it seems many of us have forgotten that we're part of a diverse community of subjects, and the rest of the natural world has shaped our bodies and minds over millions of years. We need to spend time in the biotic community nurturing it and allowing it to nurture us in return – it's a two-way ecological restoration. As Indigenous professor Robin Wall Kimmerer said, "As we work to heal the Earth, the Earth heals us."[1] Connections and reciprocity are vital.

But many of us now live in brick, metal and glass cubes, often stacked on top of one another, hundreds of feet high in the polluted ravines of the concrete jungle. Our physical, emotional and experiential ties with our co-evolved friends (plants, animals and microbes) have been severed. We've created new socio-ecological systems and desolate urban cocoons where nature's collection of holobionts is shut out. So, now more than ever, 'nature' seems like a separate entity instead of a community we are part of.

Nested Nature

Many of us – and I speak through the 'Western' lens – distance ourselves from fostering reciprocal connections with the land (or are prevented from doing so). We're consumed by the notion of convenience at all costs. At the same time, our ecosystems pay a heavy price. But it's not just our external ecosystems.

As immunologist Dr Tari Haahtela said, "We are protected by two nested layers of biodiversity".[2] He's referring to the microbes in the plants, air, water and soil, but also the microbes in and on our bodies, our 'walking ecosystems'. Indeed, the current global megatrend of biodiversity loss coincides with the rapid increase in chronic autoimmune diseases and mental health issues.[3]

Social isolation is also rising and is a significant risk factor for mortality; some experts liken it to smoking heavily every day. Scientists now link these biological, psychological and social maladies to our growing disconnection with natural environments and their biodiverse residents, including our microbial 'old friends'.

Prescribing Nature

My second PhD paper was 'Green prescriptions and their co-benefits: Integrative strategies for public and environmental health'.[4] In this paper, Martin Breed and I explored the concept of green prescribing.

Doctors can now prescribe activities that involve spending time engaging with natural environments to benefit health and wellbeing – much like in our fictional example at the beginning of this chapter. It's a way to bring people closer to the rest of nature. A green prescription might involve gardening activities known as 'therapeutic horticulture'. Alternatively, it could include volunteering for a local biodiversity conservation group, or it could be in a much simpler form: walking in a forest or park and immersing oneself in the sights, smells and sounds of nature – also known as forest bathing or *shinrin-yoku* (森林浴).

Green prescriptions can potentially contribute to both reactive (healthcare) and proactive (health-promoting) public health solutions while enhancing the natural environment. This proactive healthcare approach can reduce the likelihood and impact of disease outbreaks.

I believe holistic approaches to 'creating' health rather than the reactive model of 'let's treat the ailment with a drug' offer greater promise to reduce the likelihood and severity of infectious disease outbreaks. Regular exposure to nature likely enhances immune function.[5] It can make individuals more resilient to infections. The soothing effects of nature can significantly reduce stress levels, which, when high, can weaken the immune system and increase susceptibility to diseases.[6] Improved mental health resulting from time spent in nature encourages healthier lifestyle choices. Additionally, outdoor activities promote physical exercise, further bolstering overall health and immune function. Green prescriptions also help mitigate the risk of disease transmission in crowded indoor spaces by providing opportunities for social interaction in safer, open environments. Fostering a strong connection to nature cultivates a healthier, more resilient population that's better equipped to handle the challenges of infectious disease outbreaks.

My colleagues and I recently published a review of biodiversity and human health linkages.[7] We drew upon research by others to propose five pathways to health through nature engagement:

- Biological pathway
- Psychological pathway
- Social pathway
- Physical activity pathway
- Environmental buffering pathway

Imagine walking through a dense green forest. As you breathe in the 'fresh' air, your immune system receives a boost, thanks to the essential oils (also known as 'phytoncides') emitted by the trees. These natural chemicals enhance your body's production of white blood cells, fortifying your defences against illness. A similar thing may happen when you're exposed to the diverse assemblage of microbes in the forest. Research with children has shown that contact with forest materials can diversify the human microbiome and enhance immune function.[8] You may also be exposed to microbial 'old friends'. These are microbes that humans have co-evolved with for millennia. They play important roles in regulating our innate immune system. Without such regulation, some scientists think our immune system goes awry and starts attacking ordinarily innocuous substances like pollen and dust. In extreme cases, a faulty immune system will attack the body's own cells, manifesting as an autoimmune disorder.

Exposure to these tiny organic particles in forests is an example of the *biological pathway* at work – nature provides unseen, subtle gifts to enhance your immune system, making you more resilient to infections. It's worth re-emphasising that despite this book being about pathogens, most microbes (99.9%) are either harmless or beneficial and vital for our survival.

Now imagine sitting by a tranquil lake, feeling the breeze and listening to the gentle lapping of water against the shore. Your breathing begins to synchronise with the rhythm of the water, your mind clears and your mood lifts. This is the *psychological pathway*. The serene environment calms your nervous system, reducing stress hormones and fostering a sense of peace and wellbeing that strengthens your mental health.

There's another example we can use to show how the biological and psychological pathways are deeply interconnected. Let's go back to the forest. The fractal patterns of the leaves and flowers make you feel calm and soothed. How? Viewing such patterns can trigger a chain of biological reactions and likely mirror the natural geometry your brain has evolved to process efficiently. First, light energy is converted into electrical signals in the eye's retina. These signals are then biochemically relayed to the visual cortex, which interacts with the hypothalamus. Ultimately, these reactions lead to the release of the chemical acetylcholine. This chemical attaches to receptors in your heart and slows down your heart rate, making you feel calm and soothed.[9]

Now, consider a sunny afternoon spent in the park with friends and family. Laughter rings out and conversations flow effortlessly. These moments of connection are more than just enjoyable; they nourish your body and mind. This is the *social pathway*. Positive social interactions in natural settings can enhance mental and emotional health, creating a support network that bolsters your resilience to life's challenges. I mentioned earlier that social isolation is a major risk factor for mortality. This makes the conviviality-supporting nature of *nature* all the more valuable.

Parks, woodlands, meadows, lakes and so on are also great for stimulating physical activity. Now, envision yourself on a brisk hike up a hillside trail. Your heart pounds, muscles flex, and sweat trickles down your back as you navigate the path. This physical exertion is a boon for your body, improving cardiovascular health, building strength and boosting overall fitness. This is the *physical activity pathway*. Nature motivates you to move, transforming exercise from a chore into a joy, with benefits for your physical health.

Let's test your imagination one last time. Think of seeking refuge from the noise and air pollution of the city in a park. The air feels cleaner and cooler, and the surrounding trees act as a shield against the urban clamour. This is just one of many examples in the *environmental buffering pathway*. Nature provides gifts. We often call them 'ecosystem services'. In our example, the park acts as a sanctuary, protecting you from environmental stressors, reducing exposure to harmful pollutants and providing a calming backdrop that enhances your overall health.

Direct evidence linking green prescriptions and nature engagement to a reduction in infectious diseases is limited. However, there are several ways – via the five pathways – in which these practices can contribute to overall health and potentially reduce the risk and severity of such diseases.

How Nature Engagement might Help Protect against Infectious Diseases

Enhanced immune function

Spending time in nature can boost the immune system. Studies suggest that exposure to natural environments can increase the number and activity of natural killer (NK) cells, which play a crucial role in the body's defence against infections.[10] Exposure to diverse natural environments can help build a more robust microbiome, which is vital for a healthy immune system. A diverse microbiome can protect against infections by outcompeting pathogenic bacteria (think 'there's no room at the dining table' for opportunistic pathogens) and enhancing immune function.

Reduced stress and anxiety levels

Chronic stress can weaken the immune system, making the body more susceptible to infections. Nature engagement can reduce stress, which in turn can enhance immune function and lower the risk of illness. Meta-analyses have shown that increased nature exposure is associated with decreased salivary cortisol, anxiety, self-reported stress, systolic blood pressure and diastolic blood pressure, and with increased restorative outcomes.[11] Better mental health, fostered by nature engagement, can lead to healthier lifestyle choices and better adherence to preventive measures. Positive mental health is linked to lower levels of inflammation and improved immune response.

Increased social interactions and physical activity

As mentioned, green prescriptions often encourage outdoor activities like walking, cycling or gardening. Regular physical activity strengthens the immune system, reducing the likelihood of severe infections and improving overall health. For instance, physical activities can benefit the response to viral communicable diseases.[12] The amount of green space in your neighbourhood is positively associated with physical activity.[13] Moreover, engaging in outdoor activities provides opportunities for social interaction, which improves mental health and provides emotional support, further bolstering the immune system.

Caveats

We shouldn't view green prescriptions as a low-cost alternative to conventional treatments. To be effective, they still need investment and resources. If the concept is to be successful in the long run, governments must commit to scaling up while addressing systemic social issues. All this will take time. If this holistic approach is not adopted, then people in crisis with more immediate priorities will be less likely to engage with (the rest of) nature. Green prescribing must also be considered part of a holistic health promotion strategy based on planetary health principles. To care for ourselves, we also need to care for our environment.

However, it's important to say that green prescriptions aren't a one-size-fits-all solution. Some people may experience anxiety about outdoor spaces, making these activities more stressful than soothing. Additionally, certain disorders might be exacerbated by outdoor settings rather than alleviated. And have you heard of 'time poverty'? This is when long work hours or life demands leave little room for rest, leisure or self-care. Time poverty plus unequal access to green spaces mean that many people are unable to fully benefit from nature-based prescriptions.

Lower-income communities often have less ready access to safe, biodiverse, well-maintained natural areas. Moreover, promoting increased interaction with nature could inadvertently harm biodiversity without careful planning. Increased foot traffic and human activity can damage delicate ecosystems, leading to unintended negative consequences. Green prescriptions hold great promise. But we must approach them thoughtfully, ensuring they are accessible, equitable and sustainable for both people and the planet.

During my PhD, I spent many hours working on this concept with Sheffield's Greener Practice group, led by Aarti Bansal.[14] They're a group of GPs who recognise that the health of humans and our planet are deeply interconnected. As GP Gillian Orrow often says to me, "We need to create the conditions for health to flourish." This requires a community-centred approach whilst allowing biodiversity – including our invisible friends – to reach the core of our society and remain there as an integral part of our community. Our children must have the opportunity to mingle with other life forms and develop a sense of reciprocity with the land from an early age. Otherwise, we will continue to see ourselves separate from nature, and our health and ecosystems will continue to suffer.

Protect our habitats, walk through them, climb trees and touch the soil and leaves. As Margaret Atwood wrote, "In the spring, at the end of the day, you should smell like dirt".[15]

What About During a Pandemic (Coping) and after a Pandemic (Recovery)?

I once completed a series of microbiome and geospatial studies. One of the latter involved assessing the impact of nature exposure (to parks, woodlands, meadows, lakes, gardens and so on) on supporting people's health during the COVID-19 pandemic.[16] The pandemic reshaped our lives in ways we never imagined. It brought a surge in mental health issues as people grappled with lockdowns and social restrictions.

At its peak, natural environments emerged as a sanctuary for many, offering solace and a much-needed escape. I remember the lockdowns in the UK were quite brutal. At one point, MPs told us we could only exercise for 30 minutes outside. My partner and I were cycling through the Peak District, near where we lived, when the police pulled us over. They asked us where we lived to gauge whether we'd journeyed 'too far' outside our local area. It was mayhem. It was stressful. The last thing we needed was a restriction on how long we could spend outside engaging with nature.

In our study, we created a web-based questionnaire to gather stories and data from over 1,100 participants, mainly from England. We wanted to understand how often people visited natural spaces, what they felt during these visits, and how their access to nature might have influenced their wellbeing. We also used mapping tools to look at the different types of green and blue spaces around their homes. The results were fascinating. As the pandemic confined us indoors, people flocked to parks, forests and gardens more frequently and for extended periods. Many shared that these moments in nature were a lifeline, helping them cope with stress and anxiety. We discovered that those who lived near greener areas – with lush trees, gardens and allotments – reported higher levels of mental wellbeing. It wasn't just the presence of nature but the quality and accessibility of these green spaces that made a difference.

Our study highlighted an important lesson: nurturing our connection with nature is vital for our mental health, especially in times of crisis. I think the same is true for recovery. Conserving and enhancing our natural environments can build resilient communities and promote overall wellbeing. The pandemic showed us that our bond with the natural world is not just a luxury but a necessity for a healthier, happier life. I say, 'not just a luxury', but as I mentioned earlier, access to quality natural environments *is* a luxury for many – they're not equally distributed.

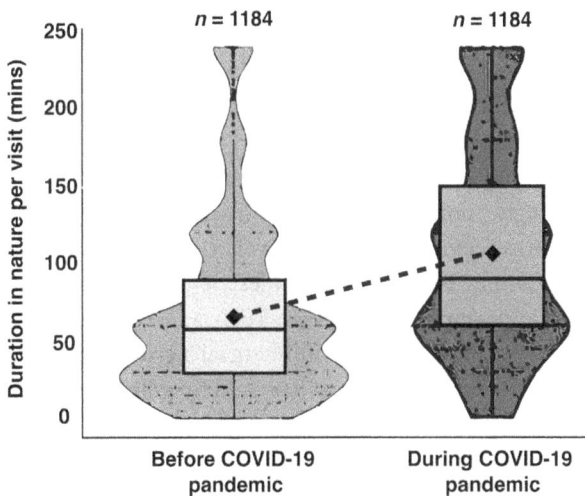

People spent significantly more time in nature as a COVID-19 coping mechanism

Could Chronic, Non-Infectious Diseases be Considered Pandemics?

Most of the diseases I've spoken about thus far are known as 'communicable' diseases. This means they can be transmitted from one person to another, either directly or indirectly – they're infectious. But what about non-communicable diseases or 'NCDs'?

These are conditions like heart disease, diabetes, mental health conditions, cancer and chronic lung conditions. These diseases are responsible for over 70% of all deaths globally, impacting millions and placing a considerable burden on healthcare systems and economies. The number of people suffering from these diseases is skyrocketing due to factors like unhealthy diets, lack of exercise and an ageing population.

Non-communicable diseases are non-infectious in the traditional sense, but their risk factors and consequences do not respect borders. Some scientists say that their widespread and growing impact on our lives makes them a pandemic of a different kind. Others say that we should classify them as 'communicable', just with non-traditional vectors. They argue that changing the terminology will have positive implications for public health.

These non-communicable diseases, once primarily seen in high-income countries, now affect low- and high-income nations. Despite their devastating impact, the response to non-communicable diseases has been slow and underfunded compared to the response to infectious diseases like HIV/AIDS, Ebola and Zika. A paper by Allen in 2017 highlights that non-communicable diseases are not solely due to individual lifestyle choices. They're also driven by broader social, political, environmental and economic factors such as urbanisation, globalisation and the marketing of unhealthy products.[17]

Interestingly, non-communicable diseases can spread through social networks, environments and even viral transmission in some cases. For example, behaviours and lifestyle choices that increase the risk of non-communicable diseases, such as poor diet and lack of physical activity, often spread within social networks as friends and family members influence each other. Additionally, living in environments with limited access to healthy foods or safe spaces for exercise can increase the prevalence of these diseases within a community. Moreover, certain non-communicable diseases have direct links to viral infections. Hepatitis B and C viruses can lead to liver cancer, while the human papillomavirus (HPV) is a major cause of cervical cancer. This interconnectedness challenges the notion that non-communicable diseases are entirely non-infectious.

Allen argues that viewing non-communicable diseases through the lens of a pandemic can provide valuable insights and drive a more coordinated global response. Like infectious diseases, non-communicable diseases have widespread, severe impacts.

Green prescriptions and engaging with natural environments are known to alleviate the symptoms of some non-communicable diseases. In a world facing a double burden of disease, from contagious outbreaks to chronic illnesses, reconnecting with the natural world may offer a powerful form of relief.

CHAPTER 17

Black-market Biodiversity

There are more tigers in American backyards than in the wild.

—World Wildlife Fund

Hidden in the forest's underbrush, a group of men wait. One spots something and gestures to the others. In the moonlit shadows, a magnificent tiger appears, unaware of the danger lurking nearby. The crack of a gunshot shatters the night's tranquillity, and the tiger collapses, its life extinguished in an instant.

In a city market, a different kind of transaction takes place. Exotic birds squawk from cramped cages. Reptiles, packed into small containers, lie motionless. A dealer negotiates prices with buyers, exchanging wads of cash for endangered species destined for private collections or traditional medicine markets. This is where our fallen tiger – or its body parts – will end up. The tiger's bones might be ground into powder and used in traditional medicine for their supposed ability to treat pain, arthritis and other ailments. Its claws and teeth might be used as amulets believed to bring protection and strength. The tiger's genitals are believed to have aphrodisiac properties, so they're consumed in tonics. There's no scientific evidence for any of these purported benefits. You might as well tell people to spread fairy dust on their morning cereal.

The scenes of animal cruelty and exploitation are driven by a dark network, a black market. From the poachers who risk everything for a meagre profit to the go-betweens who smuggle animals across borders, the illegal wildlife trade involves many players. At the top of the chain are the wealthy collectors and consumers who drive demand. It's a multi-billion-pound industry. Some say it's worth up to 258 billion US dollars annually – the fourth largest transnational organised crime.[1]

The reach of this clandestine trade extends across continents, and its impact on global biodiversity is devastating. People disrupt entire ecosystems by removing key species from their natural habitats.

But we're here to talk about diseases, so why is all this relevant? Well, the illegal wildlife trade and the illicit importation of bushmeat pose significant public health risks due to the spread of zoonotic diseases from exporting regions.

As an example, Paris Charles de Gaulle airport intercepts substantial amounts of illegal bushmeat – approximately 5.25 tonnes per week.[2] This trade mainly originates from African countries and brings with it a heightened risk of zoonotic infections. Individuals directly involved – such as poachers, local market sellers and consumers – face increased dangers of contracting these infections. The improper handling and consumption of wildlife exacerbate these risks. Wildlife meat markets are notorious for lacking hygiene, having inadequate hand-washing facilities, mixing wildlife products with other fresh goods and using unsanitised structures. These practices create fertile ground for wildlife-associated pathogens to thrive and spread. The markets become hotbeds of zoonotic disease transmission.

Remember, scientists have linked the illegal wildlife trade to the emergence and spread of numerous zoonotic diseases. One of the most infamous examples is severe acute respiratory syndrome (SARS), which originated from bats and was transmitted to humans via civet cats sold in wildlife markets. The 2002–2003 SARS outbreak resulted in over 8,000 cases and nearly 800 deaths worldwide – a 10% case fatality rate.[3] Similarly, the Ebola virus, believed to be transmitted through contact with infected wildlife such as fruit bats and primates, has caused multiple deadly outbreaks in Africa, with fatality rates sometimes exceeding 50%.[4] Another significant example is the H5N1 avian influenza, which spread from birds to humans, primarily through markets selling live poultry. This led to numerous human infections and deaths, particularly in Asia. Moreover, the COVID-19 pandemic, caused by the SARS-CoV-2 virus, is widely believed to have originated from a wildlife market in Wuhan, China. Some say various species were kept in close quarters, aiding the cross-species transmission of the virus (although a lab leak hasn't been unanimously ruled out).

Have scientists studied these markets? Yes indeed. For instance, a study conducted in Laos explored the potential transmission of zoonotic pathogens from wildlife traded in local markets. The researchers identified 36 zoonotic agents that were transmitted.[5] The illegal wildlife trade not only poses significant risks to human health but also has dire consequences for other wildlife and ecosystems. Such diseases can decimate native species and sensitive ecosystems. One striking example is the chytrid fungus *B. dendrobatidis*, which, as we've discussed in Chapter 11, has spread through the global trade of amphibians.

Then, we have the illicit trade of plants. This is equally pernicious. Diseases transported through trade can devastate agriculture, native plant populations and entire ecosystems. A notable example is *Phytophthora ramorum*, the pathogen responsible for sudden oak death. It's led to significant tree mortality in California and Oregon, affecting forest ecosystems and biodiversity. Another major plant pathogen is *Xylella fastidiosa*, known for causing 'olive quick decline syndrome' in Italy and diseases in grapevines, citrus trees and ornamental plants.

As a child, I remember crouching beside the tall grasses, looking at frothy spittle nests. The froth is made by mixing plant sap, which the nymphs of spittlebugs or 'froghoppers' feed on, with a special fluid they secrete. They use their abdomens to whip air into this mixture, creating bubbles that form a frothy mass around them. In a way, these froghopper insects are the mosquitoes of the plant world – they hop from tree to tree, feeding on the sap but inadvertently inoculating the tree with the *Xylella* bacterium.

The illegal and unregulated trade of infected plants has exacerbated the spread of diseases such as *Xylella* and *Phytophthora*, causing widespread economic damage.[6]

Online and Traditional Postal Services

Nowadays, we can get just about anything delivered to our doors with a few clicks of a button. Need a tiny screwdriver to fix your spectacles? No problem; click, click, done. Need some bananas, coffee, washing liquid or clothes? No problem; click, click, done.

The internet has made it easy to get everyday items without leaving the house. However, it's also made it easy for criminals to sell illegally acquired wildlife products to consumers across the globe. In fact, you can find live, endangered species for sale online in just a few seconds. Each year, people traffic hundreds of thousands of pangolins for their scales, exploit tens of thousands of sea turtles and poach thousands of elephants for their tusks.

The illegal wildlife trade has found itself a dark and thriving marketplace online. Traffickers exploit the anonymity and global reach of the internet to conduct their illicit business. Social media platforms, e-commerce sites and even encrypted messaging apps have become hubs for buying and selling endangered species, their parts and their products. From exotic pets to ivory and traditional medicines to rare animal skins, these items are advertised and sold with alarming ease. They're often hidden behind code words, encrypted apps or innocent-seeming listings. Buyers and sellers can connect across borders. This makes it difficult for authorities to track transactions and enforce laws.

The speed and scale at which these illegal trades happen online have made it a challenge for conservationists and law enforcement agencies alike. From a disease perspective, the rapid and often unregulated movement of wildlife through these digital channels creates ideal conditions for spreading diseases. The online trade only amplifies the risk of new outbreaks. As animals are captured, transported and sold inhumanely, the likelihood of disease transmission increases.

You only need to go to Facebook to find wild animals for sale. I looked and found countless reptiles on the market – some will be from legitimate sellers, but others probably will not.

I did a quick search on Google and found live cheetahs (*Acinonyx jubatus*) for sale. I clicked on the link, and it recognised my location. I received the message, "Sorry – locals only." I thought to myself, "Is this supposed to be comforting?" I guess it means this particular organisation is doing something to prevent cheetahs from being sold abroad.

Their advert is certainly *not* comforting:

> **Cheetah cubs for sale**
> Wildlife South Africa Classifieds
> 17 Dec 2021 – All our Cheetah cubs are bottle-fed
> and raised in our home as home pets, so they are
> perfectly socialized and will make very good pets.

Should cheetahs be kept as pets? (Spoiler alert: in my view, no!)

Apparently, South Africa is the only country with CITES-registered captive-breeding facilities that legally allow the international trade of live specimens, primarily for zoos and safari parks. This means the facilities are signed up to the global agreement that regulates international trade in wild animals and plants to ensure it doesn't threaten their survival. However, it seems the country's authorities are also fine with people breeding the cheetahs as pets, locally at least.

A report by the wildlife trade monitoring network Traffic suggests that around 70% of the trade in live cheetahs is unfolding on social media.[7] Over just six months, two hundred and twenty-two unique URLs were found, which shows an alarming trend in the online trade of live cheetahs. The top five countries with URLs selling cheetahs as pets were the United Arab Emirates, Saudi Arabia, Kuwait, South Africa and the United States.

The illegal wildlife trade exploits traditional postal routes, making it a stealthy conduit of potential disease outbreaks. Traffickers use the postal system to smuggle live animals and products derived from endangered species. They often hide them in ordinary-looking packages that slip through standard inspections. These shipments can carry deadly pathogens, including those responsible for zoonotic diseases. As these packages travel across borders, passing through multiple hands and regions, they create ample opportunities for spreading pathogens.

I'm acquainted with someone who visited Australia Post's international gateway facility to explore their anti-wildlife trafficking technology. Here, a high-tech arsenal is at work to combat the illegal wildlife trade. As packages arrive from around the world, they pass through sophisticated X-ray scanners that can detect hidden animals, animal parts or other illicit goods. The 2-D scanners require specially trained humans to notice when a package seems unusual. Trained sniffer dogs also patrol the facility. The dogs' noses are keenly attuned to the scent of wildlife and alert handlers to anything suspicious.

But it's an ongoing arms race between wildlife traffickers and those trying to stop them. Criminals constantly evolve tactics to evade detection at facilities like Australia Post's international gateway. For example:

- Smugglers often go to great lengths to hide illegal wildlife and their parts within seemingly innocuous items. They might embed animal products in everyday objects, conceal them within layers of packaging or even disguise live animals in stuffed toys, electronics or other everyday items. By doing this, they hope to avoid detection by X-ray machines and other scanning technologies.
- Instead of sending large shipments that might attract attention, criminals often send smaller packages more frequently.

This method is known as 'smurfing'. It helps them avoid the suspicion that comes with bulk shipments and reduces the risk of losing a large quantity of goods in a single interception.

- Smugglers often attempt to mask the scent of illegal wildlife products by packaging them alongside strong-smelling substances like coffee, spices, perfumes or cleaning agents. The hope is that these odours will confuse or overwhelm the dogs' senses.

My contact shared some fascinating insights into the cutting-edge technology being deployed at Australia Post to combat the illegal wildlife trade. At the heart of this effort are AI and machine-learning algorithms. They've changed how packages are scanned and inspected. Unlike traditional 2-D imaging machines, which can miss subtle details, these advanced systems use 3-D imaging to create a comprehensive, multi-dimensional view of the contents inside each package.

As packages pass through the scanner, the AI software analyses the 3-D images in real time, breaking down the shapes, densities and textures of everything inside. The system has been trained on vast datasets of legal and illegal items. It learns to recognise the telltale signs of smuggled wildlife or other suspicious objects. This training involves feeding the AI thousands of images of different types of animals, animal parts and other illegal items that are hidden in various ways. Over time, the AI becomes incredibly adept at identifying even the most well-concealed contraband.

This technology can detect anomalies that human inspectors or less sophisticated machines might miss. For example, the AI can flag a package that appears to contain a stuffed toy but whose internal structure doesn't match the typical materials or shapes expected in a toy. It might recognise the subtle difference in texture or density that suggests an object inside isn't what it seems. The flagged package is then set aside for further manual inspection, where trained officers can take a closer look.

Moreover, the AI system continuously improves as it processes more and more data. Each time it analyses a package, it learns something new. This learning process refines its ability to distinguish between legitimate and illicit items. This means that the longer the system is in operation, the more effective it becomes at identifying smuggled wildlife, making it several times more effective than standard imaging technology. However, remember that it's an arms race. This means criminals are probably working hard to develop new ways of evading such technology. It reminds me of the Red Queen hypothesis mentioned back in Chapter 5: "it takes all the running you can do to keep in the same place."[8]

'Is the Illegal Wildlife Trade the Most Serious form of Trafficking?'

This was the title of a recent paper.[9] The authors discuss how wildlife trafficking has long been seen as a lesser crime, overshadowed by the global focus on human, drugs and weapons trafficking. However, this perception fails to account for the true gravity of the situation. Their paper illustrates that the illegal wildlife trade is also a ticking time bomb for public health.

The trafficking of species like pangolins and bats is driven by demand for traditional medicine and exotic delicacies. This trade creates a direct pathway for zoonotic diseases to leap from wild animals to humans. And it isn't a hypothetical risk – it's already happening, as evidenced by the emergence of seven new coronaviruses from animals since the 1960s. The paper argues that the societal and economic costs of wildlife trafficking are vastly underestimated. COVID-19 alone has cost the global economy trillions of US dollars and caused millions of deaths. Financial gains from this illicit trade pale in comparison to the human and economic devastation it can unleash.

The paper also highlights the need to shift how we approach wildlife trafficking. It's now about survival. The continued exploitation of wildlife poses a clear and present danger to global health. The conditions created by illegal wildlife markets are breeding grounds for the next pandemic. The paper calls for a rethinking of policies, stronger international cooperation and a more informed public that understands the true cost of wildlife trafficking: lost species but also shattered human lives and economies.

To conclude, the authors make a compelling case that we should no longer view wildlife trafficking as a secondary crime. Stopping the illegal wildlife trade should be a high priority for governments based on animal welfare and ecological consequences alone. However, the authors of this paper argue that its potential to cause pandemics like COVID-19 also elevates it to one of the most serious global threats we face today. The world can no longer afford to be complacent. We must curb wildlife trafficking, not just for the sake of biodiversity, but to protect humanity from future catastrophic pandemics.

We've discussed some high-tech solutions to help tackle the illegal wildlife trade. But isn't this just sticking a plaster on a deep and ever-growing wound? The root causes – rampant demand, cultural practices and economic pressures – remain largely unaddressed. While technology can help intercept illegal shipments and track down traffickers, it's only a *part* of the solution. To truly stop the illegal wildlife trade, we need to change the underlying factors that drive it.

So, what are these factors?

The factors driving the illegal wildlife trade are complex and deeply rooted in various social, economic and cultural contexts. To get an idea of the scale and complexity of the issue, I journeyed into the city to meet with an illegal wildlife trade expert. He wishes to be unnamed, so let's call him John. It was an ordinary day in Adelaide; the sun was shining, and the streets were lively with the usual hustle and bustle. Yet amidst this normality, I couldn't help but think how, at any given moment, someone was out there capturing wild animals for illicit trading on the internet.

This is a huge problem in Australia, particularly for reptiles. The iconic blue-tongued skinks (*Tiliquas* spp.), for instance, are regular victims. Our scaly friends are captured, individually bound with tape, stuffed into dodgy packaging, like an old sock, and put into electrical appliances, like a rice cooker or a hollowed-out DVD player. Shingleback lizards (*T. rugosa*) are a heavily trafficked species here. The lizards can sell for thousands of dollars abroad thanks to their distinctive shape and patterns. Organised traffickers sent to Australia by overseas cartels hire campervans and disappear into the outback with a shopping list of reptiles to capture.

The shingleback lizard can host a highly contagious coronavirus known as the shingleback flu, which first emerged in reptiles in the 1990s but is becoming more widespread.[10] This coronavirus means that trafficking the lizards abroad can spread infectious diseases to other ecosystems.

John told me that the core of the problem is rampant demand for wildlife products, whether for use in traditional medicine, as luxury items or as exotic pets. A lack of awareness about the devastating impact on species and ecosystems fuels this demand, as does the mistaken belief in the supposed benefits of these products. But often, people know of all this and simply don't seem to care.

Cultural practices also play a significant role, particularly in regions where people have used wildlife products for centuries in traditional medicine or as status symbols. In some cultures, consuming or owning rare wildlife is seen as a sign of wealth and power, making these practices hard to change. Unless these deeply ingrained cultural beliefs are addressed, the demand will persist.

Economic pressures also drive the illegal wildlife trade, especially in poorer regions where people see wildlife trafficking as a lucrative way to make a living. John said, "In many cases, the lack of viable economic alternatives leaves individuals and communities with little choice but to engage in or tolerate illegal activities. Until we provide sustainable economic opportunities that can replace the income generated by wildlife trafficking, it will be challenging to break this cycle."

Global inequalities contribute to the problem, as wealthy consumers in developed countries often drive the demand, which leads to exploitation in less economically stable regions. Addressing these inequalities, promoting global cooperation and ensuring that both source and demand countries are committed to ending the trade are essential for long-term success.

Weak law enforcement and corruption in certain areas work against addressing the issue. This enables traffickers to operate with relative impunity. In places where officials are easily bribed or laws are poorly enforced, the risks of engaging in wildlife trafficking are low compared to the potential rewards. Strengthening governance, enforcing laws more rigorously and reducing corruption are crucial. But this is incredibly hard as these issues are deeply rooted in broader social, economic and cultural systems.

John told me that we must keep raising awareness of illegal wildlife trafficking and lobby to tackle the root causes – the social and cultural drivers. Otherwise, we're merely treating the symptoms, not curing the disease.

Seven Things You Can Do

The Zoological Society of London (ZSL) is on a mission to protect the most endangered species and drive changes in behaviour and attitudes that will reduce the demand for taking these creatures from their native habitats. Efforts like theirs are crucial in helping these species recover and stopping the illegal wildlife trade. But real change happens when we all come together. We can all play a vital role in protecting pangolins, tigers, sharks and countless other species from the threat of illegal wildlife trade. The ZSL outlines seven simple ways you can make a difference, which I've paraphrased below:[11]

Be a Mindful Traveller

When travelling overseas, be mindful of animal-related experiences that could be linked to wildlife trafficking. Think twice before taking photos with animals like tigers, lorises, monkeys, parrots, leopard cats, orangutans or bears, as they may have been illegally captured from the wild. By steering clear of these interactions, you help protect wildlife and reduce demand for exploitative practices.

Choose Sustainably Sourced Products

As a responsible consumer, choose products with transparent and ethical supply chains. Look for wood certified by the Forest Stewardship Council (FSC) to avoid supporting the illegal timber trade, which drives deforestation and biodiversity loss. Be mindful of palm oil, opting for products certified by the Roundtable on Sustainable Palm Oil (RSPO). When buying fish, prioritise sustainably sourced options with traceable supply chains—tools like the Good Fish Guide[12] and the Giki app[13] can help you make informed choices.

Say No to Exotic Animal Products

Stay informed and avoid buying items made from wild animals or plants – whether online or in person – even if they appear legal locally. Products like exotic bird feathers, tortoiseshell, ivory, big cat teeth or claws, coral jewellery, cacti, wild orchids and sea turtle shells are often linked to the illegal wildlife trade. This trade is vast and not always obvious. When unsure, ask about the item's origin and request clear documentation to ensure it's legal and sustainably sourced.

Steer Clear of the Unusual

While travelling, it can be tempting to try something unusual for the experience or the story – but think twice. Avoid dishes like shark fin soup, bird's nest soup, or drinks infused with animal parts, such as tiger bone wine. Steer clear of exotic wild meats, including salamanders, snakes, songbirds, monkeys, turtles or tortoises. Skip wild-harvested eggs – especially sea turtle eggs – and traditional medicines made from endangered species, like bear bile, wild ginseng, caterpillar fungus or tiger parts. These choices can drive species toward extinction, harm ecosystems and increase the risk of emerging diseases.

Choose the Right Pet

When choosing a pet, keep in mind that demand in Europe and the USA fuels much of the illegal wildlife trade. This trade threatens wild populations of amphibians, reptiles, birds, insects and more. Always buy from reputable sources that provide proof of captive breeding, and even then, consider the conservation status of the species – some may be endangered in the wild despite being available in captivity.

Report the Crime

If you suspect illegal wildlife trade, report it to the National Wildlife Crime Unit (NWCU) right away. Avoid buying any items you believe may be illegal – doing so not only risks legal consequences for you but also drives demand and helps sustain the cycle of wildlife exploitation.

Support Charities Paving the Way to Recovery

As a global conservation charity, ZSL works with local communities, NGOs, governments and law enforcement agencies to tackle the illegal wildlife trade. With your support, they can build effective pathways for wildlife recovery and long-term protection.

What about Plants?

I once gave a talk to the Native Orchid Society of South Australia (NOSSA). Most members of the society were at retirement age and had a wealth of knowledge. I learnt about a critically endangered *underground* orchid. The orchid can't photosynthesise as it's wholly subterranean and leafless, so it acquires its energy by hooking up to a mycorrhizal fungus. The fungus itself forms an association with a *Melaleuca* broom bush plant. In essence, the *Melaleuca* plant is photosynthesising for the underground orchid, presumably benefiting from the fungus in return. It's a beautiful kind of symbiosis.

One of the members told me how they thought an underground orchid existed in South Australia. He said, "If I ever find it, I won't even tell other NOSSA members!" He told me some crazy stories about orchid hunters who steal endangered orchids from the wild and trade them. Sometimes, they merely collect them to keep like football cards. It's the buzz of knowing only they have one of the few specimens in existence, even if it means threatening their survival as a species.

The illegal trade of plants extends far beyond orchids. Their movement from the wild can introduce pathogens to new areas where native plants have no natural resistance. Just like with wildlife, the illegal trade in plants threatens individual species and the broader environmental health of regions. It increases the vulnerability of native plants to diseases and other stresses.

Ending the illegal wildlife and plant trade (although some, like the late ecologist Oliver Rackham, would consider plants also to be wildlife) isn't solely about saving individuals and species, although that's a priority. By shutting down this global black market, we reduce the risk of the next pandemic, panzootic and 'phyto-pandemic'. We also safeguard the future for all life on Earth.

CHAPTER 18

Sowing the Seeds of Change

The hopeless don't revolt, because revolution is an act of hope.
—Peter Kropotkin, *The Study of Revolution*

Revolution! Revolution!
The word 'revolution' is thrown around so frequently in our everyday vernacular that it's easy to lose sight of its true meaning. Originally reserved for moments of meaningful change – whether in politics, society or science – we've now stretched it to cover everything from minor innovations to the most trivial trends. I can't help but think there's a direct correlation between the increasing frequency of its use and the growing triviality of what it describes.

There's the selfie revolution, describing the rising popularity of posing in front of a smartphone. There's the skincare revolution, referring to new trends in face creams and the like. We even use it in business names. I remember a coffee shop called Coffee Revolution in Sheffield during my university days. The coffee was... fine.

We've diluted the word to the point where it's become a catch-all for anything slightly new or improved, regardless of its actual impact.

The irony is that by using 'revolution' to describe everything, we risk diminishing the significance of true revolutions – those moments of profound, far-reaching change that genuinely alter the course of history. The digital revolution, for instance, transformed how we communicate, work and live in ways still unfolding today. The agricultural revolution began thousands of years ago and laid the foundation for modern civilisation by transitioning many societies from hunting and gathering to farming and settlement. These are the kinds of changes that merit the term 'revolution'.

So, while it's easy to get caught up in the excitement of the latest 'revolution' being advertised or discussed, it's worth pausing to ask how we might start reserving the term for moments that truly deserve it. I've already mentioned the agricultural revolution, a true revolution that unfolded in various cultures at different times across the globe. This was not just a slight shift in how we grew our food; it was a seismic transformation that altered the course of human history. It changed humanity itself. It led to the development of megacities and nations (for better or worse). But the impact of the agricultural revolution didn't stop with humanity. It also reshaped the natural world. It led to dramatic changes in biodiversity. As humans began cultivating land, we transformed wild landscapes into farmlands and cleared forests. When industrial agriculture began (arguably another 'real' revolution), it drove countless species to extinction or near extinction. The diversity of life on Earth was permanently altered as many human societies began to dominate ecosystems, often to the detriment of other species.

The agricultural revolution also set in motion changes to the climate we're still grappling with today. As a systems thinking law goes: "Today's problems come from yesterday's 'solutions.'"[1]

Large-scale deforestation, land use changes and the subsequent rise of industrial agriculture have contributed to greenhouse gas emissions and ecosystem degradation. What began as a means of feeding a growing population has evolved into a significant driver of climate change, with far-reaching consequences for the planet's future.

Moreover, the agricultural revolution had a far-reaching impact on disease dynamics. As humans settled in closer quarters and lived in larger, denser populations, the spread of infectious diseases became more common.[2] Domesticated animals lived close to humans and became reservoirs for diseases that could jump to human hosts. This increased interaction between human crowds, animals and the environment led to increased zoonotic disease outbreaks. In many ways, the agricultural revolution set the stage for the types of pandemics we see today. (Though it's worth noting that it also helped fuel the amazing healthcare facilities we developed.)

This kind of sweeping, multifaceted transformation genuinely deserves to be called a revolution. It's amusing to compare this to the so-called 'revolutions' in skincare, selfies and 'okay' coffee. The agricultural revolution fundamentally altered the life trajectory on Earth in ways that continue to shape our world thousands of years later. I think we need another revolution.

We need another *agricultural* revolution. A *real* one, not just a slight shift in food production trends, but an overhaul of the food industry on a global scale. We need this kind of revolution for many reasons: to restore ecosystems, reduce the impacts of climate change and reduce poverty. But we also need another agricultural revolution to reduce diseases in humans, wildlife and domesticated plants and animals. Currently, just three crops – wheat, maize (corn) and rice – account for more than 50% of the world's food energy intake.[3] We've seen in previous chapters that cultivating vast swathes of monoculture annual crops destroys natural habitats, which brings us closer to novel pathogens. It also kick-starts a chain reaction. In this reaction, we have three components: (1) an evolutionary barrage of plant diseases, (2) an arms race to stop them, and, in doing so, (3) a selection pressure for resistance and cross-resistance to clinical drugs.

We also know that high-intensity animal agriculture creates a hotbed for zoonotic diseases. The close quarters in which animals are kept, combined with the routine use of antibiotics to prevent disease outbreaks, contribute to the emergence of antibiotic-resistant bacteria. These conditions also facilitate the spread of viruses between animals and humans, increasing the risk of pandemics. The practice of raising large numbers of genetically similar animals further exacerbates this risk, as a single pathogen can easily spread through a population with little resistance. In plant and animal agriculture, the lack of genetic diversity weakens resilience and increases vulnerability to disease. And thus, we have a dangerous cycle that threatens human and planetary health.

So, What can We do About it?

People are doing some remarkable things, but we haven't reached the critical mass for that 'revolution' just yet. *Diversity* is a common denominator of most solutions to most problems I have encountered in my life and career. For instance, in my experience of landscape architecture, incorporating *diverse* features to stimulate our multiple senses (think colours, patterns, scents, natural sounds, things you can touch) can maximise human wellbeing. In the 'think tanks' I have been a member of, having a *diversity* of minds from people of different ages, genders, and backgrounds has invariably led to more creative and impactful solutions. Having a *diverse* gut microbiome can outcompete opportunistic pathogens and keep us healthy. But to sustain this diversity, we also need to consume *diverse* foods – especially those that feed the 'friendly' microbes – as neurobiologist John F. Cryan would say, "You are what your microbes eat!"[4]

The same goes for agriculture. Diversity is key. Just imagine walking through a farm, and the fields aren't just a monotonous expanse of a single crop. In each direction, you see a rich array of different plants. You might bump into the farmer, who explains this is 'syntropic agroforestry' in action. Crop rotation and trees growing alongside the crops create a resilient system that promotes biodiversity and naturally fends off pests and diseases.

Syntropic agroforestry draws upon principles of ecology – strata, lifecycle and natural succession.[5] The farmer no longer relies on monocultures; the land thrives, supporting the farmer and the surrounding wildlife. No harsh chemicals are used, so the selection pressure for disease resistance also diminishes. Some plants are annuals. But these annuals are planted alongside perennial crops in a carefully stratified arrangement. The perennials live for many years. This reduces the need to till the soil and allows the land to sequester more carbon. The farmer talks excitedly about how these crops are helping to restore the soil and reduce their carbon footprint, all while providing a reliable food source.

I've written before (in *Treewilding*) about meeting such a farmer who lives in Aotearoa/New Zealand. Klaus Lotz and his family run a commercial syntropic agroforestry farm on the North Island. They restored a plot of land and now grow considerable amounts of vegetables, fruits, nuts and mushrooms, surrounded by a thriving ecosystem. Klaus once told me that the farm is such a wildlife haven that they literally trip over the endangered kiwi birds. He also said a particular plant might yield 5–10% less produce than if it were in a monoculture system with synthetic fertiliser and pesticide; however, when factoring in all the different crops they grow, the overall yield of produce is considerably more. Klaus and his family are living proof that you can make a decent living while caring for the land, supporting biodiversity and minimising disease risks. Doing syntropic agroforestry at scale seems to pose a challenge. Yet, Klaus is convinced we can practise this farming in most biomes and at scale. The family runs courses and inspires others to start their own farms.

Researchers are testing ways to mechanise syntropic agroforestry in thousands of hectares. The Research Center for Syntropic Agriculture says scaling up this form of agriculture "will not only transform the rural landscape of the planet but also guarantee the future for the human species, because, if there will be a future for us, it is intrinsically connected to the restoration of the entire Earth".[6]

Regen AG

Meanwhile, on a regenerative farm, you might see low-density cattle grazing in a way that mimics natural patterns. The farmer might describe how rotational grazing is about raising animals and rebuilding the land. The limited number of cows move from pasture to pasture. The land recovers and regenerates, increasing biodiversity and capturing more carbon in the soil. I visited a regenerative farm back in the UK. It was a living example of how we can reduce our reliance on high-intensity animal agriculture while restoring the health of our ecosystems.

As with syntropic agroforestry, running a regenerative agriculture farm at scale will be challenging, but the rewards are worthwhile. The transition from conventional farming to regenerative practices requires a deep understanding of soil health, biodiversity and ecosystem management. This can be complex and resource-intensive. Scaling up these practices often means overcoming financial hurdles, adapting to new techniques and managing the unpredictability of nature. However, the path to success may lie in embracing innovation and collaboration. Farmers are turning to technology, such as precision agriculture tools, to optimise their practices and make them more efficient.

I caught up with Kym Kruse, founder of RegenAG in Australia. I've known Kym since 2023. He contacted me because he was interested in some of my research on stimulating a plant growth-promoting fungus, *Trichoderma harzianum*, with sound waves. Kym helps establish collaborative regional partnerships and farmer-to-farmer networks and convenes regenerative agriculture events around Australia. He said networking with other regenerative farmers and participating in cooperative ventures can help farmers scale up, spread the financial risk and share valuable knowledge. He noted that consumer demand for sustainably produced food is growing, offering regenerative farmers new markets and economic incentives. It's all about staying resilient,

252 The Nature of Pandemics

seeking support and continuously innovating. By doing so, regenerative farms can thrive on a larger scale and transform agriculture from the ground up.

Syntropic agroforestry and regenerative farming promote healthier soils teeming with beneficial microbes that can outcompete harmful pathogens. Diverse crops break the cycles that allow pests and diseases to thrive. Moreover, by reducing the need for chemical inputs like pesticides and antibiotics, which can drive resistance and disrupt ecological processes, these farming practices can help mitigate the emergence of new, hard-to-treat diseases. Additionally, by promoting resilient and more diverse ecosystems, regenerative practices can reduce the spread of diseases, offering a healthier alternative to the pathogen hotbeds of intensive animal farming.

There's an important caveat to consider. Blending food production with natural habitats, like in syntropic agroforestry, isn't without risks. By bringing wildlife and livestock closer together, we might set the stage for more interactions that could raise the odds of diseases jumping from animals to humans. These systems might inadvertently create new pathways for pathogens to spill over from wildlife to domestic animals and, eventually, humans.

Establishing buffer zones, such as vegetative barriers or managed land strips, can be a protective measure. These zones could reduce direct contact between wildlife and livestock, minimising the risk of disease transmission while still supporting biodiversity enhancement efforts. The key is to strike a balance that protects both ecosystems and public health. Another option is to remove livestock from the agroforestry system. This is a controversial suggestion. Those who live on plant-based diets would argue otherwise.

Lab-Grown Meat

Let's leave the farm behind and head into a lab. Here, a new kind of protein is taking shape. In a sleek, climate-controlled facility, scientists are growing meat by cultivating cells. No feathers, no hooves, no fields. Just muscle tissue, grown in bioreactors.

I spoke with the founder of one such startup, who explained how both plant-based and cultured proteins are far gentler on animals. But he also suggested that they dramatically reduce the environmental costs of meat production. The future of some foods is being shaped by innovation. I'm not completely sold on this idea, but I'm also aware that sticking with the status quo has its own set of problems – environmental, ethical and otherwise.

It's controversial but might offer a sustainable alternative to traditional, intensive livestock farming. By cultivating meat in a controlled environment, scientists eliminate the need for large-scale and dense animal farming, which, as we know, is often a breeding ground for zoonotic diseases. We reduce our reliance on these high-risk farming practices by shifting to cultured and plant-based proteins. This may, therefore, lower the chances of new disease outbreaks.

These innovative proteins can also be produced with fewer environmental inputs, reducing the strain on ecosystems and lowering the incidence of diseases linked to ecosystem degradation. But this is all speculation, hence the 'may's' and 'mights'.

Lab-grown meat and plant-based proteins might be hailed as game-changers for reducing disease risks and antibiotic resistance, but it's essential to consider their broader impact. These innovations aren't without their challenges. The industrial-scale production of cultured meat relies heavily on technology and energy-intensive processes. This could offset some environmental and health benefits if not managed sustainably. Additionally, the widespread adoption of lab-grown and plant-based proteins may disrupt traditional farming communities, potentially leading to economic and social challenges that could indirectly affect public health. Moreover, as these products enter the market, there may be unforeseen health implications from consuming ultra-processed foods, even those derived from plants or cultured cells. Thus, while these innovations hold potential, it's crucial to approach their development carefully while considering the broader ecological, social and health impacts.

Other technologies are playing their part in transforming agriculture. Drones now hover over fields, gathering data on soil moisture and plant health. Farmers use this information to apply water, fertilisers and pesticides with precision, reducing waste and maximising yields. This is precision agriculture, where AI and sensors make farming more efficient. Whether it's sustainable or not is another question. It might end up perpetuating our obsession with annual monocultures! I've even heard of robotic pollinators. Crikey… I think we need to keep reminding ourselves of that systems thinking law: "Today's problems come from yesterday's 'solutions'."

Scientists also work with CRISPR technology, editing the genes of crops to make them more resistant to diseases. One aim is to reduce the need for chemical inputs. The potential of these innovations is vast, though there's a need for caution to avoid unintended ecological consequences. Moreover, this doesn't promote the type of holistic approach I think is needed for a healthy revolution.

The Role of Governments

But change isn't just happening on the farm. Governments are beginning to recognise the need for broader shifts. Policymakers are exploring ways to shift agricultural subsidies from large-scale monocultures to more sustainable practices. The conversation is turning towards backing small-scale farmers, boosting local food systems and supporting organic farming. The goal? To create a more diverse and resilient agricultural landscape that moves away from the grip of industrial farming.

It's not happening fast enough, but it is happening.

Across the globe, governments are increasingly turning their attention to agroecology and organic farming, recognising these approaches as keys to a sustainable future. In Argentina, a bold initiative was launched in 2020, during peak COVID-19 pandemic times, to weave agroecology into the fabric of the nation's agricultural practices.[7]

Ghana is taking a different approach. The government's Investing for Food and Jobs plan, now in its second phase, subtly integrates organic practices into the country's agricultural policies.[8] The aim is to promote organic fertilisers and sustainable soil management. The latter is vital. Remember, 75% of the world's soils are currently damaged. This could rise to 90% by 2050 unless we change how we treat soil.[8]

Mexico has enacted a constitutional ban on the cultivation of transgenic maize and is working towards phasing out glyphosate use by 2027. While the government promotes sustainable agricultural practices, including agroecology, the scale and specifics of support for farmers transitioning to these methods are not fully detailed. If this works, it should boost local production and healthier food options.[9]

Saudi Arabia has made significant investments in its organic agriculture sector, aiming to increase organic farming by 300% through substantial funding and support for farmers transitioning to organic practices.[10] Meanwhile, Tanzania is implementing its National Ecological Organic Agriculture Strategy (2023–2030), focusing on promoting organic farming to protect natural resources, boost exports and improve livelihoods.[11]

These global stories highlight a growing recognition that sustainable agriculture is a necessity. As more governments invest in organic and agroecological practices, the hope is that these efforts will lead to a healthier planet and more resilient food systems that reduce the risks of emerging diseases.

However, with stories of hope, caution is always needed. In the 2020s, Sri Lanka grappled with a severe food crisis. The streets were filled with frustration as families struggled to put food on the table. Skyrocketing prices amplified their worries. And market forces weren't solely to blame. It was deeply rooted in an ambitious yet hasty governmental decision.

In a bold move to champion environmental sustainability, the Sri Lankan government decided to ban chemical fertilisers nationwide, pushing for a shift to organic farming. The vision was to lead the world in eco-friendly agriculture, but the reality quickly became a nightmare. Crop yields plummeted, and the lush green landscapes gave way to fields struggling to produce enough to feed the nation. As food production dropped, the cost of essentials soared. The things that were once readily available became a luxury. The policy, though now abandoned, had already taken its toll. There's a lesson here. The dream of a greener future, when rushed and unplanned, can come at a high cost.

As they say, 'The road to hell is paved with good intentions.' The intentions may be great, but without equally great planning and execution, they risk fading into wishful thinking. Still, this shouldn't prevent us from pursuing a food system 'revolution' – but bold visions require meticulous planning to become lasting realities.

I don't think 'infectious disease' prevention is top of the list. But consumers are becoming more aware of their food choices. I see families at the supermarket reading labels carefully, often choosing products that are locally produced or certified by sustainability standards. They've learned about the impact of their food choices and are part of a growing movement to demand sustainably produced food. They're also more mindful of food waste, planning meals carefully to ensure nothing goes to waste and knowing that reducing food waste is essential to building a more sustainable food system. However, just like we need social policies to ensure people from all backgrounds have equitable access to natural environments (see Chapter 16), the same goes for healthy, sustainable foods.

Growing Food in the Concrete Jungle

In the concrete jungle, urban agriculture is taking root. Rooftop gardens, edible bus stops and vertical farms are turning grey concrete spaces into lush green areas producing fresh food. Urban foraging can take several forms, from harvesting the fruits of street trees and bushes to community gardening in food forests.

In the UK, there's a community group called the Sheffield Abundance Project, a volunteer-led initiative that harvests surplus urban fruit across the city and redistributes it to food banks and local communities.[12] Urban foraging projects may need to adapt to the complexities of urban life; for instance, the ownership of urban land regularly changes. However, innovations are helping to address this issue. For example, mobile allotments, such as those created by the Avant Gardening project[13], can be installed on vacant lots to provide communities with a foraging hub and can be easily moved if the land status changes. They're like vegetable plots on wheels.

Community gardening can promote other benefits, including physical exercise, mental wellbeing, nature connectedness and social cohesion. So, it's a win for human health – it could help strengthen people's immune systems while contributing towards stemming the non-communicable disease 'pandemic'.

All these efforts are promising, but they face significant challenges. Transitioning from entrenched agricultural practices to more sustainable ones is complex, and large-scale policy shifts are needed to support these changes. The economic realities of global food markets also present obstacles, but there's a growing awareness that the status quo is unsustainable. The increasing availability of innovative and holistic alternatives offers hope that we can build a food system that is not only more sustainable but also more resilient to the challenge posed by emerging infectious diseases.

Revolution! Revolution!

Let's not use the term lightly. It should be reserved for true transformation – moments when the world shifts on its axis and everything changes. That moment is overdue for agriculture; the sooner it arrives, the better.

CHAPTER 19

Vaccinations for Humans *and* Wildlife?

Life or death for a young child too often depends
on whether he [or she] is born in a country where
vaccines are available or not.

—Nelson Mandela, *Vaccine Fund Board address*

The journey to East Africa began long before I booked the plane ticket. It started in a sterile clinic in the small village of Somercotes, in Derbyshire, England. The sharp scent of alcohol wipes and the obligatory flick of the syringe marked the first feeling of 'it's happening!'.

Preparing to work in the dense rainforests required a series of jabs to shield against the invisible dangers that lurked in the jungle, some of which we've covered in this book. As the nurse calmly explained the vaccines – yellow fever, typhoid, hepatitis and rabies – the sting of each needle reminded me that the wild beauty of the rainforest comes with risks. The reality of the journey ahead became tangible.

I had to go back for a rabies booster. The only silver lining was that the government covered the cost because I was classed as a 'bat worker'. I also took a course of malaria prophylaxis – a decision I was thankful for when a colleague who skipped it ended up contracting malaria during the trip. Fortunately, she was okay after a couple of weeks of hellish symptoms.

And thus, we have vaccinations.

Some of you will know this story, but it's worth sharing for those who don't, as it's quite the historical event. For those who know the story, there may be a twist you haven't encountered before.

The practice of vaccinating supposedly began in the late eighteenth century with the pioneering work of English physician Edward Jenner. The story goes that in 1796, Jenner observed that milkmaids who had contracted cowpox, a relatively mild disease, seemed to be immune to smallpox, a deadly and widespread disease at the time.[1] To test this, he took material from a cowpox sore on a milkmaid's hand and inoculated it into an eight-year-old boy named James Phipps. Later, Jenner exposed the boy to smallpox (which would obviously be an ethical nightmare today!), and remarkably, the boy did not develop the disease.

This experiment led to the development of the first smallpox vaccine, and the term 'vaccine' itself comes from *vacca*, the Latin word for cow, in honour of the cowpox virus used in Jenner's early experiments. Jenner's work laid the foundation for the field of immunology and set the stage for developing vaccines against many other infectious diseases.

However, there's a white lie at the heart of vaccine history. This story is found in medical textbooks and reputable sources worldwide, but it's largely a myth crafted by Jenner's biographer to elevate his legacy. In reality, the link between cowpox and smallpox immunity was discovered by John Fewster years earlier when he noticed that some patients who didn't respond to smallpox inoculations had previously contracted cowpox.[2] Fewster's observations were discussed among local physicians, including a young Edward Jenner. It was Jenner, building on these insights, who later formalised and popularised the use of cowpox as a vaccine, though he wasn't the first to experiment with it.

This myth likely persists because it simplifies a complex history into a narrative of a lone genius. It's a pattern that's repeated in science. The true story of vaccination is one of collaboration, with contributions from various doctors, farmers and even cultures around the world that used early methods of smallpox prevention long before Jenner's time. While Jenner's work was undeniably important, the invention of vaccines was a collective effort built on centuries of global knowledge. The legend of the milkmaid might be a good story, but it's time to give credit to the many unsung heroes behind one of humanity's greatest medical achievements.

As alluded to, it's also important to point out that many cultures practised various forms of inoculation or traditional immunisation techniques for hundreds, if not thousands, of years that bore some resemblance to vaccination. For example, in parts of Africa, Asia and the Middle East, people practised *variolation* to protect against smallpox long before Edward Jenner developed the vaccine.[3] Variolation involves deliberately exposing a person to material from a smallpox sore, usually by inhalation or rubbing it into a small scratch on the skin. This can induce a mild form of the disease that would confer immunity.

In addition to these forms of early inoculation, many Indigenous cultures had their own methods for promoting health and preventing disease, often including herbal medicines.[4] While these methods were not 'vaccinations' as we define them today, they reflect an understanding of the importance of disease prevention and health maintenance within Indigenous societies. So, the concept of vaccination as we know it today – using a weakened or inactivated pathogen to stimulate the immune system – was a development of Western medicine. Still, the roots of understanding immunity and disease prevention are found in many cultures worldwide.

Vaccines have been one of the most powerful tools in public health. They've transformed the world in ways that are almost hard to imagine. For instance, the smallpox vaccine eradicated the disease entirely by 1980, saving an estimated five million lives every year.[5] Polio, once a global scourge paralysing over 350,000 children annually, has seen a 99.9% reduction in cases since 1988 thanks to widespread vaccination efforts.[6]

In recent history, the measles vaccine has been a game-changer, reportedly preventing over 23 million deaths between 2000 and 2018 alone.[7] This translates to an 80% drop in measles deaths worldwide. The HPV vaccine, introduced in 2006, is on track to virtually eliminate cervical cancer in countries with high vaccination coverage, potentially saving hundreds of thousands of lives every year.[8] The introduction of the Haemophilus influenzae type b (Hib) vaccine in the 1990s has led to a more than 90% reduction in Hib-related illnesses like meningitis in many countries.[9] This vaccine has prevented countless cases of severe disability and death, particularly in children under five. The rotavirus vaccine, introduced in 2006, has dramatically reduced cases of severe diarrhoea in children – a leading cause of child mortality in low-income countries. Hospitalisations were cut by up to 90% in countries with high vaccination coverage.[10]

The pneumococcal conjugate vaccine (PCV) has been successful, preventing an estimated 500,000 deaths annually from pneumonia and other pneumococcal diseases worldwide.[11] Incidentally, my friend Marja Roslund, a Finnish environmental scientist and immunologist, recently found that rubbing hands in microbe-rich soils might help strengthen pneumococcal vaccine effectiveness by enhancing immune functions.[12] This sounds utterly kooky, right? But it all goes back to the 'old friends' hypothesis and the notion that exposure to a diverse range of microbes has an important role in immune regulation.

Remember, most microbial species are either harmless or beneficial and vital for our survival! Marja's team have done some other interesting experiments exploring how increasing biodiversity in urban environments could influence children's immune systems by altering their microbiomes. In one study, over 28 days, daycare yards were transformed with natural materials like forest floor materials, providing a richer microbial environment.[13] Researchers found that children exposed to these biodiverse settings experienced increased skin and gut bacterial diversity, particularly in beneficial Proteobacteria. The researchers linked this microbial shift to a rise in immune-regulating markers like TGF-β1 and regulatory T cells, which help prevent immune-mediated diseases.

The children's microbiomes became more similar to those in nature-based daycare centres where kids regularly explore forests. The study supports the idea that reconnecting children with natural environments could strengthen their immune systems. It potentially offers a natural way to reduce the risk of conditions like allergies and autoimmune diseases in urban populations. In a way, exposure to diverse microbes from natural environments and enhancing the microbiomes of your 'walking ecosystems' could be viewed as a natural vaccination against immune disorders.

We don't just measure the impact of vaccines in terms of lives saved, but also in how they prevent suffering and protect future generations. The data is clear: vaccines have not only changed the course of history; they've redefined what's possible for humanity.

Some vaccine critics believe that natural immunity, gained through exposure to the disease, is better than immunity acquired through vaccination. There's some sense in this with some diseases. However, it also overlooks that many vaccine-preventable diseases can cause severe illness, disability or death, often making vaccination the much safer option. Vaccine critics also frequently cite misinformation, such as the debunked claim that vaccines cause autism, which originated from a fraudulent study in the 1990s.[14] Despite being thoroughly discredited, this claim still circulates widely and fuels vaccine hesitancy.

Moreover, the argument for natural immunity ignores a critical equity issue: not everyone has the same access to healthcare and resources needed to recover from serious infections. Vaccines provide a level of protection that transcends socioeconomic boundaries. They ensure that vulnerable populations, including the immune-compromised, the elderly and those in low-resource settings, are shielded from preventable diseases. Relying solely on natural immunity would disproportionately harm these groups.

There's no doubt that vaccines have saved countless lives. But the discourse thus far has been about humans. What about vaccinations for wildlife?

Back to Bats with White Noses

Remember David Blehert from Chapter 10? I asked him: What's the latest on vaccine development and rollout efforts to protect bats from white-nose syndrome? He said, "The laboratory within my branch, led by Tonie Rocke, has developed and registered (for field application) an orally ingestible vaccine that protects bats against severe white-nose syndrome." The vaccine was developed using the raccoonpox virus and engineered to express fungal antigens, priming the bat's immune system to fend off the deadly infection.

The raccoonpox virus is a member of the poxvirus family. In this instance, the team engineered the virus to carry specific antigens from the fungus that causes white-nose syndrome. Raccoonpox is a virus that primarily infects raccoons but is not harmful to other species. It has been repurposed in this case as a viral vector – a delivery system that introduces fungal proteins to a bat's immune system without causing disease. When presented with these proteins, the immune system recognises them as foreign. It mounts a defence, creating memory cells that can respond more effectively if the bat later encounters the actual white-nose syndrome fungus, *P. destructans*.

The process works similarly to how many other vaccines function. The viral vector carries the fungal antigens into the bat's body, where they stimulate an immune response without triggering an infection. This preparation gives the bat's immune system a 'head start' in fighting off the real fungal invader. Administering the vaccine is relatively straightforward. A small liquid dose is dropped into the bat's mouth while it's held gently. This method minimises stress on the bat, and only a single dose is required to be effective.

The choice of raccoonpox as the viral vector is significant because it's well suited to wildlife vaccination. The virus has been used previously in successful vaccination campaigns against rabies in vampire bats. This is one of those moments to pause, reflect and marvel. If you're like me, you may be thinking, "Wow, we're treating viral diseases with viruses!" It's also been used to treat plague in black-footed ferrets, showcasing its versatility in different species and different pathogens.

Tonie Rocke's team has field-tested the vaccine on little brown bats in several states. The trials have shown that vaccinated bats experience lower levels of fungal infection than those given a placebo.[15] The vaccine has proven particularly promising in regions like Wisconsin, where the disease has taken hold, and in Idaho's Minnetonka Cave, where researchers first detected the fungus in 2021, the vaccine has been a ray of hope. Four years in, bats show no signs of the disease.

Now, researchers are expanding trials to see how well the vaccine works across different bat populations and under varying conditions. This includes whether it can protect bats before exposure to the fungus or immediately after infection begins. The next step is to develop delivery methods to simultaneously reach large numbers of bats. Ideas include aerosol sprays applied to roosting sites or edible vaccines that bats would ingest as they groom. These developments aim to reduce the need for handling individual bats, making the vaccination process more scalable and less stressful for the animals. If successful, this vaccine could be a game-changer in the fight against white-nose syndrome. Fingers crossed.

But the innovation doesn't stop there. Other teams are adapting pandemic-era technology, like UV-C light, used to sanitise hibernation sites before bats return. It's an approach that mirrors sterilisation practices in hospitals and grocery stores, with early tests showing a significant reduction in fungal loads on treated bats. However, the uneven surfaces and complex structures of caves pose challenges in eradicating the fungus, so researchers are refining methods to target roosting spots more effectively.

There's also a probiotic angle. In Washington and British Columbia, scientists dust bat roosts with a clay powder mixed with beneficial bacteria.[16] This probiotic treatment, derived from healthy bats in the region, aims to bolster the bats' natural defences by enriching their skin microbiomes. Although the initiative is still in its early stages, initial studies suggest it could significantly improve survival rates through winter hibernation. This reminds me of my friend Brendan Daisley in Canada, who developed a probiotic called 'BioPatty'. He essentially puts these probiotic patties in beehives, and it enriches the bees' microbiome and immune defences.[17]

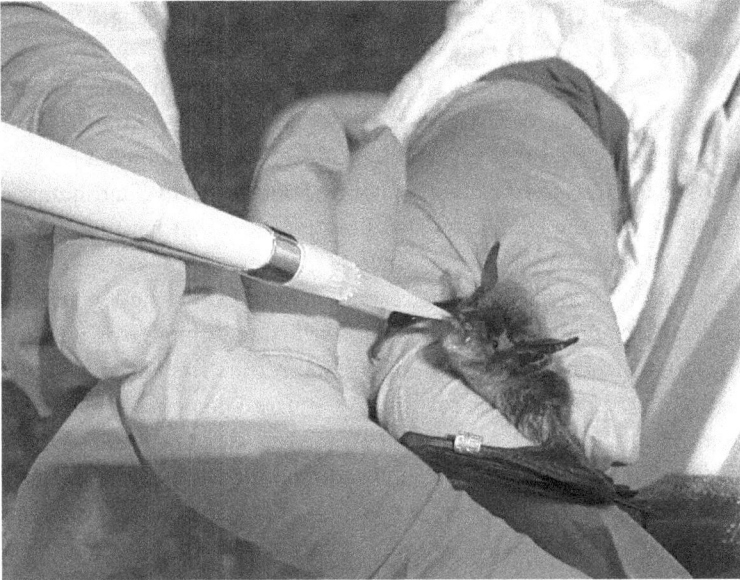

Bat receiving the white-nose syndrome vaccine (USGS: public domain)

In Georgia, researchers are taking a different approach, using volatile organic compounds like those found in orange peels and wild pineapples.[18] Scientists hope to slow fungal growth without disturbing the bats by fogging hibernation sites with these compounds. After years of trials, they've seen the bat population in one test tunnel slowly recover, though fine-tuning is needed to balance efficacy with preserving other essential cave microbes.

Together, these methods form a multi-pronged defence against a seemingly unstoppable enemy – and I for one love multi-pronged approaches! Each approach tackles a different aspect of the fungus's lifecycle, from disrupting its DNA with light to boosting bat immunity through innovative vaccines and probiotics.

The race is on to save these beautiful creatures before it's too late, but the combination of cutting-edge science and creative problem-solving offers a glimmer of hope in the dark recesses of the bat caves.

Badgers and Tuberculosis

For as long as I can remember, the debate over whether to cull badgers (*Meles meles*) in the UK has been a constant back and forth. Arguments and passionate views ricochet from conservationists on one side and farmers on the other. It all boils down to badgers carrying a form of tuberculosis that can be fatal to cattle.

In Cornwall, a pioneering project is giving hope to cattle farmers struggling with the devastation caused by bovine tuberculosis (bTB). A recent study shows that vaccinating badgers – often blamed for trans-mitting the disease to cattle – could be a more effective and humane solution than the controversial culling practices that have led to the deaths of over 210,000 badgers since 2013.[19]

Led by the Zoological Society of London (ZSL) alongside farmers and conservationists, this project was the first of its kind to be initiated by farmers themselves. Over four years, researchers vaccinated 265 badgers across 12 farms, covering 11 square kilometres. The results were encouraging. Despite initial scepticism about whether enough badgers could be reached, the study showed that 74% of local badgers received the vaccine, and the proportion of infected badgers fell from 16% to zero. This result caught the attention of scientists and farmers alike, signalling a potential shift in how we tackle bovine tuberculosis.

There's little to no evidence that culling badgers effectively reduces bovine tuberculosis in cattle. It's worth emphasising this point. Although badgers can play a role in bovine tuberculosis transmission, scientists estimate that around 94% of infections in cattle come from other cows.[20] I remember my undergraduate supervisor, Richard Yarnell, who also served as the chief executive officer of the Badgers Trust, shared with me his involvement in earlier trials. This early research indicated that a significant number of infections in cattle could be attributed to poor husbandry practices and the stresses associated with cattle transportation.

Badgers are used as a scapegoat. This only prevents holistic action. Again, it's like putting a plaster on a wound rather than addressing the root cause of the symptoms. Many experts agree that the continued spread of bovine tuberculosis in cattle is primarily due to issues related to the movement of cattle rather than transmission from wildlife like badgers. The transportation of infected cattle between farms, inadequate biosecurity measures and ineffective management of cattle herds are all major contributors to its spread.

Therefore, improving cattle management practices, such as reducing the movement of infected animals, implementing better quarantine protocols and enhancing biosecurity measures, could be more effective in controlling bovine tuberculosis than wildlife culling. To me, the spread of bovine tuberculosis is just another symptom of our unsustainable agricultural practices. It's another reason we need a revolution in farming and land management (see Chapter 18).

It remains to be seen whether this vaccination strategy will directly impact bovine tuberculosis rates in cattle. However, the study marks a promising step towards a more balanced approach to disease control that brings together science, farming and conservation. In addition, the vaccination programme may reduce disease outbreaks and symptoms in wild badger populations (bovine tuberculosis can lead to severe health issues for them, including weight loss, coughing and eventually death). The project team calls for more research and government support to scale these efforts. This could pave the way for badger vaccination to be a key tool in eradicating bovine tuberculosis while conserving the badger, an iconic member of the UK's biotic community.

Black-Footed Ferrets

Cast your mind back to Chapter 10. Remember how black-footed ferret populations are endangered and threatened by sylvatic plague? Spread by fleas, sylvatic plague devastates prairie dog populations, both the primary food source and preferred shelter providers for ferrets. There is a vaccine for this disease. However, delivering the vaccine to prairie dog colonies over vast areas of prairie habitat is time-consuming and labour-intensive. Conservationists needed to find a way to deliver the vaccine to the ferrets more efficiently. Cue the flying robots!

A partnership involving the World Wildlife Fund, the US Fish and Wildlife Service, Model Avionics and other collaborators led to the development of an innovative solution: drones equipped with custom-designed bait dispensers.[21] Drones whizzing overhead drop vaccine baits, like tiny treats, across vast stretches of prairie. Amusingly, prairie dogs love the smell and taste of one of my favourite cupboard staples: peanut butter! This is one of the components of the vaccine baits. Drawn to the irresistible scent, they munch away, unknowingly boosting their defences against sylvatic plague.

It's hoped that this high-tech approach can help transform ferret recovery efforts, allowing precise, GPS-guided distribution of vaccines over thousands of acres. The introduction of 'triple-shooter' mechanisms, which fire baits in three directions at once, enhances vaccine distribution. This innovation boosts efficiency. It's made it possible to cover large areas quickly – something that once seemed out of reach. These advancements in drone technology may help to stabilise prairie dog populations. This will ensure a sustainable food source for black-footed ferrets and contribute to broader conservation goals.

As an ecologist, I've used drones to map habitats, collect water samples, survey parasites (using tick drags) and create 3-D models of trees, so I can testify to their versatility. Here's hoping they can help our slinky mustelid friends in the States.

Devil Tumour Vaccines

In a groundbreaking move to protect Tasmanian devils from the deadly, contagious cancer, researchers are preparing to test a new vaccine that could be a lifeline for this iconic species.[22] As discussed in Chapter 11, devil facial tumour disease has decimated up to 80% of the devil population in the past few decades, posing a severe threat to their survival. Inspired by the success of COVID-19 vaccines, scientists have developed a similar approach using an adenovirus to carry the vaccine into the devil's cells. The goal is to make the tumour cells more recognisable to the immune system by encouraging them to express specific proteins, thereby triggering a defensive response. The initial trials will focus on testing the safety and effectiveness of the vaccine in 22 captive devils. If successful, the team plans to roll out a practical distribution method involving AI-driven bait dispensers. This would target devils in the wild.

This effort isn't the first attempt at creating a devil facial tumour disease vaccine, but the current project represents the most promising step forward. Even if the vaccine doesn't provide complete immunity, researchers hope it can at least extend the devils' lifespan, giving them more chances to breed and stabilise the population. By combining cutting-edge technology with conservation science, this project could be the breakthrough to save one of Australia's most endangered marsupials.

Frog Vaccines and… Saunas?

Tiny droplets with chytrid fungus trickle over the smooth skin of a tadpole, slipping between the folds where hundreds of bacterial species cling. As we saw back in Chapter 11, for years, this fungus has wreaked havoc on amphibian populations worldwide, decimating species. But now, instead of succumbing, the microbiome on this tadpole's skin starts to shift, and bacteria produce antifungal compounds. The microbiome fortifies the tadpole's natural defences without introducing any foreign microbes.

This probiotic attempt is orchestrated by a carefully crafted vaccine. In a groundbreaking study, researchers at Penn State uncovered how a vaccine designed to protect frogs from the deadly chytrid fungus can also shift the composition of their skin microbiomes, enhancing the frogs' resilience against future infections. The researchers applied a non-lethal dose of the fungus's metabolic product to tadpoles and observed significant changes in the microbial community on their skin.[23] The shift favoured bacteria known to produce antifungal compounds. This finding suggests that vaccines can do more than just trigger an immune response – they can also help reshape the microbiome in ways that boost natural defences.

This approach contrasts with traditional methods, such as adding probiotics, which struggle to establish themselves in the ever-changing microbiome. Instead of introducing new bacteria, this technique gently nudges the microbial community towards a more protective composition. It's a strategy the researchers call 'microbiome memory'. The microbial community reorganises so that it holds its ground against one of the deadliest pathogens amphibians have ever faced. Notably, the overall diversity of the microbiome remained stable, which is key to maintaining frog health (once again, remember the 'no room at the dining table' effect). This innovative work could pave the way for new strategies in conservation and broader vaccine development. The results of this study seem like a major victory in the fight to protect vulnerable amphibians. I say this with cautious optimism, but here's hoping the vaccine influences the microbial landscape in wonderfully beneficial ways.

Let's pause again here. I won't pretend to read your mind, but let me guess: somewhere in the last few paragraphs, you thought the recent success in chytrid vaccinations was great. But then you remembered the subheading for this section, and that one word caught your eye: 'SAUNAS'. Am I right? Either way, let's get to it.

In a surprising breakthrough, Australian scientists have turned to 'frog saunas' as a potential solution to combat the chytrid fungus. The idea is simple and innovative: provide a warm and controlled refuge for the frogs where they can naturally fend off the infection. While not a vaccination, it could be a complementary strategy for some species. Researchers at Macquarie University constructed greenhouse 'saunas' using painted bricks as shelters.[24] When green and golden bell frogs (*Litoria aurea*), an endangered species, were placed in these saunas, the warmer conditions allowed them to clear the fungus from their skin – a rare victory against the pathogen that has driven at least 90 amphibian species to extinction. Past attempts to heat-treat other frog species had failed. However, these bell frogs responded positively. The secret lies in how closely the saunas mimic the frogs' preferred microhabitats. This helps them regain the natural defence mechanisms undermined by the fungus.

Anthony Waddle, who led the research, highlighted the significance of this discovery. For a species like the green and golden bell frog, which is clinging to survival, this method gives them a fighting chance. Traditional approaches have failed, but these controlled, heated environments show that even simple interventions can have powerful results. Moreover, many of the treated frogs resisted reinfection. This raises hopes that the sauna approach could help stabilise their populations in the wild.

Despite this success, Waddle and other experts caution that the technique might be species-specific. The effectiveness of these saunas for other amphibians remains uncertain. Still, the breakthrough offers a glimpse of hope in a field where solutions have been hard to come by. Mini saunas could offer a creative and sustainable strategy to protect some of the world's most vulnerable frogs. I'm not sure how it would work at scale, though!

Vaccinating Plants?

Plants don't get vaccines like humans do. However, scientists have developed methods that essentially 'vaccinate' them against diseases. Exposing plants to harmless versions of pathogens or beneficial microbes can stimulate the plant's immune system. It teaches the plant to recognise and fight off more serious threats in the future. This is known as 'induced resistance'. It's a bit like giving plants a heads-up before the real attack comes.

Human dependence on monoculture for food production undermines the plants' natural defence mechanisms. In a diverse ecosystem, plants can develop a range of resistance genes, allowing them to better adapt to evolving threats. However, we limit this natural evolutionary process by cultivating genetically uniform crops. If all plants in a field are genetically identical, a disease that affects one plant can easily spread to the rest. Plant pathogens rapidly evolve. Yet, we grow crops with the same resistance genes, leaving them vulnerable. As a result, we increasingly depend on pesticides to protect our crops.

Researchers are exploring RNA-based treatments, like mRNA vaccines in humans, that could prepare plants to defend against specific viruses.[25] These 'vaccines' might not involve needles or syringes, but they're proving powerful tools in protecting crops and boosting global food security. However, I think this is just another plaster on the wound, without treating the root (pardon the pun) cause.

Who would have thought you could vaccinate a tree?

I recently watched some talks by Glynn Percival, an academic at the University of Reading, UK. He revealed that trees can be boosted, much like our immune systems – they can essentially be 'vaccinated' to enhance their natural defences. With pests and diseases on the rise due to climate change, this approach might be crucial for the future of species like oak, ash and horse chestnut (*Aesculus hippocastanum*), which face serious threats. Percival's research uses natural soil amendments like chitin and willow mulch, which slowly release salicylic acid (the principal chemical in aspirin) as it decomposes, encouraging trees to produce defence compounds.[26]

In one talk, Percival said: "Once a tree is infected, it's really hard to manage that disease … we're also starting to see a build-up in tolerance in pests and diseases to pesticides … So, we need a different way of looking at how we will manage these problems."[27]

"And how do we control human diseases?" he asked the audience.

"Vaccination!" one person shouted out.

Percival pointed out that vaccinating plants against diseases is not novel. The idea of inducing resistance was recognised in the 1920s when heat-treated *Botrytis cinerea* was exposed to Begonia plants. Instead of causing infection, it resulted in the plants developing resistance. It's much like a flu vaccine, where we're given a live-attenuated (weakened) version of the pathogen to trigger an immune response. However, Percival said that humans need a specific vaccine for a specific disease (a flu vaccine for the flu, a typhoid vaccine for typhoid and so on), and yet once you switch on the defence system in trees, you don't just switch on one system; you switch on about thirteen. You get an increase in antifungal compounds, leaves become thicker, resin production is enhanced, and you get resistance to various pathogens – viruses, fungi, bacteria and so on.

Percival has found that willow mulch switches on the tree's defences by interacting with the root system. His research team have found that it's good but not quite as good as spraying fungicides. They're still fine-tuning the system and hope to produce a refined treatment in the future.

In the Netherlands, more than half a million urban elms (*Ulmus* × *hollandica*) have been vaccinated with a fungus that enhances resistance to Dutch elm disease, resulting in just a 0.1% infection rate.[28] These plant vaccines work broadly, bolstering defences against multiple diseases by thickening leaves and boosting antimicrobial compounds. Developing this type of vaccination for other tree diseases may be possible.

We're grappling with rising threats to our ecosystems. But try imagining a future where we arm our crops and forests with a biological defence system fine-tuned over millennia. There is hope; good people are working on this. Imagine reducing our dependence on harmful pesticides. Imagine embracing a more resilient and holistic approach to treating our ecological and agricultural systems. With nature's own defences as our guide, here's hoping we can strike a balance where people and the planet are better protected.

CHAPTER 20

Staying Ahead
of the Curve

*We must read the ripples before the wave hits and
find clarity in the uncertainty.*

Tension was high in the crowded airport terminal. As passengers
shuffled through the security lines, they watched the oversized
screens flashing updates – another outbreak. Somewhere across
the globe, a single mutation in a renegade virion had turned a contained
virus into a potential global threat. In a lab, epidemiologists stared at
their monitors, watching the digital threads of a pathogen's movement
light up across the map. It's a routine drill now – data streams from
satellite tracking, wastewater sampling and genomic sequencing labs.
The warning systems have grown more sophisticated, but they're only
as good as the weakest link in the chain.

This is the front line of pandemic prevention. We don't really see
what goes on as we're busy living our lives – working, commuting,
socialising, sleeping. But things are happening. Some people dedicate
their lives to detection, rapid response and adaptation plans. These will
all determine how close we come to another global crisis.

But what does it really take to stay ahead of the next outbreak? As
we scan for microbial signals and monitor hotspots, the future might
depend on these people and the foresight they bring to a world that's
often reactive, not prepared.

In the background of a pandemic like COVID-19, there's organised
mayhem of data streaming, epidemiologists darting around, and
genomic sequencing happening. When COVID-19 first emerged,
global health systems and surveillance networks scrambled to respond.
The world was caught off guard. Despite sophisticated tracking systems

and international health protocols, the virus's rapid spread revealed how unprepared we were for a pandemic of this scale. The World Health Organization was alerted to unusual pneumonia cases in Wuhan, China, in late December 2019. It quickly activated its International Health Regulations (IHR) framework, notifying countries and coordinating an international response. However, by the time the virus's genetic sequence was shared via platforms like GISAID in early January 2020, SARS-CoV-2 was already spreading across borders.

There are many groups and labs across the world that form part of a pandemic early-warning system. Let's imagine a European lab where scientists at GISAID pore over shared genetic data, tracing mutations and identifying variants. Over in Africa, field epidemiologists at the Centres for Disease Control and Prevention (CDC) might receive an alert and coordinate ground-level investigations. The gears of the global pandemic-surveillance machine grind into action – a collaboration of research scientists, public health officials and number-crunching gurus. Each link is crucial in the chain of pandemic prevention.

In the race against emerging pathogens, timing is everything – remember, there are also 'unknown unknowns' that we somehow need to make known. These surveillance networks and early-warning systems, layered across continents and driven by cutting-edge technology and local expertise, currently represent humanity's first line of defence. But really, our first line of defence should be more holistic; our first line should be to protect nature and let nature's shield protect us.

Beneath the structured response, we're still learning how to predict, prevent and contain pandemics. We now live in a world where ecological destruction and urban sprawl continue to push us closer to new outbreaks. So, we need to learn fast. As the world changes, so must these systems – constantly evolving to address the next unknown threat.

Mpox

As I write these words, a recent surge in mpox cases has triggered global concern, especially in Africa, where outbreaks have become more severe and widespread. Mpox, formerly known as monkeypox, is a viral zoonotic disease caused by the mpox virus, which belongs to the Orthopoxvirus genus, the same family as the smallpox virus.

Before we go too deep into what mpox is all about, let's briefly touch on why it was renamed. The disease was renamed to 'mpox' by the World Health Organization in 2022 in response to concerns that the original name was both inaccurate and stigmatising. The term 'monkeypox' led to misconceptions, as monkeys are not the primary reservoir of the virus. Additionally, the name fuelled negative stereotypes and stigmatisation, especially in African regions where the disease was endemic.

Mpox was first identified in laboratory monkeys in 1958 (hence the original name).[1] The disease was later found to be endemic in certain wild animal populations in Central and West Africa. The first human case was recorded in 1970 in the Democratic Republic of the Congo (DRC).[2] The virus has since been associated with occasional outbreaks in various African countries.

Mpox presents symptoms like those for smallpox, although it's generally less severe. The illness typically begins with flu-like symptoms such as fever, headache, muscle aches and swollen lymph nodes. A few days after these first symptoms, a characteristic rash appears. It progresses through stages of macules, papules, vesicles, pustules and scabs before eventually healing (unless you're unlucky). The rash is often concentrated on the face, palms and soles but can spread to other body parts, including the genital area.

Human-to-human transmission occurs primarily through close contact with infectious lesions, bodily fluids, respiratory droplets and contaminated materials like bedding. We can also contract the disease through contact with infected animals, particularly rodents, through bites, scratches or handling. While the virus has been circulating in certain regions for decades, recent outbreaks have highlighted its potential for wider global spread, largely due to increased travel and human encroachment on wildlife habitats.

The Democratic Republic of the Congo is at the centre of the recent crisis, reporting over 54,000 cases and 1,250 deaths in 2024.[3] What makes this wave particularly alarming is the emergence of a strain, Clade Ib, that may be deadlier than the earlier Clade II (but probably less deadly than Clade Ia). It has pushed the case fatality rate up to 3–4%, much higher than the strain from the global outbreak in 2022 (Clade II). This latest resurgence of mpox is not confined to the DRC. Several neighbouring countries, such as Rwanda and Uganda, have detected their first-ever cases. Public health experts believe this is only the 'tip of the iceberg'. There are weaknesses in the surveillance system across the region, particularly in conflict zones and areas with poor healthcare infrastructure. Children under 15 are especially vulnerable. They make up 70% of the reported cases and 85% of deaths in the DRC.

In response, the World Health Organization and the Africa CDC declared this outbreak a public health emergency. This declaration is a call to action, urging countries to move from a reactive to a proactive stance in fighting the disease. Experts emphasise the need for coordinated international efforts, with the Africa CDC highlighting that this crisis is "a fight for all Africans".[4]

It will be interesting to see how the world will respond to mpox. Will the response be different, given the recency of the COVID-19 pandemic? We can perhaps get an idea of how countries might respond by looking to Scandinavia. Sweden is vigilant after detecting the newer strain of mpox, Clade Ib. The case involved a patient who had recently travelled to parts of Africa where this virulent strain is spreading. Swedish health officials activated contact tracing and prepared containment strategies, even though there was no indication of a major outbreak... yet. Swift action is essential to prevent the spread of this virus. Key measures include rapid diagnosis, isolation and contact tracing. And global support, particularly in African regions where the virus originated, is crucial for stopping its spread.

At the time of writing, Sweden is one of 15 non-African countries with a confirmed case of this new strain (the others being Australia, Germany, India, Thailand, the UK, Zambia, Zimbabwe, the USA, Canada, Pakistan, Oman, the UAE, Belgium and China). These are all travel-related cases. However, officials are urging healthcare providers to stay alert as more cases could emerge globally. The discovery of the new strain has raised awareness across Europe, with agencies warning of potential new cases and advising precautions for travellers heading to affected regions. So, this level of vigilance seems to be promising. But vigilance alone is like watching the flood approach; early action is what keeps the flood at bay.

Technology

So, what about getting ahead of the curve? The book thus far has emphasised the need to 'future-proof' against pandemics through One Health approaches. But can we also adapt and detect invisible threats before they become disasters? That's where cutting-edge surveillance technology comes in. It might soon keep us one step ahead of the next pandemic. Our defence against emerging diseases has often been reactive: waiting until infections spread, then rushing around for solutions. But soon, a new wave of tech might change the game. It may allow us to catch pathogens in real time, perhaps even before they spill over into human populations.

Let's take environmental biosensors, for example. Scientists are developing these devices to deploy in high-risk areas such as live animal markets, wastewater systems and wildlife reserves. The sensors could detect tiny traces of environmental pathogens, like viruses in the air or bacteria in water, long before infections become widespread.

Biosensors are being explored and developed for detecting bird flu. These sensors offer the potential for rapid, on-site detection of viruses like H5N1, which could be game-changing in managing outbreaks in poultry and wildlife. Recent advancements focus on making these biosensors more efficient, portable and applicable in field settings, but their use in real-world scenarios is still emerging. While the current state of biosensors is promising, they're mainly in the research and development phase, and large-scale deployment is still limited. This technology is evolving. In the future, we could see biosensors playing a pivotal role in monitoring bird flu and other zoonotic diseases in a way that allows for more proactive interventions. It'll be like having a digital nose constantly sniffing out danger. If a pathogen is detected, alerts will be sent to health authorities instantly, giving us a critical head start.

A couple of years ago, I founded an initiative called the Aerobiome Innovation and Research Hub (AIR Hub).[5] We're looking at improving air-sampling devices for collecting microbes. This is going to sound a little crazy, but we're also investigating whether vacant spider webs could be an innovative, low-cost and environmentally friendly alternative to traditional air-sampling methods. By analysing the microbes collected on the webs, we aim to determine if they reflect the microbial composition of the surrounding air, thus offering insights into the environment's health. This approach could change how we monitor air quality and microbial diversity, providing critical data on how airborne microbes interact with factors like pollution, climate change and human activity. This isn't a completely novel and crazy idea. One study has already explored using spider webs to collect airborne viruses, and another used the webs to detect the DNA of animals.[6, 7] Given that we're in Australia, spiders can be a little challenging to work with, to put it lightly. We've teamed up with local spider expert Bruno Alves Buzatto to ensure the webs are vacant and safe to sample! Our work is somewhat speculative now. Yet, we may see this webby affair evolve into a nature-based infectious-disease-detection system in the future.

Then there's genomic sequencing. It was once a cumbersome lab process, but now it's a nimble tool in disease surveillance. Genomic sequencing allows scientists to read the genetic code of viruses, bacteria and other organisms. It gives us deep insights into how these pathogens behave and evolve. The technology was critical in tracking the different variants of COVID-19, helping public health officials understand how the virus was mutating and spreading across regions. Portable sequencers, like Oxford Nanopore's MinION, are changing fieldwork.[8] They enable researchers to perform rapid sequencing even in remote areas. This means that scientists can quickly identify the presence of a pathogen, offering clues about its origin and how it might spread. For example, during the 2014–2016 Ebola outbreak in West Africa, scientists used portable sequencers to monitor viral mutations in real time.[9] As we progress, genomic surveillance could be used routinely at key points like border controls, monitoring humans and non-human animals to catch emerging threats early.

Artificial intelligence (AI) and machine learning are also key players. We can use these to sift through vast oceans of data – from social media chatter to emergency room visits – spotting unusual patterns and predicting outbreaks. For instance, algorithms can flag a spike in flu-like symptoms in a community before it's officially reported. AI can even monitor wildlife migration patterns and climate data to anticipate where a zoonotic disease might emerge next.

In 2019, the Canadian health-monitoring platform BlueDot flagged the early signs of COVID-19 several days before official alerts were issued.[10] By analysing data from sources like news reports, airline ticketing and official health announcements, BlueDot's AI predicted the spread of the virus to other countries. At the time, BlueDot's Founder, Kamran Khan, said, "We did not know this would become the next pandemic. But we did know that there were echoes of the SARS outbreak, which we should be paying attention to."[11] In the future, we could use such systems globally, creating a digital early-warning system that spots outbreaks in real time.

What about wearable tech? In today's modern world, a common sight is a jogger warming up and looking at their watch. Wearable health tech like smartwatches that track your vitals could soon be connected to broader health networks. If we detect unusual spikes in illness across many users, it could signal the start of an outbreak. The data collected should be anonymous, but it could clearly show where interventions are needed most. During the COVID-19 pandemic, researchers found wearables could detect signs of infection days before symptoms, indicating that someone might be contagious. In the future, we could link these devices to larger health networks, where anonymised data is pooled and analysed for trends. For instance, if multiple wearables detect rising body temperatures or abnormal vitals in a community, it could alert health officials. This would allow for rapid interventions like testing, isolation or targeted vaccination efforts before the virus spreads widely.

There's an argument that these technologies shouldn't work in isolation, but form part of an interconnected web of surveillance that spans from local communities to global networks. By combining environmental monitoring, rapid diagnostics, AI-driven insights and real-time data sharing, we could catch emerging pathogens before they have a chance to cause global chaos. The goal is to shift from a reactive stance – scrambling to contain outbreaks after they've already spread – to a proactive, data-driven approach that spots and stops threats at their source.

This technology sounds promising, but my ethics alarm bells are also ringing – perhaps I've been in academia too long. We mustn't ignore the ethical questions when using these powerful surveillance technologies. Imagine living in a world where your health data, environmental conditions and even daily movements are constantly tracked – all to prevent the next pandemic. While this might help catch outbreaks early, it raises major privacy concerns. Who controls this data? Can we trust that it won't be misused? There's also the issue of fairness. What happens if only wealthy nations can afford these advanced tools while poorer regions are vulnerable? A similar story unfolded during the COVID-19 pandemic with vaccines and treatments. If we aim to protect everyone, we must use these technologies responsibly, ensuring they're accessible and that privacy and equity are prioritised. Balancing innovation with ethical considerations will be key to building a smarter, safer future.

Staying ahead of the curve means predicting tomorrow's challenges today. In a world of fast-moving threats, *proactive* beats *reactive* every time.

CONCLUSION

Looking Back, Moving Forward

As we conclude this journey into infectious diseases and the vital role of protecting biodiversity in preventing outbreaks, it's clear that our relationship with the natural world is at a critical crossroads. Recall the Introduction and the story of Eyam's resilience in the face of the bubonic plague. The lessons in this story are timeless. Just as those villagers sacrificed for the greater good, we must now confront our broken relationship with nature to mitigate future pandemics.

We first explored the groundwork for shifting perspectives. Pathogens spilling across species boundaries isn't nature's 'fault'; it mirrors our fragmented landscapes, our fragmented ontologies, our fragmented thinking. Our relentless transformation of the natural world propels the frequency and diversity of infectious diseases. We're *Homo sapiens*; this is what we do. We change landscapes. We always have and likely always will. But it's *how* we change these landscapes and view and interact with the other lifeforms on the planet that matters the most.

It's entirely possible to do this in a way that's more in sync with natural processes and *promotes* reverence and reciprocity rather than *opposes* them. Reverence and reciprocity mean recognising that we're not separate from nature – we're part of it. By rethinking how we interact with other species and their habitats, we can reconfigure our relationship with the natural world to mitigate risks rather than exacerbate them. This may seem abstract or even idealistic, but the approach is within our reach. We can achieve this by aligning our actions with the rhythms of nature rather than forcing our will upon it. Plenty of people are doing this, from restoring ecosystems to reshaping farming practices.

We explored the dilution effect and found that *diversity* is often vital. This is true for many things in life. Ecosystems with richer biodiversity can (variably) reduce disease risk. When diverse species are present, they can act as buffers by diluting the pools of highly efficient disease carriers. However, biodiversity's protective effect is fragile. We unravel these natural defences when we fragment habitats or eliminate key species. Instead of reducing disease, we inadvertently amplify risks.

Remember the vultures? This is probably my favourite story in the book. It's truly jaw-dropping to think that the decline in one or two species can lead to such dramatic consequences for human health. The cascading effects that span both the ecological and social realms are fascinating – from a lack of scavenging and a boom in dog populations to a rise in rabies cases and massive impacts on people's lives and economies.

And what about bats? From the forests and skies to the damp caves, we explored their role as vital ecosystem players and misunderstood scapegoats. While bats carry viruses, their unique immune systems allow them to coexist with pathogens that would devastate other species. The real danger isn't the bats themselves but how our actions – deforestation, urbanisation and pollution – create conditions ripe for disease spillovers. Rather than blaming bats, we must recognise that disrupting natural systems heightens the risk of pandemics. As we push these creatures into corners, we stress them, sparking viral shedding that fuels the very fears we cast upon them. In this section, I hoped to challenge some to look beyond simplistic blame and embrace the intertwined complexities that govern contagion. If we shift our perspectives – seeing ecosystems as allies rather than threats – we might fend off future pandemics and discover pathways to a healthier coexistence.

Next, we explored the idea of Disease X, which embodies our ongoing struggle against invisible threats. This unknown pathogen lurks, whether it emerges from a marketplace, a lab accident or a factory farm, waiting to exploit our globalised world. Humanity's actions – deforestation, wildlife trade and industrial farming – set the stage for the next pandemic. As species constantly evolve to survive, we must adapt our approaches to anticipate the next crisis. But which pathogen is most likely to cause the next pandemic? Viruses, bacteria, protozoa and fungi all have the potential to cause death and disruption on unimaginable scales. In fact, they already are. In antimicrobial-resistant bacteria, we see a rising tide of untreatable diseases – some even consider this resistance a type of pandemic.

While the fictional idea of zombie-inducing fungi infecting humans is unlikely, fungal pathogens like *Aspergillus* and *Candida* are already unleashing turmoil, especially among populations with weakened immune systems. Despite their deadly impact, fungal diseases remain underfunded and under-researched compared to bacterial and viral pandemics. With climate change and human activities driving fungal evolution, proactive measures are crucial to preventing these overlooked threats from escalating into worse crises.

Personally, I think the pathogen most likely to cause something akin to or worse than the recent COVID-19 pandemic is avian influenza (H5N1). Unlike many pathogens, H5N1 has a high fatality rate, killing over 50% of those it infects. While it mainly spreads between birds and occasionally from birds to humans, if mutations enable it to transmit efficiently between humans, it could then lead to a highly lethal global outbreak.

And while we're talking about a disease that also affects other animals, this is a key takeaway from the book. While we all wonder when there will be another pandemic that affects humans, 'pandemics' are occurring everywhere, as we speak, in wildlife. From the millions of birds and mammals recently lost to avian flu to the plague ravaging ferret populations, and from chytrid fungus decimating amphibians worldwide to white-nose syndrome blighting North American bats, the impact of diseases on wildlife is widespread and severe. And let's not forget prion diseases. These are some of the most perplexing and destructive pathogens known to us. We also have the numerous ailments afflicting often overlooked plants and fungi. Many of these devastating outbreaks are, at least in part, driven by human actions.

Moving forward, we must embrace the principles of One Health. The health of people, (other) animals and ecosystems is interconnected – protecting one protects all. The future of humankind lies in understanding that safeguarding nature is about protecting ecosystems for their intrinsic and instrumental value. It's also about securing our own survival. Ultimately, we need to build bridges and respect boundaries. Building bridges to mindfully connect with nature while guarding the boundaries between species is key to securing our shared future.

Notes

Introduction: Invisible Foes, Planetary Woes

1. If you're ever in the Peak District, Eyam is well worth a visit: https://www. eyamvillage.org.uk. Go to the small Eyam Museum and wander around the grave sites and old cottages for the full experience. There's a nice stone outcrop with a natural arch, called Cucklet Delf, and it is said that the village rector, William Mompesson, could preach into the valley from here and be heard by his congregation on the opposite hillside. There's also a story of two lovers, Emmott Sydall and Rowland Torre (one living in Eyam and one in nearby Stoney Middleton), who were separated during the plague outbreak. The story, while likely embellished over time, tells of their secret meetings at a stream near to Cucklet Delf, where they would shout and wave to each other, unable to physically touch due to the plague quarantine. The story suggests that Emmott eventually died of the plague, with Rowland continuing to go to their meeting place in vain, hoping for her return. See https://www.bbc.com/news/uk-england-35064071 for Thompson's comments.

2. See Stephens et al., 2021 for a comprehensive analysis of the world's largest zoonotic outbreaks, offering valuable insights into their drivers, impacts and patterns – an essential resource for understanding how and why diseases spill over from animals to humans.

3. See Jones et al., 2008a. This landmark study was among the first to map emerging infectious disease events globally, identifying hotspots and key drivers – including land use change, wildlife trade and human population density – that contribute to spillover risk.

4. See Acharya et al., 2020. This paper underscores the importance of the One Health approach – recognising the interconnectedness of human, animal and environmental health – in preventing and responding to pandemics. It calls for better integration of veterinary, medical and ecological expertise to detect and control outbreaks more effectively.

5. See Agnelli and Capua, 2022. This brief but thought-provoking article challenges the narrow framing of SARS-CoV-2 as purely a human pandemic, proposing the term *panzootic* to better capture its wide host range and ecological implications. It highlights the need for language that reflects the multi-species nature of disease emergence.

6. See Assefa and Gilks, 2020. This paper explores two possible futures for HIV: eradication versus long-term endemicity. It emphasises that ending the epidemic will depend not just on biomedical breakthroughs, but on strengthening health systems and addressing social inequalities – lessons that resonate far beyond HIV.

7. See Gelband et al., 2004. This influential report examines the economic and policy dimensions of antimalarial drug resistance, underscoring the urgent need for global coordination, equitable access to effective treatments and investment in research – all of which remain crucial in managing infectious disease threats.

8. See Whitfield, 2002.

9. Pomeroy (2019). This article critically examines the oft-repeated claim that malaria has been responsible for half of all human deaths throughout history. Pomeroy traces the origins of this assertion and, through expert interviews and

historical analysis, concludes that while malaria has been a significant killer, the 'half of everyone' figure is likely an overstatement lacking solid empirical support.

10. See Didelot, 2016. Didelot explores whether the villagers' decision to quarantine truly saved lives – or may have cost more than it spared – offering a nuanced look at community sacrifice, contagion and the ethics of containment.

11. See Dean et al., 2018. This study reshapes our understanding of historical disease transmission and highlights the complexity of epidemic dynamics.

12. See Robinson, Cameron and Jorgensen, 2021. In this paper, we explore how people's exposure to and understanding of biodiversity may influence their attitudes towards microbes. We suggest that a deeper connection with the natural world could reduce germaphobia and foster more nuanced views of the microbial world, with an understanding that many microbes are essential to life and health.

13. See Denning, 2024. The rise of fungal diseases. This news release underscores the rising yet often overlooked burden of fungal pathogens and calls for greater global investment in fungal disease research and surveillance.

14. See the following article with Diana Bell: https://phys.org/news/2024-03-pandemic-earth-wildlife-biologist.html

15. See Rackham, 2016.

Chapter 1: A Trip Down Spillover Lane

1. See Sansonetti, 2006. In this foundational paper, Sansonetti explores the biological and ecological factors that shape species barriers. His analysis provides critical context for understanding zoonotic spillover and the emergence of new infectious diseases.

2. See IPBES, 2020. This landmark report, authored by 22 global experts, argues that pandemics are becoming more frequent due to human activities such as land-use change, agricultural expansion and wildlife trade. It emphasises that prevention – through biodiversity conservation and systemic change – is far more effective and economical than reacting to outbreaks after they occur.

3. See Milbank and Vira, 2022. This paper examines the complex socio-economic and cultural factors influencing wild meat consumption and its role in zoonotic spillover. The authors argue that simplistic policy responses, such as blanket bans, may not be effective and could harm communities reliant on wild meat for sustenance and income. They advocate for nuanced, context-specific strategies that balance public health concerns with the livelihoods and cultural practices of local populations.

4. See Kaddumukasa et al., 2014. This study documents the diversity and distribution of mosquito species in the Zika Forest. It provides valuable insight into the complex ecology of mosquito-borne diseases and the importance of long-term surveillance in regions where new pathogens may emerge.

5. See Musso, 2015. This short communication was among the earliest to suggest that the Zika virus may have spread from French Polynesia to Brazil, potentially during international sporting events. It highlights the role of global travel in accelerating the geographic expansion of emerging infectious diseases.

6. See Brady et al., 2019. This large-scale study provides some of the most compelling evidence linking Zika virus infection during pregnancy to cases of microcephaly in Brazil. Analysing data from over four million births, the authors demonstrate a strong spatial and temporal association, reinforcing the urgent need for surveillance and reproductive health support during viral outbreaks.

7. See Marbán-Castro et al., 2021. This review synthesises current evidence on the impacts of Zika virus infection during pregnancy, including adverse outcomes such as miscarriage, microcephaly and congenital Zika syndrome. It highlights the critical need for maternal health interventions and long-term monitoring of children exposed in utero.

8. See Hayes, 2001. This paper traces the remarkable journey of West Nile virus from its first recorded isolation in Uganda in 1937 to its unexpected emergence in New York City in 1999. It provides insight into the virus's shifting epidemiology and highlights the role of globalisation and ecological change in facilitating disease spread.

9. See CDC, 2024a. This official CDC resource provides an overview of West Nile virus, detailing its transmission, symptoms and prevention strategies. It notes that West Nile virus is primarily spread to humans through the bite of infected mosquitoes, with cases typically occurring during mosquito season, which starts in the summer and continues through to autumn.

10. See Gill Jr et al., 2005. This study documents one of the most extraordinary feats of bird migration ever recorded: a non-stop 11,000-kilometre flight by bar-tailed godwits. It provides powerful evidence of how migratory birds can connect distant ecosystems and (indirectly) potentially act as vectors in the global spread of infectious diseases.

11. See McQuiston et al., 2002. This article offers a concise overview of Q fever, which remains a relevant example of how close contact with livestock can lead to outbreaks with public health implications.

12. See Roest et al., 2011. This paper examines the origins of the Netherlands outbreak in goat farms, the delayed public health response and the lessons learned for managing future zoonotic disease outbreaks – particularly those linked to intensive animal farming.

13. See van der Hoek et al., 2012. This book chapter offers an in-depth analysis of the human impact of the Q fever epidemic in the Netherlands. It covers epidemiological trends, clinical features, and the challenges of diagnosis and treatment, reinforcing the need for integrated veterinary and human health surveillance in managing zoonotic outbreaks.

14. Ibid.

15. See Hunink et al., 2010. This report explores how environmental factors – such as wind direction, landscape features and proximity to infected farms – contributed to the spread of Q fever in the Netherlands. It illustrates how disease dynamics can be shaped not only by pathogens and hosts but also by the physical environment, underscoring the value of spatial and ecological analysis in outbreak investigations.

16. See Otterstatter and Thomson, 2008. This study investigates the risk of pathogen spillover from managed bumblebee colonies to wild pollinators. The findings raise concerns about commercial pollination practices, suggesting that diseases introduced through domesticated pollinators may pose a significant threat to biodiversity and ecosystem health, an often-overlooked dimension of emerging infectious disease ecology.

17. *See Goodwin*, 2021. This article offers an accessible overview of COVID-19, covering its symptoms, modes of transmission, prevention strategies, and the current state of treatments and vaccines. It serves as a valuable resource for understanding the evolving nature of the pandemic and the measures individuals can take to protect themselves and others. Also see Business Standard, 2025.

18. See Choo et al., 2020. This paper critically assesses the evidence linking pangolins to the origins of SARS-CoV-2. While coronaviruses related to SARS-CoV-2 have been detected in pangolins, the authors argue that evidence of direct transmission to humans remains weak. The study urges caution in targeting pangolins and emphasises the need to avoid fuelling misinformed narratives that could undermine conservation efforts.

19. See Jin et al., 2021. This systematic review assessed the clinical evidence for the use of pangolin scales (*Squama Manitis*) in traditional medicine. After evaluating 15 studies – including randomised controlled trials, case reports and case series – the authors found no reliable evidence supporting the therapeutic efficacy of pangolin scales. The review concludes that the removal of *Squama Manitis* from the Chinese

Pharmacopoeia is scientifically justified, both to protect endangered species and to uphold standards of clinical evidence.

20. See Storr, 2020. Storr explores how the human brain is wired for narrative, drawing on insights from psychology and neuroscience to explain why storytelling is such a powerful tool for understanding the world. His work offers valuable lessons for science communicators seeking to convey complex issues – like pandemics – in ways that resonate with the public.

21. Robinson, 2024. This book explores the deep connections between humans, biodiversity and forests – past, present and future. I weave science, history and personal narrative to show how restoring our relationship with trees and woodland ecosystems can play a vital role in creating a healthier, more resilient world.

22. See Hoang and Kanemoto, 2021. This study employs remote sensing data and a multi-region input–output model to quantify and map global deforestation footprints over a 15-year period (2001–2015) at a 30-metre resolution. The authors find that while many developed countries, China and India have achieved net forest gains domestically, they have concurrently increased the deforestation embodied in their imports, with tropical forests being the most threatened biome. The study highlights the need for robust transnational efforts, improved supply chain transparency and financial support for tropical regions to form zero-deforestation policies.

Chapter 2: What Have Vultures Got to Do with It?

1. See Burton, 2014. The 95% decline in vultures is underreported and underappreciated. Also see *Interconnected* Episode 3 (podcast) for a discussion of this major decline and the ecological and health implications.

2. See Prakash, 1999. This study documents a greater than 95% decline in key vulture species in India's Keoladeo National Park during the 1990s, likely due to high adult mortality and breeding failure. Urgent conservation action was recommended. Also see Jalihal et al., 2022. This paper highlights how vultures contribute to public health by efficiently disposing of carcasses, reducing disease risk. It also notes the limited attention given to vultures in India's public health policies.

3. See Pain et al., 2003. This paper discusses the temporal and spatial declines of *Gyps* vultures in Asia, summarising population crashes and proposing several possible causes (including habitat loss, food shortage, persecution and poisoning), but *not* diclofenac. That hypothesis was first clearly put forward in Oaks et al. (2004), which provided evidence linking diclofenac residues in livestock carcasses to kidney failure in vultures.

4. See Santangeli et al., 2024. The authors surveyed global experts who confirmed that vultures provide key ecosystem services – especially waste removal and disease control – supporting health and sustainability goals.

5. See Sault, 2016. The author explores how hummingbirds and vultures symbolise the duality of life and death in Latin American cultures. Hummingbirds are linked to vitality and renewal, while vultures represent death and transformation. These birds are seen as mediators between the earthly and spiritual realms, embodying the cyclical nature of existence. The study draws on examples from Mexico, Costa Rica and Peru to illustrate these symbolic roles.

6. See Stapleton, 2005. This paper explores how the Rockefeller Foundation's wartime 'Louse Laboratory' helped pioneer DDT-based typhus-control methods, influencing global disease-prevention strategies during and after WWII.

7. See Carson, 1962. In this third instalment of *Silent Spring*, Carson highlights the ecological and health consequences of pesticide overuse, laying groundwork for the modern environmental movement.

8. See Green et al., 2004. This study provides compelling evidence that diclofenac is the primary cause of catastrophic Gyps vulture population declines in the Indian subcontinent.

9. See Markandya et al., 2008. This study quantifies the economic and health impacts of vulture population declines in India, linking reduced scavenging to increased feral dog numbers and higher rabies incidence, and estimating significant associated costs.

10. See Burfield and Bowden, 2022, for details of the impact of the vulture decline.

11. See Frank and Sudarshan, 2024. This study quantifies the human health and economic impacts of vulture population declines in India, attributing over 500,000 excess deaths between 2000 and 2005 to the loss of these scavengers. The authors estimate annual damages at $69.4 billion, underscoring the critical ecosystem services provided by vultures.

12. See Duffy, 2018. This article examines the exceptionally high mutation rates of RNA viruses. The study suggests that high mutation rates may result from selection for rapid replication rather than being inherently advantageous.

13. See SNF, 2025. This profile chronicles Dr Vibhu Prakash's journey from a young nature enthusiast in Meerut to a leading figure in vulture conservation in India. As principal scientist at the Bombay Natural History Society, he spearheaded the Vulture Conservation Programme, establishing breeding centres and advocating against diclofenac use, significantly aiding the recovery of Gyps vulture species.

14. See Mackenzie, 2004. 'The mysterious mass die-off of vultures solved'.

15. See Nambirajan et al., 2018. This study indicated ongoing illegal or unintended exposure and underscored the need for stricter enforcement of the veterinary diclofenac ban.

16. See Espunyes et al., 2022. This study found that Eurasian griffon vultures in north-eastern Spain harbour antimicrobial-resistant strains of salmonella and campylobacter, including types shared with humans and livestock. These vultures, feeding on livestock carcasses and landfill waste, may act as environmental sentinels and vectors for resistant bacteria, highlighting concerns about wildlife's role in the spread of antimicrobial resistance.

17. See Morales-Reyes et al., 2015. This study quantifies the environmental and economic costs of replacing natural scavenger services with artificial carcass disposal in Spain. Following regulations enacted after the bovine spongiform encephalopathy crisis, livestock carcasses were collected and processed in authorised plants, leading to annual emissions of 77,344 metric tons of carbon dioxide equivalents and approximately $50 million in additional costs. The authors argue that allowing vultures and other scavengers to naturally dispose of carcasses would mitigate these emissions and expenses.

18. See Bullock, 1956. This brief note discusses observations from a Chilean sheep farm where vultures feeding on anthrax-infected carcasses were suspected of spreading the disease. Experiments demonstrated that vultures' excrement contained viable anthrax spores, suggesting they could act as vectors for the disease.

19. See Van Den Heever et al., 2021. This review highlights the critical role of African vultures in disease regulation by rapidly consuming carcasses, thereby limiting pathogen spread. Their decline disrupts scavenger guilds, potentially increasing disease transmission risks at the human–wildlife–livestock interface. The authors advocate for integrating vulture conservation into One Health strategies to bolster ecosystem and public health resilience.

20. See Mudur, 2001. This article reports that wildlife scientists in India suspect a link between the resurgence of human anthrax cases and the sharp decline in vulture populations.

21. See Subramanian, 2008. This article explores the impact of vulture population declines on Zoroastrian sky burial practices in India. The piece highlights the intersection of cultural traditions and ecological changes, emphasising the need for conservation efforts to preserve both biodiversity and cultural practices.

22. See Neale, 2013. This article explores the remarkable contributions of the Parsi community in India, despite their small population size. It highlights their

significant roles in various sectors, including shipbuilding, industry, science, and the arts. The piece also discusses the challenges faced by the community, such as declining numbers and the impact of vulture population collapse on traditional funerary practices.

Chapter 3: Why Bats Get Bad Press

1. See Low et al., 2021. This study reviews cultural perceptions of bats across 60 cultures in 24 Asia–Pacific countries. Findings reveal that 62% of these cultures hold exclusively positive views of bats, 10% exclusively negative and the remainder a mix. The authors advocate for culturally informed conservation strategies that leverage positive local beliefs to promote bat conservation.

2. See Greenspoon et al., 2023. This study estimates the total global biomass of wild terrestrial mammals at about 20 million tonnes and marine mammals at about 40 million tonnes. In comparison, livestock and humans together exceed one billion tonnes, illustrating the vast dominance of human-associated biomass over wild species.

3. See Frick et al., 2020. This review identifies key threats to global bat populations, including habitat loss, hunting, climate change and disease. It emphasises the need for improved data on bat species, as many are understudied, and calls for integrated conservation efforts involving research, monitoring and collaboration across regions to address these challenges.

4. See Wilkinson and South, 2002. This study examines why bats often live significantly longer than other mammals of similar size. Analysing 64 bat species, the authors found that factors like hibernation, larger body size and lower reproductive rates are associated with increased longevity. They suggest that reduced exposure to external threats and seasonal energy conservation contribute to bats' exceptional lifespans.

5. See Halley et al., 2022. This study presents the first motor cortex map of a bat species, the Egyptian fruit bat (*Rousettus aegyptiacus*), revealing an enlarged tongue representation linked to their unique tongue-based echolocation. Additionally, it identifies coordinated forelimb and hindlimb motor areas supporting flight, suggesting that the motor cortex of bats has co-evolved with their specialised behaviours.

6. See Boonman et al., 2014. This study reveals that certain Old World fruit bats, previously considered non-echolocating, emit audible clicks using their wings to navigate in darkness. While this wing-generated echolocation is rudimentary compared to vocal-based systems, it suggests that echolocation may have evolved multiple times in bats.

7. See Deshpande et al., 2022. This study assesses the socio-economic benefits and costs of fruit bat seed dispersal in India's Western Ghats, focusing on crops like cashew and areca. Findings indicate that bat-mediated seed clumping enhances crop regeneration, aligning with positive local perceptions. Despite concerns about zoonotic risks, no direct link between perceived benefits and disease risk was found, supporting the conservation of bats for their ecosystem services.

8. See BCI, 2020. This article highlights the Mexican long-nosed bat, an endangered species. Researchers are studying their mating habits and migration patterns, and developing conservation strategies to protect vital habitats like the Cueva del Diablo, the only known mating site for the species.

9. See Nuding, 2024. In this episode Swift, president of the Texas Pecan Growers Association, discusses his transition to regenerative pecan farming at Swift River Pecans. He highlights practices such as enhancing soil health, using cover crops and integrating beneficial wildlife like bats for natural pest control.

10. See Frank, 2024. This study quantifies the economic and health impacts of declining bat populations due to white-nose syndrome in North America. The

findings underscore the significant costs of substituting natural pest control with chemical alternatives.

11. See Sieradzki and Mikkola, 2021. This chapter explores the role of bats as bioindicators of ecosystem health. It discusses how bats' sensitivity to environmental changes, such as habitat degradation and pollution, makes them valuable for monitoring biodiversity and ecological integrity. The authors highlight various studies that demonstrate bats' responses to environmental stressors and advocate for their inclusion in conservation and monitoring programmes.

12. See Low et al., 2021.

13. See López-Baucells et al., 2023. This study analysed 1,095 bat-related articles from 15 major newspapers across Western Europe (1956–2019). It found that 17% focused on bats and diseases, with 80% of these portraying bats as threats to human health. Articles framing bats negatively received more reader engagement. The authors advocate for media to emphasise the ecological roles of bats to support conservation efforts.

14. See Horror Show, 2024. Bats evolution in cinema: Scary bats in 60 movies by size

15. See Voigt and Kingston, 2016. This comprehensive volume examines the multifaceted threats bats face in the Anthropocene, including habitat loss, urbanisation, disease and human–wildlife conflicts. It underscores the importance of bats in ecosystems and advocates for integrated conservation strategies that address both ecological and socio-cultural dimensions.

16. See Rose, 2001. This illustrated encyclopedia catalogues mythical creatures from global folklore, including giants, dragons and other legendary beings. It provides cultural context and cross-cultural comparisons, highlighting common themes and variations across traditions.

17. See Eklöf and Rydell, 2021. This paper documents attitudes towards bats in Swedish history – including their association with witchcraft (but also how they've been viewed in a positive light). Also see Lunney and Moon, 2011. This chapter examines how cultural myths and negative media portrayals have led to public indifference or hostility towards bats in Australia. Despite comprising a significant portion of the country's mammal fauna, bats are often misunderstood and overlooked. The authors argue that education and outreach are essential to counteract misconceptions and promote bat conservation.

18. See Stoker, 1897.

19. See Temmam et al., 2022. This study identified bat coronaviruses in Laos closely related to SARS-CoV-2. These viruses can bind to human ACE2 receptors and infect human cells, though they lack the furin cleavage site found in SARS-CoV-2. The findings suggest a potential for direct bat-to-human transmission.

20. See Epstein et al., 2020. This study investigates the dynamics of Nipah virus in *Pteropus medius* bats in Bangladesh. It reveals that Nipah transmission in bats is influenced by factors such as population density, waning immunity and viral recrudescence. These dynamics lead to sporadic outbreaks in bat populations, which can result in spillover events to humans, particularly through the consumption of date palm sap contaminated with bat excreta. The research underscores the importance of understanding bat ecology and behaviour to predict and prevent future Nipah outbreaks in human populations.

21. See O'shea, et al., 2014. This article explores the hypothesis that bats' ability to fly – and the associated elevated body temperatures during flight – may enhance their immune responses, allowing them to harbour a diverse array of viruses without exhibiting disease symptoms. This adaptation could explain why bats are reservoirs for numerous zoonotic viruses, such as SARS-like coronaviruses, Ebola and Nipah, which can be highly pathogenic in other mammals.

22. See Fu et al., 2023. This study investigates the role of the STING pathway in the innate immune response of bats, specifically the Brazilian free-tailed bat (*Tadarida brasiliensis*), against RNA viruses. Researchers cloned the bat STING

(BatSTING) gene and demonstrated that its overexpression in bat lung cells led to increased production of interferon-beta (IFN-β) and interferon-stimulated genes (ISGs), effectively inhibiting RNA virus replication. Conversely, knocking down BatSTING impaired the IFN-β response, highlighting its essential role in antiviral defence. The findings suggest that BatSTING is crucial for mediating IFN-β responses and controlling RNA virus infections in bats.

23. See Hess et al., 2011. This article reviews the current understanding of Hendra virus. It discusses transmission pathways and clinical manifestations, and outlines public health and veterinary responses to outbreaks. The authors emphasise the importance of surveillance, infection control measures and the development of preventive strategies, including vaccination, to mitigate the impact of the virus.

24. See Leligdowicz et al., 2016. This review analyses the clinical progression and management of Ebola during the 2013–2016 West African outbreak. It highlights that Ebola often begins with nonspecific symptoms like fatigue and fever, progressing to gastrointestinal issues and, in severe cases, multiorgan failure. The study emphasises that while no specific antiviral treatments were available, supportive intensive care – such as fluid resuscitation, electrolyte correction and organ support – significantly reduced mortality rates in well-resourced settings.

25. See McFarlane et al., 2011. This study analysed Hendra virus spillover events in Australia from 1994 to 2010, identifying a significant association with the dry season. Postal areas with flying fox roosts were found to be approximately 40 times more likely to experience spillover events, independent of horse density. The findings suggest that seasonal and ecological factors, such as bat behaviour and land use changes, play a crucial role in virus transmission.

26. See McMichael et al., 2017. This study examined the relationship between physiological stress and Hendra virus excretion in wild flying foxes in Australia. Researchers measured urinary cortisol levels – a stress indicator – in two flying fox species and assessed Hendra virus presence via PCR. Findings revealed a small but statistically significant association between elevated cortisol levels and Hendra virus excretion, suggesting that physiological stress may influence viral shedding. However, the study did not establish a direct causal link, indicating that stress is one of multiple factors affecting Hendra virus dynamics in flying fox populations.

Chapter 4: Biodiversity and the Dilution Effect

1. See Steer et al., 1986. This article provides a historical overview of Lyme disease, tracing its recognition from the 1970s outbreak of arthritis in Lyme, Connecticut, to the identification of *Borrelia burgdorferi* as the causative agent. It discusses the clinical manifestations, epidemiology and early research efforts that established Lyme disease as a distinct tick-borne illness.

2. See Walter et al., 2017. By sequencing 146 genomes from ticks collected between 1984 and 2013, researchers found ancient and widespread genetic diversity. The recent rise in human Lyme disease cases is attributed to ecological changes – such as forest fragmentation, increased deer populations and climate change – rather than new bacterial mutations. Birds likely facilitated the bacterium's long-distance spread, while small mammals contributed to its local transmission.

3. See Hook et al., 2022. It provides the 500,000 figure (and other references to it). It also shows that 64% of surveyed residents in high-incidence US states were willing to receive a Lyme disease vaccine. Unwillingness was linked to lower education, non-White race, vaccine safety concerns and low perceived risk. The authors suggest targeted communication to address hesitancy.

4. See Sonenshine, 2018. This review examines the northward expansion of four medically significant tick species in North America – *Dermacentor variabilis*, *Amblyomma americanum*, *Amblyomma maculatum* and *Ixodes scapularis* – driven

largely by climate change. It highlights the public health risks posed by these shifts, as ticks transmit a wide array of pathogens. The study discusses both environmental and biological factors influencing tick dispersal and establishment in new regions.

5. See Carroll et al., 1995. Black-legged ticks (*Ixodes scapularis*) use chemical cues from white-tailed deer – especially from does – to choose ambush sites, improving their chances of finding a host.

6. See Salomon et al., 2020. This paper presents a standardised protocol for collecting questing hard ticks using the drag cloth method. It provides detailed instructions on constructing a one-square-metre white drag cloth, establishing linear transects to sample at least 750 square metres and collecting ticks at regular intervals. The protocol aims to enhance consistency in tick surveillance, facilitating reliable comparisons across studies and regions.

7. See Bunnell et al., 2011. This study found that sick hedgehogs emit higher levels of indole in their faeces, attracting more *Ixodes hexagonus* ticks. Ticks preferred the scent of sick hedgehog faeces, suggesting they use health-related odours to select hosts. Dr Bunnell and her study inspired me during my parasitology days.

8. See Long et al., 2023. This study found that *Ixodes pacificus* ticks are more strongly attracted to a combination of carbon dioxide and volatile compounds produced by deer-associated microbes than to either cue alone. The synergy between carbon dioxide and microbial odours enhances the ticks' host-seeking behaviour.

9. See Nalage et al., 2023. This review examines how environmental toxins – such as pesticides, heavy metals and industrial pollutants – disrupt the gut microbiome in animals. Such disruptions can impair digestion, weaken immunity and increase disease susceptibility. The authors emphasise the need for further research to understand these impacts and develop strategies to mitigate them.

10. See Fackelmann et al., 2021. This study of spiny rats found that human disturbances—such as contact with domesticated animals and exposure to agricultural environments – led to reduced microbial diversity and increased presence of potentially harmful bacteria in the rats. These changes could compromise wildlife health and elevate the risk of zoonotic disease transmission.

11. See Coutinho-Abreu et al., 2024. This study identified specific compounds produced by human skin bacteria that influence mosquito behaviour. Notably, 2- and 3-methyl butyric acids, along with geraniol, were found to significantly reduce mosquito landing rates. Conversely, lactic acid, commonly produced by skin microbes, enhances mosquito attraction. These findings suggest that modifying the skin microbiome to decrease lactic acid production or increase repellent compounds could be a novel strategy to prevent mosquito bites.

12. See Couret et al., 2017. This study found that larval black-legged ticks with higher engorgement weights were more likely to acquire the bacterium that causes Lyme disease. The results suggest that the extent of blood intake during feeding influences the likelihood of infection in tick larvae.

13. See Van Buskirk and Ostfeld, 1995.

14. See Ostfeld and Keesing, 2000.

15. See Allan, Keesing and Ostfeld, 2003. This study found that smaller forest patches (less than two hectares) in southeastern New York had higher densities of *Borrelia burgdorferi*-infected nymphal ticks compared to larger patches. The increased Lyme disease risk in fragmented forests is attributed to reduced biodiversity and higher populations of white-footed mice. These findings support the dilution effect hypothesis, suggesting that greater biodiversity can mitigate disease transmission.

16. See Keesing and Ostfeld, 2021. This comprehensive review explores how biodiversity influences disease transmission through the dilution effect. The authors discuss mechanisms such as encounter reduction, vector regulation and the role of low-competence hosts in disrupting transmission cycles. They also address

challenges in studying these effects in natural ecosystems and emphasise the importance of biodiversity conservation in managing disease risks.

17. See Liu et al., 2020. This study examines how plant diversity influences the prevalence of plant diseases across different latitudes. The authors found that the dilution effect is more pronounced in temperate regions compared to tropical ones. The strength of this effect depends on specific ecological contexts, including host–pathogen interactions and environmental conditions.

18. See Khalil et al., 2016.

19. See Kouba et al., 2020. This study analysed 46 years (1973–2018) of data on Tengmalm's owls in western Finland, revealing a significant population decline linked to reduced reproductive success. Key factors include decreased availability of primary prey (bank voles) due to habitat loss from clear-cutting of mature forests, and climate-related changes such as milder winters and deeper snow cover delaying breeding. These conditions led to increased nestling starvation and lower fledgling survival, contributing to the long-term decline of the owl population.

20. See Halliday et al., 2017. This study demonstrates that increasing host diversity doesn't always reduce disease risk. While microbial parasites showed no significant response to host diversity, insect parasites nearly doubled in abundance in more diverse plant communities. This amplification was linked to changes in host composition and parasite transmission modes, highlighting the complex interplay between biodiversity and disease dynamics.

21. See Carías Domínguez et al., 2025. This scoping review synthesises recent literature on intestinal dysbiosis, highlighting its broad and nonspecific symptomatology, which complicates timely diagnosis and treatment. The authors emphasise the need for standardised diagnostic criteria and explore the role of probiotics as adjunctive therapy in managing dysbiosis-related conditions. The review underscores the importance of further research to better understand the clinical parameters for early detection and effective intervention strategies.

22. See Lee and Hwang, 2025. This study investigated how different pulse species affect the human gut microbiome using an in vitro digestion and fermentation model. Red bean, mung bean and Heunguseul were found to promote beneficial bacteria like *Bacteroides*, *Eubacterium* and *Akkermansia*, while reducing harmful *Escherichia-Shigella*. These pulses also increased microbial diversity, despite having lower nutrient contents. The findings suggest that a balanced macronutrient profile in foods supports gut microbial eubiosis.

23. See Spragge et al., 2023. This study demonstrates that diverse gut microbiome communities can protect against pathogens like *Klebsiella pneumoniae* and *Salmonella enterica* by consuming nutrients essential for pathogen growth. The researchers found that individual bacterial species had minimal protective effects, but when combined into diverse communities, they significantly inhibited pathogen colonisation. The protective effect was linked to the community's ability to outcompete pathogens for shared nutrients, a mechanism termed 'nutrient blocking'. These findings suggest that enhancing microbiome diversity could be a strategy to prevent infections by limiting pathogen access to nutrients.

Chapter 5: Disease X – It's Coming

1. See Van Kerkhove et al., 2021. This article emphasises the urgent need for global preparedness against Disease X. The authors highlight that factors such as climate change, ecosystem alterations and increased urbanisation elevate the risk of emerging infectious diseases. They advocate for proactive measures, including enhanced surveillance systems, accelerated research and development of medical countermeasures, and equitable access to healthcare resources, to mitigate the impact of unforeseen health threats.

2. See Chr, 1979. This paper examines the Red Queen Hypothesis and discusses how this constant adaptation contributes to extinction patterns observed in the fossil record, emphasising that extinction is often driven by biotic interactions rather than solely by environmental catastrophes.
3. See Carroll, 1871. This book explores themes of logic, identity and inversion, famously introducing the concept of the mirror world and characters like the Red Queen.
4. See Carroll, 1865.
5. See Davey Smith, 2016. Davey Smith draws parallels between Donald Rumsfeld's 'known unknowns' framework and cardiologist Sir James Mackenzie's 1919 insights on medical ignorance, emphasising the importance of recognising uncertainties to advance evidence-based medicine.
6. See Ingham and Luft, 1955. The Johari Window emphasises the importance of feedback and self-disclosure in personal development and group dynamics.
7. As of April 2024, most countries have stopped recording COVID-19 cases and deaths. Therefore, the Coronavirus Tracker (https://www.worldometers.info/coronavirus/) is no longer being updated for feasibility reasons.
8. See Marani et al., 2021. This study estimates the annual chance of a COVID-19–scale pandemic, with risks potentially tripling due to environmental changes.
9. See UNMC, 2023. Airfinity's modelling estimates that rapid vaccine deployment within 100 days could reduce the risk to 8.1%.
10. See Williams et al., 2023. This article emphasises the need for proactive pandemic preparedness, advocating for strengthened surveillance, investment in early-stage research and sustained vaccine-manufacturing capacity. It underscores that global coordination and collaboration are essential to respond effectively to future pandemics.
11. See Gautam, 2022. Gautam warns that antimicrobial resistance is a looming global threat, potentially surpassing COVID-19 in impact. The article emphasises the urgency of addressing antimicrobial resistance through improved stewardship and education, particularly among medical professionals.
12. Robinson, 2023.
13. See Waglechner et al., 2019. This study reveals that glycopeptide antibiotic biosynthesis and resistance co-evolved in *Actinobacteria* approximately 150–400 million years ago, indicating that resistance mechanisms are ancient and intrinsic to antibiotic-producing organisms.
14. See Smith, 2003. This comprehensive review explores how *M. tuberculosis* evades host defences, focusing on virulence factors like cord factor and its ability to inhibit phagosome–lysosome fusion. It also discusses how the bacterium's genetic adaptability contributes to its persistence and resistance.
15. For snippets of Fleming's Nobel Prize speech, see De la Fuente-Nunez, 2019.
16. See Cillóniz et al., 2012. This study found that bacterial co-infections, particularly with *Streptococcus pneumoniae* and *Staphylococcus aureus*, were common in patients hospitalised with H1N1-related community-acquired pneumonia, leading to increased severity and mortality.
17. See Costa et al., 2022. In a Brazilian study of 191 COVID-19 patients in ICU, 30% developed secondary infections – primarily ventilator-associated pneumonia – caused by multidrug-resistant gram-negative bacteria like *Acinetobacter baumannii* and *Pseudomonas aeruginosa*. These infections were linked to longer ICU stays, extended mechanical ventilation and higher mortality.

Chapter 6: Going Viral

1. See Hill, 2006. This article discusses how cyanophages – viruses that infect marine cyanobacteria – carry and express Photosystem II genes (psbA and psbD) acquired from their hosts. By doing so, they enhance the host's photosynthetic

machinery during infection, boosting viral replication. The study highlights the role of cyanophages in horizontal gene transfer and their impact on marine photosynthesis.

2. See Dance, 2021. This article explores the vast and largely uncharted diversity of viruses, highlighting their roles in ecosystems and evolution and the potential for future discoveries. It emphasises that while viruses are often associated with disease, many are integral to ecological balance and genetic innovation.

3. See Ma, 2023. Ma introduces the virome comparison (VC) method, featuring two metrics – virus species specificity (VS) and virome specificity diversity (VSD) – to efficiently compare complex viromes and identify unique or enriched viral species.

4. See Chua, 2003. Chua details the 1998–1999 Nipah virus outbreak in Malaysia, which resulted in 265 human encephalitis cases and 105 deaths. The outbreak was linked to close contact with infected pigs, with fruit bats identified as the natural reservoir. The study highlights the importance of accurate diagnosis and swift public health responses in managing zoonotic disease outbreaks.

5. See Khan et al., 2024. This review examines Nipah virus outbreaks over 25 years in Southeast Asia, highlighting the virus's high fatality rates and its emergence as a significant zoonotic threat.

6. See Epstein et al., 2006. This paper outlines Nipah virus emergence, with fruit bats as the reservoir and human outbreaks linked to pig farming and bat–human contact. It highlights deforestation and land-use change as key drivers.

7. See Mwangilwa et al., 2025. This study found that during the COVID-19 pandemic, measles mortality in Zambia increased by 220%, largely due to disrupted immunisation services. Children aged zero to four, unvaccinated individuals and females were at higher risk, while vaccination significantly reduced mortality odds. The authors recommend prioritising routine immunisation and improving measles case management to prevent further outbreaks.

8. See UNICEF, 2022. In early 2022, global measles cases surged by 79% compared to the same period in 2021. UNICEF and WHO attribute this rise to pandemic-related disruptions in immunisation services, increasing inequalities in vaccine access and the diversion of resources from routine immunisation. They warn that these factors create a 'perfect storm' for measles outbreaks, particularly affecting children.

9. See Fischer et al., 2014. Influenza causes significant cardiopulmonary illness in low-income regions, exacerbated by limited healthcare access, comorbidities like HIV, and scarce resources such as vaccines and antivirals. Enhanced surveillance and support are essential to mitigate its impact.

10. See Abbasi, 2023. Reports of H5N1 transmission among mink and seals raise concerns about mammal-to-mammal spread and potential human spillover.

11. See Eisfeld et al., 2024. This study found that bovine H5N1 virus infects mice and ferrets, with limited airborne transmission. The virus binds to human-type receptors, raisin

12. g concerns over potential adaptation. The authors emphasise the need for surveillance to prevent human spread.

13. See Mason and Haddow, 1957. This study confirmed Chikungunya virus as the cause of a 1952–1953 epidemic in Tanganyika (now Tanzania) by isolating the virus from acute-phase sera and detecting antibodies in convalescent samples.

14. See Powers, 2011. This review examines how genetic mutations, particularly the E1-A226V substitution, enhanced Chikungunya virus adaptation to *Aedes albopictus* mosquitoes, facilitating its spread during the 2004–2011 Indian Ocean outbreak. It underscores the importance of genomic surveillance in understanding viral evolution and guiding vaccine development.

15. See Chippaux and Chippaux, 2018.

16. See Tyagi and Vythilingam, 2025. This chapter traces mosquito research from ancient Indian texts to modern vector control, highlighting key discoveries like Ronald Ross's malaria findings in India.
17. See Chaves-Carballo, 2005.
18. See Clements and Harbach, 2017. This article revisits the historical discovery that mosquitoes transmit yellow fever, highlighting the contributions of Carlos Finlay, Henry Rose Carter and Walter Reed. It critically examines primary sources, revealing that earlier accounts may have overstated certain individuals' roles, and underscores the collaborative nature of this significant medical breakthrough.
19. See Monath, 2001.
20. See Prasad, 2023. This chapter discusses the challenges of urban sustainability, highlighting issues like pollution, resource overuse and social inequality. It advocates for integrated planning and long-term strategies to balance economic growth with environmental and social well-being.
21. See Kiss et al., 2015. The authors envision a self-sufficient urban model for 2050, integrating renewable energy, hydroponic agriculture and efficient resource use to reduce environmental impact.
22. See Maquart et al., 2022. The authors advocate for integrating plastic pollution control into public health strategies.
23. Ibid.
24. See Ferronato et al., 2024. This review examines circular economy initiatives for plastic waste in seven developing countries. It highlights challenges such as open dumping and burning, and emphasises the need for improved policies, infrastructure and community engagement to align with sustainable development goals.
25. See Tan et al., 2024. The authors explore converting plastic waste into energy via pyrolysis and gasification, noting significant carbon dioxide emissions as a challenge. They advocate for integrating carbon capture technologies to enhance sustainability, especially in developing nations.
26. See Wilder-Smith and Massad, 2018. Eleven unvaccinated Chinese workers contracted the disease, with cases imported into China. Modelling indicated that nearly all the 259,000 Chinese workers in Angola were unvaccinated, despite international health regulations. The study underscores the need for stricter vaccination enforcement to prevent disease spread.
27. See Science Direct, 2025. Yellow Fever virus.
28. Some suggest that 18.2 million people died from COVID-19 during the peak period. See Larkin, 2022.
29. See Maxmen and Mallapaty, 2021. This article examines the ongoing debate over COVID-19's origins, highlighting that while a natural spillover from animals remains the leading theory, a lab leak cannot be definitively ruled out due to limited evidence. The authors call for further investigation to clarify the virus's source.
30. See Liu et al., 2021. This entry provides an overview of these four human coronaviruses, covering their virology, transmission and clinical significance. It highlights their roles in respiratory infections and their relevance to public health.
31. See Peeri et al., 2020. This article compares SARS, MERS and COVID-19, highlighting how globalisation and inadequate early responses contributed to the rapid spread of COVID-19. It stresses the need for improved preparedness and the potential role of technologies like the Internet of Things in managing future outbreaks.
32. See Bamford, 2020. 'The original Sars virus disappeared – here's why coronavirus won't do the same'
33. See Ramadan and Shaib, 2019.

34. See Hui et al., 2021. This chapter provides an overview of SARS, MERS and COVID-19, detailing their virology, transmission and clinical features. It underscores the importance of surveillance and preparedness for emerging respiratory viruses.

Chapter 7: Can We ESKAPE this Bacterial Blitzkrieg?

1. See Rice, 2008. Rice highlights that ESKAPE pathogens – *Enterococcus faecium, Staphylococcus aureus, Klebsiella pneumoniae, Acinetobacter baumannii, Pseudomonas aeruginosa* and *Enterobacter* spp. – are leading causes of hospital-acquired infections and exhibit high antibiotic resistance. He criticises the lack of targeted federal research funding to combat these specific threats.

2. See CNN, 2016.

3. See Klevens et al., 2007. The study highlights the significant burden of MRSA and underscores the need for effective prevention strategies.

4. See Blot et al., 2002. Critically ill patients with MRSA bacteraemia had a significantly higher mortality rate (64%) compared to those with MSSA - methicillin-susceptible Staphylococcus aureus (24%). The study highlights the increased risk associated with MRSA infections in ICU settings.

5. See Washio et al., 1997. This study found that elderly patients with MRSA infections in a Japanese geriatric hospital had a higher case fatality rate compared to those with methicillin-sensitive strains. Key risk factors included poor nutritional status (hypoalbuminaemia), fever and reduced activities of daily living. Patients admitted from nursing homes or other hospitals showed higher MRSA isolation rates than those from their own homes.

6. See Infectious Diseases Society of America (IDSA), 2011. IDSA outlines urgent actions to address antimicrobial resistance, including incentives for new antibiotic development, enhanced surveillance and improved stewardship programmes. The report highlights the significant health and economic burdens posed by resistant infections.

7. See ECDC, 2024 for patient stories. The ECDC coordinates European Antibiotic Awareness Day (EAAD), an annual initiative held on the 18th of November to raise awareness about antimicrobial resistance and promote prudent antibiotic use. EAAD supports national campaigns across Europe and collaborates with global partners to combat antimicrobial resistance.

8. See Murray et al., 2022. Antimicrobial resistance caused 1.27 million deaths in 2019, with the highest burden in low-income regions. Urgent action is needed on surveillance and new treatments.

9. See Anthony et al., 2023. This study estimates that approximately 59% of Earth's species inhabit soil, making it the most biodiverse habitat. The authors highlight the critical role of soil organisms in ecosystem functions and advocate for their conservation. See also Blakemore, 2025. Blakemore argues that over 99.9% of Earth's species reside in soil, predominantly microbes. He critiques lower estimates and warns that soil biodiversity faces severe threats from erosion and pollution.

10. See Brady et al., 2017. This is a concise overview of the *Acinetobacter* genus.

11. See Zhang et al., 2020. This study identified that inpatients with *Pseudomonas aeruginosa* bacteraemia in China had higher mortality rates when infected with multi-drug-resistant strains. Key risk factors included advanced age, ICU admission and prior antibiotic use.

12. See ASM, 2023. The fascinating story of a dangerous eye infection from tainted eye drops.

13. See Bhagirath et al., 2016. This review explores how the altered lung environment in cystic fibrosis patients promotes chronic *Pseudomonas aeruginosa* infections. It highlights the complex interactions between host factors, microbial communities

and environmental conditions, suggesting that a multifaceted therapeutic approach is essential for effective management.

14. See Robinson, Breed and Beckett, 2024. This article introduces the concept of 'probiotic cities', advocating for urban designs that incorporate microbial ecology knowledge in green and blue infrastructure and biointegrated materials. Such designs aim to enhance biodiversity, improve human health and encourage sustainable urban ecosystems. We also run a Probiotic Cities Symposium every two years – the next should be in 2026. Sign up for my newsletter at www.jakemrobinson.com to stay updated.

15. See Otto, 2023. This paper reviews efforts to block *S. aureus* virulence via quorum quenching. The efforts show limited success so far, with risks like more biofilm.

16. See WHO, 2024. In 2023, TB surpassed COVID-19 as the leading infectious disease killer, with 8.2 million new diagnoses and 1.25 million deaths. The disease remains prevalent in low- and middle-income countries, notably India, Indonesia, China, the Philippines and Pakistan. Challenges include underfunding and the rise of drug-resistant TB strains.

17. See Ravikoti et al., 2025. The article reviews recent TB treatment advancements, including WHO's recommended six-month BPaLM regimen for MDR/RR-TB, which has achieved 89% success. It also highlights improved diagnostics and 16 vaccine candidates in trials.

Chapter 8: A Fungal Frenzy – Could This Be 'The Last of Us'?

1. See Robinson, 2023. *Invisible Friends: How Microbes Shape Our Lives and the World Around Us.*

2. See Hassett et al., 2015. Mushroom spores, rich in hygroscopic sugars like mannitol, attract water droplets in humid air, facilitating cloud formation and potentially enhancing rainfall, especially over forests.

3. See Loreto and Hughes, 2019.

4. See de Bekker et al., 2021. This review explores how certain fungi manipulate insect behaviour to enhance spore dispersal. Proposed mechanisms include secretion of effectors, interference with host signalling pathways and circadian timing. The study highlights the need for further research into these complex host–pathogen interactions.

5. See Will et al., 2023. This study integrates metabolomic, transcriptomic and genomic data to explore how *Ophiocordyceps camponoti-floridani* manipulates *Camponotus floridanus* ants. Findings suggest that the fungus alters neurotransmitter levels and suppresses immune pathways, leading to the ants' summit-climbing behaviour that facilitates fungal spore dispersal.

6. See van Roosmalen and de Bekker, 2024. This review examines how *Ophiocordyceps* fungi manipulate ant behaviour, potentially through secreted proteins and circadian rhythm interference. It discusses whether such mechanisms are unique or have evolved independently across parasitic fungi.

7. See Beckerson et al., 2023. The study suggests that aflatrem-like compounds produced by *Ophiocordyceps* fungi may induce the 'zombie ant' behaviour by affecting the ant's neuromuscular and sensory functions. Injection of aflatrem into healthy ants replicated symptoms like reduced activity and staggering, aligning with gene expression changes observed in naturally infected ants.

8. See ISID, 2024. Fungal infections cause over 3.8 million deaths annually. Awareness is urgently needed.

9. See Nwachukwu et al., 2023. This review compares the efficacy of ultraviolet-C (UVC) radiation and medicinal plant extracts in combating *Candida auris*. While UVC shows effectiveness, its application is limited by factors like exposure time

and safety concerns. In contrast, plant-based treatments, particularly essential oils, offer a safer and more accessible alternative for disinfection.

10. See RMIT, 2021. Researchers at the Royal Melbourne Institute of Technology developed an ultra-thin black phosphorus coating that kills over 99% of drug-resistant bacteria and fungi within two hours, without harming human cells. The coating degrades naturally, making it suitable for use on medical implants and wound dressings.

11. See SMH, 2024. The article exposes how Australia's pursuit of 'perfect' strawberries relies heavily on toxic pesticides, risking farmworker health and harming the environment. It calls for more sustainable practices and highlights organic alternatives as safer options.

12. See the University of Sydney, 2022. The WHO published its first fungal priority pathogens list, identifying 19 fungi that posing significant public health threats. The list aims to guide research and policy, addressing challenges like antifungal resistance, limited diagnostics and the impact of climate change on fungal disease spread.

13. Ibid.

Chapter 9: Protozoan Perils: Should We Eradicate Mosquitoes?

1. See Shannon et al., 2024. The authors review how mosquitoes, often seen only as pests, also visit flowers and contribute to pollination. They highlight the ecological role of mosquitoes in plant-pollinator networks and call for more research into this underappreciated behaviour.

2. See Allan, Budge and Sauskojus, 2023. The authors explore how strategies used to control *Aedes* mosquitoes, which spread dengue, could inform responses to invasive *Anopheles* species. Despite the insects' behavioural similarities, the paper stresses that misapplying *Aedes*-focused tactics could undermine efforts to control malaria.

3. See Poinar, 2021. The author reviews fossil evidence of ancient viruses, parasitic bacteria and protozoa, showing that parasitism has deep evolutionary roots. Amber fossils reveal early host–pathogen relationships, highlighting the long co-evolution of microbes and their hosts.

4. See Galaway et al., 2019. The authors investigated how *P. falciparum* evolved to infect humans by reconstructing an ancestral RH5 protein. This ancestral protein could bind both gorilla and human red blood cell receptors, suggesting a cross-species transmission. A subsequent single amino acid change in RH5 led to human-specific infection, offering insights into malaria's zoonotic origin.

5. See Liu et al., 2014. Genetic analyses of parasites from wild chimpanzees and gorillas reveal a diverse lineage closely related to human *P. vivax*. This suggests that the parasite crossed from apes to humans in Africa, and that the widespread Duffy-negative mutation in African populations likely evolved to resist this ancestral infection.

6. See Rougeron et al., 2022. The authors review the evolutionary history of *P. falciparum* and *P. vivax*, focusing on their origins, global spread and adaptation to humans. Through population genetics and comparative genomics, they highlight host-switch events from African apes and the role of human migration in shaping parasite diversity. The study underscores how genetic insights inform malaria control strategies.

7. See Oladipo et al., 2022. The authors highlight escalating challenges in malaria control across Sub-Saharan Africa. The study identifies growing resistance to artemisinin-based therapies and insecticides, as well as limited vaccine efficacy, as significant obstacles. The authors advocate for increased investment in local

vaccine development and a reassessment of public health strategies to address these pressing issues.

8. See UNICEF, 2024. Shocking malaria statistics in children under 5 years old.
9. See Rawal, 2020. The author summarises the long history of human malaria, its causative *Plasmodium* species and key human genetic adaptations. The paper also reflects on early beliefs and milestones in malaria treatment.
10. Cox, 2002; Neill, 2011.
11. See Brier, 2004. The author reviews evidence of infectious diseases in ancient Egypt, including malaria, schistosomiasis and tuberculosis, and highlights early medical responses recorded in ancient texts and mummies.
12. See Jones, 1907. The author posits that malaria significantly contributed to the decline of ancient Greece, particularly in Attica. He draws upon classical texts and medical writings to support this hypothesis. This interdisciplinary approach, combining historical analysis with medical insight, was pioneering for its time. See also Cunha & Cunha, 2008.
13. See Carroll, 2001. Interesting article asking the question: Did malaria bring Rome to its knees?
14. See Nonvignon et al., 2016. The authors assessed the economic impact of malaria on Ghanaian businesses. They found that in 2014, companies lost approximately US$6.58 million due to malaria, primarily from treatment costs and employee absenteeism. The study advocates for increased private sector investment in malaria control to mitigate these losses.
15. See Gelband et al. 2004.
16. See Becker et al., 2012. This paper discusses the global spread of invasive mosquitoes like *A. aegypti* and *A. albopictus*, facilitated by human activities such as trade and travel. These species' eggs can withstand desiccation, aiding their survival during transportation. Their adaptability and role as disease vectors underscore the need for vigilant monitoring and control strategies. See Judson, 2003. Judson argues that eradicating malaria-carrying mosquitoes could be a justified public health measure. She suggests that if current control efforts fail, eliminating these specific mosquito species might be necessary to combat malaria effectively.
17. See Bates, 2016. Bates examines the ethical and ecological implications of eradicating disease-carrying mosquitoes like *A. aegypti*.
18. Hall and Tamïr, 2022. This book explores ethical, ecological and public health perspectives on mosquito control and coexistence, questioning whether eradication is desirable or feasible.
19. See Oxitec, 2025. In Africa, Oxitec first released its modified male mosquitoes in Djibouti in 2024.
20. See Wise and Borry, 2022. The authors examine the ethical considerations of using CRISPR-based gene drives to eliminate *A. gambiae*. They argue that while the species has low moral status, environmental impact assessments and community engagement are essential before any release. The potential public health benefits must be balanced against ecological risks.
21. See Bland, 2016. The article explores the debate over eradicating disease-carrying mosquitoes like *A. aegypti* using gene-editing technologies, underscoring the tension between public health benefits and environmental considerations.
22. See Wise and Borry, 2022.
23. Ibid.
24. See Soulé, 1985.
25. See CDC, 2024b. The CDC's DPDx malaria page offers detailed information on the *Plasmodium* species infecting humans, their lifecycles, clinical features and diagnostic methods. It also provides resources for laboratory identification and training.
26. See Gates, 2019. Gates discusses gene-editing strategies to combat malaria by targeting mosquito reproduction. Techniques like the 'X-shredder' skew offspring

towards males, while editing the *doublesex* gene renders females sterile and unable to bite. These methods aim to suppress malaria-transmitting mosquito populations in specific regions. Laboratory trials have shown promising results, but further testing is needed to assess ecological impacts and scalability.

27. See Bates, 2016.
28. See Quammen, 1996.
29. See Dunning, 2022. The scientists have engineered *A. gambiae* mosquitoes that produce antimicrobial peptides in their guts, delaying malaria parasite development. This reduces the chances of transmission, as most mosquitoes die before the parasite becomes infectious.

Chapter 10: Ring-a-Ring o' Roses and Bats with White Noses

1. See Ali, 2025. Ali reports that researchers have identified a second fungal species capable of causing white-nose syndrome in bats. While only one species has reached North America, the emergence of this second species raises concerns about a potential new wave of bat die-offs if it spreads to the continent.
2. See Minnis and Lindner, 2013. The authors show that *P. destructans* is genetically distinct and likely invasive, not native to North American bat caves.
3. See Charles, 2008.
4. See Hicks et al., 2023. Hicks and colleagues demonstrated that *P. destructans* can infect and kill little brown bats solely through environmental exposure. This finding confirms that contaminated hibernacula can serve as reservoirs for the disease, even in the absence of infected bats.
5. See Zukal et al., 2016. The authors found that *P. destructans* is widespread in Europe and Asia but doesn't cause mass bat die-offs there, unlike in North America. This suggests that Palearctic bats have developed a tolerance to the fungus, possibly due to long-term co-evolution.
6. See Puechmaille et al., 2011.
7. See Campana et al., 2017.
8. See Drees et al., 2017. Drees and colleagues conducted a phylogenetic study of *P. destructans*. Their findings indicate that the North American outbreak likely originated from a European source.
9. See Meteyer et al., 2022.
10. Ibid.
11. See Johnson et al., 2014. The authors document the estimated death toll of bats between 2006–2011. They also found that female little brown bats with higher fat reserves hibernating at four degrees Celsius had better survival rates when exposed to *P. destructans*. Even low fungal doses (500 conidia) could be fatal, highlighting the importance of host and environmental factors in white-nose syndrome mortality.
12. See Weller et al., 2018. The authors explain that *Myotis* bats in the western USA hibernate in small, scattered groups, making white-nose syndrome surveillance challenging. Townsend's big-eared bats may serve as a better sentinel species.
13. See WNS, 2024. An interactive map detailing affected regions is available on the website.
14. Personal communications with David Blehert.
15. See Clark, 1976. Due to limited knowledge about the elusive black-footed ferret, Clark advocates for the use of computer simulations to inform conservation strategies.
16. See Ashe, 2019. The article reports that the number of black-footed ferrets has rebounded to around 700, thanks to captive breeding, reintroductions and conservation partnerships across North America.

Chapter 11: The Devil's Work

1. See Strain, 2011. The author reports on a study estimating that Earth is home to approximately 8.7 million eukaryotic species, with 6.5 million on land and 2.2 million in oceans. This figure, derived from a novel analytical method, significantly narrows previous estimates ranging from three to a hundred million. The study also highlights that about 86% of land species and 91% of marine species remain undiscovered, underscoring the vast unknown biodiversity on our planet.

2. See Wake and Koo, 2018. The authors provide an overview of amphibian diversity, biology and conservation. As of 2018, approximately 7,900 species were known, with frogs comprising the majority. Despite their evolutionary resilience, amphibians face significant threats from habitat loss, climate change and diseases like chytridiomycosis, leading to widespread population declines.

3. See Martel et al., 2013. The authors identified *Bsal*, first isolated from declining fire salamander populations in the Netherlands, which induces severe skin lesions, leading to rapid mortality. Its emergence raises significant concerns for amphibian conservation globally.

4. See Patel, 1991. Patel emphasises that while breathing is typically involuntary, it can be consciously regulated to alleviate stress. By adopting controlled breathing techniques, individuals can influence their autonomic nervous system, promoting relaxation and reducing stress levels.

5. See Gibson et al., 2012. The authors found that British adults' average total water intake was 2.5 litres for men and 2 litres for women, aligning with European guidelines. However, 33% of men and 23% of women consumed less than recommended. Beverages contributed 75% of total water intake, with consumption peaking in the morning and evening. Replacing calorific drinks with non-calorific alternatives was associated with reduced energy intake.

6. See Reuell, 2019. Harvard researchers have discovered that lungless salamanders express a lung-specific protein, surfactant protein C, in their skin and mouth linings. This adaptation likely facilitates their cutaneous respiration, providing molecular evidence for how these amphibians breathe without lungs.

7. See APS, 2002. During rehydration, toads adopt a 'water absorption response' posture, pressing their belly against moist surfaces to maximise water absorption.

8. See Coghlan, 2015. *Bsal*, originating from Asia, poses a significant threat to the UK's great crested newt. Its arrival in the UK, likely via the pet trade, raises concerns about potential devastating impacts on native amphibian species.

9. See Cunningham et al., 2019.

10. See Skerratt et al., 2016. The authors recommend urgent actions, including habitat protection, captive breeding and enhanced biosecurity, to prevent further declines. They call for a coordinated national response and dedicated funding to support these conservation efforts.

11. See McDonald et al., 2005. McDonald and colleagues observed a decline in chytridiomycosis prevalence among North Queensland frogs from 1998 to 2002, suggesting the disease became endemic post-epidemic. The surviving species, *Litoria genimaculata*, rebounded to pre-decline numbers, indicating possible host adaptation or environmental factors influencing disease dynamics.

12. See Berger and Skerratt, 2013. This article reports that the deadly chytrid fungus likely originated in East Asia and spread globally via the amphibian trade, devastating frog populations worldwide.

13. See Rosenblum et al., 2013. Using genome resequencing of 29 *Batrachochytrium dendrobatidis* isolates, they identified multiple divergent lineages and evidence of recombination, suggesting that hybridisation has contributed to the pathogen's global spread and virulence. This underscores the need for genomic surveillance to manage chytridiomycosis.

14. See Cobb, 2023. Froglife warns that chytrid fungus is harming UK amphibians, especially toads, and urges the public to avoid moving amphibians or plants and to report suspected cases.
15. See Snow, 2020. The story reports that efforts to combat the chytrid fungus include monitoring outbreaks, collecting skin samples, and implementing rapid response teams to treat infected individuals and boost immunity among vulnerable species like the mountain yellow-legged frog.
16. See Voyles et al., 2018. Voyles and colleagues demonstrated that the fungus remains highly virulent in Panama, yet some amphibian species are recovering, which suggests that host adaptations, rather than pathogen attenuation, are driving shifts in disease dynamics.
17. See Simon, 2018.
18. See Wallace, 2023. The article highlights that the chytrid fungus continues to threaten Australian frogs, causing skin damage and cardiac arrest. Notably, alpine tree frogs (cousins of the Whistling tree frogs) have adapted by increasing reproductive output when infected, suggesting a potential survival strategy.
19. See McCallum et al., 2009. The authors explain that the biting mode of transmission allows the disease to persist even as devil populations decline, potentially leading to extinction. The study underscores the urgency of conservation efforts to prevent the species' loss.
20. See Wells et al., 2017. The study suggests that the disease targets the most reproductively valuable devils, potentially accelerating population decline.
21. See Stammnitz et al., 2023.
22. Ibid.
23. Ibid.
24. Ibid.
25. See Jones et al., 2004. The authors found that Tasmanian devils exhibit low genetic diversity, likely due to historical bottlenecks and island isolation. Gene flow is extensive within continuous habitats up to 50 kilometres, but limited across unsuitable terrains, leading to distinct eastern and northwestern populations. This genetic structure has implications for disease spread and conservation strategies.
26. See Jones et al., 2008b.
27. See Flies et al., 2020. Flies and colleagues propose an oral bait vaccine in a strategy that aims to deliver vaccines efficiently to wild populations, enhancing disease control efforts.

Chapter 12: It's Just the Flu, Right?

1. See Bell, 2024. Diana argues that the next pandemic is already underway: avian influenza H5N1 has caused mass die-offs among birds and mammals, including endangered species. Its spread to Antarctica threatens previously untouched ecosystems. The virus's expansion is linked to intensive poultry farming and global trade, highlighting the need for overhauling poultry production to prevent further biodiversity loss.
2. Ibid.
3. Ibid. Also see Roberton et al., 2006. The authors report fatal H5N1 infection in Owston's civets in Vietnam, highlighting the virus's threat to mammalian biodiversity in Southeast Asia.
4. See Plaza et al., 2024. The study found a significant rise in H5N1 infections among mammals during the 2020–2023 panzootic. Evidence suggests possible mammal-to-mammal transmission and viral mutations enhancing replication in mammals, underscoring the need for vigilant surveillance.
5. See Wille, 2024. The article warns that H5N1 is approaching Australia, necessitating heightened surveillance and biosecurity measures.
6. See Leguia and Nelson, 2023.

7. Ibid.
8. Ibid.
9. See Tomás et al., 2024. The authors report that H5N1 clade 2.3.4.4b has caused mass die-offs of sea lions, fur seals and terns along Uruguay's coast. Genetic analysis indicates mammal-to-mammal and mammal-to-bird transmission, with mutations suggesting adaptation to mammalian hosts. This highlights a new transmission route and increased risk to wildlife.
10. See Adams, 2002. The study examined genetic variation in the captive California condor population, revealing low mitochondrial DNA diversity due to a severe bottleneck. The study identified four maternal haplotypes among founders, with one unique to a male named Topatopa, whose lineage will not persist. These findings underscore the need for careful genetic management to maintain diversity and prevent inbreeding.
11. See Wilbur and Kiff, 1980. The authors documented the historical presence of California condors in Baja California, Mexico, noting their decline by the 1930s. The study emphasised the importance of this region for condor conservation efforts.
12. See Walters et al., 2010. The authors review the California condor's recovery, noting a rise to over 350 individuals by 2009, with 180 in the wild.
13. See Robinson et al. 2021.
14. See Undark, 2023.
15. See FCCWF, 2024.
16. See Bartels and Lewis, 2023.
17. See Kanaujia et al., 2022. The authors highlight the pandemic potential of avian influenza, particularly the risk of low-pathogenic strains evolving into highly pathogenic forms like H5N1 and H7N9. They discuss the virus's ability to cross species barriers, the economic impact on poultry industries and the necessity for stringent containment measures. The study underscores the importance of surveillance and rapid response to prevent widespread outbreaks.
18. See Ebersole, 2023.
19. See Karanja, 2024. The fact sheet shows that Victoria's poultry industry comprises 250 farms, producing 268,000 tonnes of chicken meat and 66 million dozen eggs annually. The sector employs 2,550 people and contributes $957 million to the state's economy. Victoria is Australia's largest poultry exporter, with exports valued at $44 million in 2022–2023.
20. See Morten et al., 2023. Morten and colleagues assess climate change impacts on Arctic terns, revealing that while their vast migratory range offers some resilience, declines in North Atlantic food availability and Southern Ocean sea ice pose significant threats. Cumulative minor changes may undermine their long-term survival.
21. See Eikenaar et al., 2023.
22. See Learn, 2023.
23. Eikenaar et al., 2023.
24. See Richard et al., 2021. The authors call for urgent action to protect avian biodiversity.

Chapter 13: Spongy Brains and Zombie Deer – a Prion Predicament

1. See Schonberger and Schonberger, 2012. The authors clarify that the term 'prion' – coined by Stanley Prusiner in 1982 – refers to a proteinaceous infectious particle responsible for diseases like scrapie.
2. See Milisav et al., 2015. Milisav and colleagues review how protein misfolding triggers cellular stress responses in Alzheimer's, Parkinson's and prion diseases, highlighting autophagy's key role in prion pathology.

3. See Guardian, 2000. This timeline of the UK's bovine spongiform encephalopathy crisis notes over 100,000 cases by the early 1990s and links to variant Creutzfeldt-Jakob disease in humans, leading to major public health and agricultural impacts.
4. See Ness et al., 2023. The authors analyse accounts of neurological diseases in sheep and deer in England and identify early descriptions of scrapie in sheep from 1693–1722 in southwestern England and the East Midlands (although the first official recording was thought to be in 1732). Simultaneously, reports from the East of England describe a similar neurological disease in deer, which the authors attribute to rabies.
5. See Brown, 1998. Brown provides a historical overview of transmissible spongiform encephalopathies, tracing their origins from the eighteenth century to the identification of prions as the causative agents. The article discusses early observations of scrapie in sheep, the emergence of bovine spongiform encephalopathy and the subsequent human cases of Creutzfeldt-Jakob disease, highlighting the evolution of scientific understanding of these diseases.
6. See Liberski and Brown, 2009. The authors review the history of kuru and how studying it advanced prion disease understanding.
7. See Bradley and Wilesmith, 1993. The authors provide an analysis of the UK bovine spongiform encephalopathy epidemic.
8. See Paisley and Willeberg, 2002. The authors discuss European strategies for monitoring transmissible spongiform encephalopathies, focusing on scrapie in sheep and bovine spongiform encephalopathy in cattle. They emphasise the importance of surveillance in preventing transmissible spongiform encephalopathies and maintaining food safety.
9. See Yokoyama and Mohri, 2008. This paper explains how new, unusual types of prion diseases have recently appeared in sheep and cows – including in animals thought to be resistant – suggesting that tiny structural changes in these proteins may lead to different forms of the disease.
10. See BBC, 2019. The article reports that kuru had incubation periods extending up to 56 years. This discovery raises concerns that variant Creutzfeldt-Jakob disease might also have long incubation periods, potentially leading to future cases decades after exposure.
11. Ibid.
12. See Our World in Data, 2022. The chart illustrates global per capita meat consumption by type from 1961 to 2021. It shows a steady increase in total meat consumption, with poultry leading, followed by pork and beef. The data highlights regional variations, with countries like the USA and Argentina having higher beef consumption, while others like China and Brazil have seen significant increases in pork and poultry consumption.
13. See CDC, 2024c.
14. See Osterholm et al. 2019.
15. See Cassmann et al., 2021. The paper shows that chronic wasting disease can be transmitted from mule deer to sheep via intracranial inoculation, with disease progression similar to classical scrapie.
16. See Greenlee et al., 2023. The authors show that white-tailed deer can contract classical sheep scrapie via oronasal exposure, with disease signs resembling chronic wasting disease, complicating diagnosis.
17. See Wirsenius et al., 2010. In their study, the authors modelled global agricultural land use scenarios for 2030 to assess the impacts of dietary changes and livestock productivity on land requirements.
18. See Ritchie, 2025. Factory farming is characterised by the intensive confinement of animals for extended periods, often exceeding 45 days, with limited space and natural behaviours. This system aims to maximise production efficiency but raises significant concerns regarding animal welfare, environmental impact and public health.
19. Ibid.

20. See Ritchie, 2023.
21. See Ritchie and Roser, 2024.
22. See Beecher, 2015. The article emphasises the growing concern among scientists about the potential risks of chronic wasting disease.
23. See Pritzkow, 2022. The author reviews chronic wasting disease transmission, strain diversity and potential human risk, stressing the need for ongoing research and monitoring.
24. See Tranulis et al., 2016. The authors report the first cases of chronic wasting disease in wild reindeer and moose in Norway, highlighting the disease's uncertain origin and environmental persistence.
25. See Benestad et al., 2016.
26. See Song et al., 2019. The study examined hunters' reactions to bans on urine-based scent lures, a measure aimed at controlling chronic wasting disease. It found that many hunters perceived these bans as unnecessary and felt that the regulations were not scientifically justified.
27. See Mysterud et al., 2024. The authors review seven years of chronic wasting disease management in Norway, noting limited contagiousness in some species, ongoing environmental risks and challenges in eradication efforts.
28. See Mysterud et al., 2020. The paper reports that reindeer in Norway engage in antler cannibalism, which may contribute to the spread of chronic wasting disease by transmitting prions.
29. Ibid.Ibid
30. Robinson, 2023.
31. Ibid.
32. See Gambín et al., 2017. The authors found that red deer consume antlers seasonally to supplement minerals like calcium and phosphorus, supporting osteophagia as a natural nutrient source.
33. See Gonzalez and Miranda-Massari, 2014. The article reviews how diet influences stress, showing that certain nutrients can modulate the stress response and improve mental health.
34. See Mathiesen et al., 2023. The authors explore how reindeer husbandry adapts to Arctic environmental changes, highlighting cultural, ecological and economic challenges faced by Indigenous herders.
35. See Turunen et al., 2009. Turunen and colleagues examined how climate change affects the availability and quality of reindeer forage in the Arctic, finding that warming alters plant composition and nutritional value.
36. See Kivinen et al., 2012. The authors studied forest fragmentation in Swedish reindeer areas, showing how landscape changes disrupt reindeer grazing and migration patterns.
37. See Gone71, 2020. The web page offers tips for finding reindeer antlers in Norway's Varanger Peninsula, highlighting key locations and the importance of respecting local rules and culture.
38. Robinson, 2024. *Treewilding* is all about our past, present and future relationship with forests and how to restore them – the science, controversies, solutions and hope.

Chapter 14: Roots, 'Shrooms and Webs

1. See Mitchell et al., 2014. The authors review the ecological impacts of ash dieback in the UK, highlighting threats to biodiversity and ecosystem services, and discuss potential management strategies to mitigate damage.
2. See Nixon, 2009. Nixon analyses Thomas Carlyle's use of Yggdrasil, exploring its symbolic significance in his writings on history and society.
3. See McMullan et al., 2018. McMullan and colleagues reveal that the European ash dieback epidemic originated from just two genetically distinct fungal individuals, explaining its rapid spread and impact across the continent.

4. See McKinney et al., 2014.
5. See Madsen et al., 2021. Madsen and colleagues show that ash dieback and *Armillaria* fungi jointly damage trees (with *Armillaria* mainly killing older ash), complicating disease management.
6. See Hovmøller et al., 2011. The article reviews the genetic diversity of wheat stripe rust, highlighting its rapid evolution and challenges for disease control.
7. Robinson, 2024.
8. See Robinson, Liddicoat, Muñoz-Rojas et al., 2024. In this paper, we discuss strategies for restoring soil biodiversity, emphasising its crucial role in ecosystem health and resilience amid global environmental change.
9. See Leplat et al., 2013. Leplat and colleagues review the survival mechanisms of *Fusarium graminearum*, highlighting its persistence in crop residues and implications for disease management.
10. See Mukherjee et al., 2022. The authors study how *Gibberella zeae* produces inoculum and survives in maize and wheat residues, informing disease spread and control strategies.
11. See Atanasova et al. 2013. The authors compare transcriptomics data to reveal different strategies of *Trichoderma* mycoparasitism.
12. See Ahmad et al. 2022. The authors discuss bacterial blotch but also highlight the global economic figures.
13. See Hawksworth and Lücking, 2017. They estimate global fungal diversity at 2.2 to 3.8 million species, highlighting the vast, largely unexplored biodiversity of fungi.
14. See Qv et al., 2021. The article reviews how gut bacteriophages influence inflammatory bowel disease development and discuss their potential in future treatments.
15. See Mansha et al., 2023. The authors studied how salinity and *Fusarium oxysporum* stress affect onion growth and yield, showing that combined stresses reduce plant health and productivity.
16. Robinson, 2023.
17. See Robinson, Barnes, Fickling et al., 2024. In this study, we explore how micro- and macro-scale food webs interact as multilayered networks, revealing complex ecological interdependencies.

Chapter 15: One Health and Healing Nature

1. See Orrow, 2021. The author advocates for an ecological approach to health, emphasising the interconnectedness of human, animal and environmental wellbeing.
2. See Meyer, 1942. Meyer examines the ecology of psittacosis and ornithosis, focusing on transmission patterns between birds and humans.
3. See Honigsbaum, 2014. Honigsbaum profiles Karl Friedrich Meyer's work investigating psittacosis outbreaks, highlighting his role as a pioneering disease detective.
4. Ibid.
5. Ibid.
6. Personal communications with Peggy Eby.
7. Ibid.
8. Ibid.
9. See Wilderness Society, 2024. The web page reveals Australia's hidden deforestation crisis, mainly driven by beef farming, which threatens biodiversity and habitats.
10. See Queensland Government, 2021. Of the 680,688 hectares cleared, 559,844 hectares (82%) were fully cleared.
11. Personal communications with Peggy Eby.
12. See Plowright et al., 2024.
13. See Eby, 2024. The Habitat Restoration Hub is an initiative by the Restoration Decade Alliance that aims to create a comprehensive national database of

ecological restoration projects across Australia. Currently, over 4,000 restoration sites are registered in the New South Wales Northern Rivers region.

14. See Gregory, 2022.
15. See Prist et al., 2023. The authors highlight that zoonotic diseases account for 75% of emerging infectious diseases, often due to human-induced landscape changes.
16. See Robinson et al., 2022.
17. See UN Decade, 2024. This global initiative is aimed at preventing, halting and reversing the degradation of ecosystems worldwide. Co-led by the United Nations Environment Programme (UNEP) and the Food and Agriculture Organization (FAO), the initiative seeks to restore ecosystems on every continent and in every ocean, contributing to the achievement of the Sustainable Development Goals and the Paris Agreement on climate change.

Chapter 16: Green Prescriptions

1. Paraphrased from the book *Braiding Sweetgrass* – see Kimmerer, 2013.
2. See Haahtela, 2019. The author proposes that biodiversity loss reduces beneficial microbial exposure, increasing allergies and inflammation, and suggests reconnecting with nature to improve health.
3. Robinson, 2023.
4. See Robinson and Breed, 2019.
5. See Roslund et al., 2020. The authors demonstrated that a biodiversity intervention in daycare settings improved immune regulation and enriched the gut microbiota of children, suggesting that increased exposure to diverse microorganisms can enhance health.
6. See Andersen et al., 2021. The authors conducted a systematic review examining how nature exposure influences immune system functioning. Their findings suggest that interactions with natural environments can enhance immune responses, potentially reducing inflammation and promoting overall health. The review underscores the importance of integrating nature exposure into public health strategies to bolster immune health.
7. See Robinson, Breed, Camargo et al., 2024. We conducted a scoping review to map the links between biodiversity and human health, highlighting underrepresented areas such as Indigenous health, urban social equity and COVID-19. We found moderate to high evidence supporting the role of biodiversity in enhancing immune function and wellbeing, but noted a lack of specific biodiversity metrics in many studies.
8. See Roslund et al., 2021. The authors conducted a long-term biodiversity intervention in urban daycare settings, introducing biodiverse materials like forest floor soil. The study found that this exposure enriched children's skin and gut microbiota and enhanced immune regulation, suggesting that integrating natural biodiversity into urban environments can promote healthier microbiomes and immune systems.
9. See Widmaier et al., 2019. The book provides a comprehensive overview of human physiology, detailing the mechanisms that regulate body functions across systems.
10. See Li, 2010. Li found that forest bathing trips significantly enhance natural killer cell activity and expression of anti-cancer proteins. These benefits persisted for over 30 days, suggesting that regular exposure to forest environments can bolster immune function.
11. See Yao et al., 2021.
12. See Da Silveira et al., 2021. The authors reviewed literature on the role of physical exercise in enhancing immune function against COVID-19. They found that regular moderate-intensity exercise boosts immune responses, reduces inflammation and may lower the severity of viral infections. However, excessive exercise can suppress immunity, highlighting the importance of balanced physical activity.

13. See Juul and Nordbø, 2023. The authors examined how green space and walkability in Norwegian neighbourhoods influence physical activity, highlighting that perceived safety moderates these effects. Their findings underscore the importance of designing environments that are both accessible and perceived as safe to encourage active lifestyles.
14. See Greener Practice, 2024. Greener Practice is the UK's primary care sustainability network, dedicated to integrating environmental considerations into general practice. It offers resources like the Green Impact for Health Toolkit to help practices reduce their carbon footprint and improve patient health.
15. See Atwood, 1982.
16. See Robinson, Brindley, Cameron et al., 2021.
17. See Allen, 2017. Allen discusses the growing global burden of non-communicable diseases, suggesting it may constitute a pandemic requiring urgent public health action.

Chapter 17: Black-market Biodiversity

1. See Hughes, 2021.
2. See Bezerra-Santos et al., 2021. The authors warn that illegal wildlife trade increases zoonotic disease risk by promoting close, unsafe human–animal contact.
3. See Heymann and Rodier, 2003.
4. See Leligdowicz et al. 2016.
5. See Greatorex et al., 2016. The authors found that wildlife markets in Lao PDR pose a high risk of zoonotic disease due to poor hygiene and close animal–human contact.
6. See Ancona et al., 2010. The authors use *Xylella fastidiosa* as a case study to highlight biosecurity risks in agriculture and the need for early detection and response.
7. See Bega, 2024.
8. See Carroll, 1865.
9. See Doody et al., 2021. The authors argue that post-COVID-19, illegal wildlife trade poses greater global risks than other trafficking crimes due to its links to zoonotic diseases.
10. See O'Reilly and Gardner, 2023. The authors found Shingleback nidovirus 1 only in Western Australia, suggesting it hasn't spread to South Australia.
11. See ZSL, 2024.
12. See Goodfish, 2019.
13. See Giki, 2024.

Chapter 18: Sowing the Seeds of Change

1. See Maani, 2016. The author advocates for systems thinking in complex decision-making, promoting inclusive, adaptive approaches to tackle 'wicked problems'.
2. See Dobson and Carper, 1996. The authors argue that urbanisation and population growth have historically driven the emergence and spread of infectious diseases.
3. See Khoury et al., 2014. The authors found that global food supplies are becoming more uniform, increasing reliance on a few crops and raising food security concerns.
4. See Cryan, 2017. Check out John Cryan's TED talk on the human gut microbiome and why you are what your microbes eat.
5. See Jacobi et al., 2025. The paper highlights the potential of syntropic farming to boost biodiversity and restore ecosystems, though challenges like labour demands and policy gaps remain.

6. See Rebello, 2019. Large-scale syntropic farming trials in Brazil show promise for agrochemical-free grain production, though mechanisation remains a challenge.
7. See Skill et al. 2022. Agroecology initiatives aim to promote sustainable food systems through agroecological practices.
8. See Robinson, Liddicoat, Muñoz-Rojas et al., 2024
9. See Williams, 2025.
10. See Prillaman, 2018.
11. See FAO, 2025.
12. See FANIS, 2025. Ghana's Investing for Food and Jobs (2018–2021) programme sought to modernise agriculture, enhance food security and create employment opportunities.
13. See Avant Gardening, 2024.
14. See Sheffield Abundance, 2017. The project was founded in 2007 by Stephen Watts and Anne-Marie Culhane.

Chapter 19: Vaccinations for Humans *and* Wildlife?

1. See Bailey, 2011.
2. See Peachey, 1929.
3. See Boylston, 2012. Boylston traces the origins of smallpox inoculation to practices in China, India, Africa and the Ottoman Empire, noting its introduction to Britain and America in the early eighteenth century.
4. See Rizvi et al., 2022. The authors advocate for integrating traditional medicine into modern healthcare to address antimicrobial resistance and emerging diseases. They highlight that many current drugs originate from natural products and call for further research into traditional remedies.
5. See Ledford, 2025. The author highlights that vaccines have saved 154 million lives over 50 years, mainly through measles immunisation. However, recent measles deaths in the USA underscore the risks of declining vaccination rates. Public health experts warn that misinformation and funding cuts threaten these gains.
6. See WHO, 2025. Polio is nearing global eradication, but vaccination remains vital to prevent resurgence.
7. See MSF, 2025.
8. See Dattani and Samborska, 2024.
9. See Slack et al., 2021. The authors report that Hib vaccines have greatly reduced disease, but non-typeable strains now pose rising risks.
10. See Burnett et al., 2020. The authors found rotavirus vaccines highly effective in low-mortality countries, with reduced efficacy in high-mortality settings. Despite this, significant reductions in rotavirus-related hospitalisations were observed globally.
11. See von Mollendorf and Ong-Lim, 2025. The authors report that pneumococcal conjugate vaccines have significantly reduced invasive disease, though non-vaccine serotypes are emerging.
12. See Roslund et al., 2024. The authors found that daily skin contact with microbially rich soil enhanced cell-mediated immunity to pneumococcal vaccine components.
13. See Roslund et al., 2020.
14. See Poland, 2011. Poland asserts that no credible evidence links the MMR vaccine to autism, and that misinformation has led to harmful vaccine hesitancy.
15. See Rocke et al., 2019.
16. See Hoyt et al., 2019. The authors found that probiotic *Pseudomonas fluorescens* increased bat survival five-fold in trials against white-nose syndrome.
17. See Daisley et al., 2020. Brendan and his colleagues found that probiotic BioPatty reduced *Paenibacillus* larvae in honeybee larvae and boosted their immunity.
18. See USFWS, 2024.

19. See Woodroffe et al., 2024.
20. See Badger Trust, 2025.
21. See WWF 2025.
22. See Conroy, 2023.
23. See Siomko et al., 2023.
24. See Stokstad, 2024.
25. See Voloudakis et al., 2022. The authors review RNA-based plant vaccines as eco-friendly alternatives to pesticides, using dsRNA to trigger antiviral RNAi.
26. See Percival et al., 2023. The authors found that chitin and chitosan soil amendments at 1% reduced apple and pear scab severity, matching fungicide efficacy in field trials. See also Percival, 2020. In this video, Percival explains plant disease treatments and preventatives.
27. Ibid.
28. See WUR, 2016.

Chapter 20: Staying Ahead of the Curve

1. See Chauhan et al., 2023. The authors highlight that real-time PCR is the primary diagnostic tool for mpox, and that animal reservoirs include rodents and primates.
2. See Mbala-Kingebeni et al., 2024. The authors urge urgent action to contain mpox in the DRC, citing rising Clade I cases and limited healthcare access.
3. See Malembi et al., 2025.
4. See Craig, 2024.
5. See AIR Hub, 2025. Our Aerobiome Innovation and Research Hub (www.aerobiome.org) studies the beneficial biological components of air and their role in human and ecosystem health.
6. See Smither et al., 2011. The authors developed a novel method using spider webs to capture and measure aerosolised filoviruses, offering a simple tool for studying airborne virus survival.
7. See Newton et al., 2024. The authors found that spider webs can collect airborne eDNA from nearby vertebrates, offering a non-invasive tool for biodiversity monitoring.
8. See Lu et al., 2016. The article reviews the MinION portable sequencer, which enables long-read, real-time genome assembly, though it can have higher error rates than other genomic approaches.
9. See Quick et al., 2016. The authors used a portable nanopore sequencer in Guinea to deliver Ebola genome data within 24 hours, aiding real-time outbreak tracking.
10. See MacIntyre et al., 2023. The authors advocate for AI-driven epidemic surveillance using open-source data to enable early outbreak detection, especially in low-resource settings.
11. See University of Toronto, 2020.

References

Abbasi, J., 2023. Bird flu has begun to spread in mammals - here's what's important to know. JAMA, 329(8), pp. 619–621. https://doi.org/10.1001/jama.2023.1317

Acharya, K.P., Subramanya, S.H. and Neupane, D., 2020. Emerging pandemics: Lessons for one-health approach. *Veterinary Medicine and Science*, 7(1), p. 273. https://doi.org/10.1002/vms3.361

Adams, M.S., 2002. Genetic Variation in the Captive Population of the California Condor (Gymnogyps Californianus). California Polytechnic State University.

Agnelli, S. and Capua, I., 2022. Pandemic or panzootic – a reflection on terminology for SARS-CoV-2 infection. Emerging Infectious Diseases, 28(12), pp. 2552. https://doi.org/10.3201/eid2812.220819

Ahmed, C.E.E.D., Ibrahim, N., Ali, D. and Abouzeid, M., 2022. Bacterial Brown Blotch and Marketing of Cultivated Table Mushroom. A Short Communication. Egyptian Journal of Pure and Applied Science, 60(2), pp. 27–35. https://doi.org/10.21608/ejaps.2022.129763.1031

AIR Hub, 2025. The Aerobiome Innovation and Researc Hub. Online. www.aerobiome.org

Ali, L. 2025. A Fungal Disease Ravaged North American Bats. Now, Researchers Found a Second Species That Suggests It Could Happen Again. Online. https://www.smithsonianmag.com/smart-news/a-fungal-disease-ravaged-north-american-bats-now-researchers-found-a-second-species-that-suggests-it-could-happen-again-180986715/

Allan, B.F., Keesing, F. and Ostfeld, R.S., 2003. Effect of forest fragmentation on Lyme disease risk. Conservation Biology, 17(1), pp. 267–272. https://doi.org/10.1046/j.1523-1739.2003.01260.x

Allan, R., Budge, S. and Sauskojus, H., 2023. What sounds like *Aedes*, acts like *Aedes*, but is not *Aedes*? Lessons from dengue virus control for the management of invasive *Anopheles*. The Lancet Global Health, 11(1), pp. e165–e169. https://doi.org/10.1016/S2214-109X(22)00454-5

Allen, L., 2017. Are we facing a noncommunicable disease pandemic? Journal of Epidemiology and Global Health, 7(1), pp. 5–9. https://doi.org/10.1016/j.jegh.2016.11.001

Ancona, V., Appel, D.N. and de Figueiredo, P., 2010. *Xylella fastidiosa*: A model for analyzing agricultural biosecurity. Biosecurity and Bioterrorism: Biodefense Strategy, Practice, and Science, 8(2), pp. 171–182. https://doi.org/10.1089/bsp.2009.0021

Andersen, L., Corazon, S.S. and Stigsdotter, U.K. (2021). Nature exposure and its effects on immune system functioning: A systematic review. International Journal of Environmental Research and Public Health, 18(4), 1416. https://doi.org/10.3390/ijerph18041416

Anthony, M.A., Bender, S.F. and van der Heijden, M.G., 2023. Enumerating soil biodiversity. Proceedings of the National Academy of Sciences, 120(33), pp. e2304663120. https://doi.org/10.1073/pnas.2304663120

APS, 2002. Drinking Through Your Pelvic Region? A Pretty Good Trick Developed By Terrestrial Adapted Toads. Online. https://www.sciencedaily.com/releases/2002/08/020827062602.htm

Ashe, D. 2019. Celebrating the battle for black-footed ferret recovery. Online. https://www.aza.org/connect-stories/stories/celebrating-the-battle-for-black-footed-ferret-recovery

ASM, 2023. A dangerous eye infection from tainted eye drops. Online. https://asm.org/press-releases/2023/may/a-dangerous-eye-infection-from-tainted-eye-drops

Assefa, Y. and Gilks, C.F., 2020. Ending the epidemic of HIV/AIDS by 2030: Will there be an endgame to HIV, or an endemic HIV requiring an integrated health systems response in many countries? International Journal of Infectious Diseases, 100, pp. 273–277. https://doi.org/10.1016/j.ijid.2020.09.011

Atanasova, L., Crom, S.L., Gruber, S., Coulpier, F., Seidl-Seiboth, V., Kubicek, C.P. and Druzhinina, I.S., 2013. Comparative transcriptomics reveals different strategies of Trichoderma mycoparasitism. BMC genomics, 14(1), p.121. https://doi.org/10.1186/1471-2164-14-121

Atwood, M. 1982. Bluebeard's Egg. Australia: Harper Collins.

Avant Gardening, 2024. Avant Gardening. Online. https://www.avantgardening.org/

Badger Trust, 2025. Badger cull facts. Online. https://www.badgertrust.org.uk/badger-cull-facts

Bailey, I., 2011. Edward Jenner, benefactor to mankind. In Plotkin, S.A., ed., History of Vaccine Development (pp. 21–25). New York, NY: Springer New York. https://doi.org/10.1007/978-1-4419-1339-5_4

Bamford, C.G.G. 2020. 'The original Sars virus disappeared – here's why coronavirus won't do the same'. Online. https://theconversation.com/the-original-sars-virus-disappeared-heres-why-coronavirus-wont-do-the-same-138177

Bartels, M. and Lewis, T. 2023. Endangered California Condors Get Bird Flu Vaccine. Online. https://www.scientificamerican.com/article/endangered-california-condors-get-bird-flu-vaccine/

Bates, C. 2016. Would it be wrong to eradicate mosquitoes? Online. https://www.bbc.com/news/magazine-35408835

BBC, 2019. Cases of vCJD still to emerge after mad cow disease scandal. Online. https://www.bbc.com/news/uk-scotland-48947232

BCI, 2020. The Mexican long-nosed bat. Online. https://www.batcon.org/international-bat-of-mystery-the-mexican-long-nosed-bat/

Becker, N., Pluskota, B., Kaiser, A. and Schaffner, F., 2012. Exotic mosquitoes conquer the world. In Arthropods as vectors of emerging diseases (pp. 31–60). Berlin, Heidelberg: Springer Berlin Heidelberg. https://doi.org/10.1007/978-3-642-28842-5_2

Beckerson, W.C., Krider, C., Mohammad, U.A. and de Bekker, C., 2023. 28 minutes later: Investigating the role of aflatrem-like compounds in Ophiocordyceps parasite manipulation of zombie ants. Animal Behaviour, 203, pp. 225–240. https://doi.org/10.1016/j.anbehav.2023.06.011

Beecher, C. 2015. 'Surprising' Discovery Made About Chronic Wasting Disease. Online. https://www.foodsafetynews.com/2015/06/researchers-make-surprising-discovery-about-spread-of-chronic-wasting-disease/

Bega, S. 2024. Social media fuelling illicit trade in cheetahs. Online. https://mg.co.za/the-green-guardian/2024-03-06-social-media-fuelling-illicit-online-trade-in-cheetahs/

Bell, D. 2023. The next pandemic? It's already here for Earth's wildlife. Online. https://theconversation.com/the-next-pandemic-its-already-here-for-earths-wildlife-222306

Benestad, S.L., Mitchell, G., Simmons, M., Ytrehus, B. and Vikøren, T., 2016. First case of chronic wasting disease in Europe in a Norwegian free-ranging reindeer. Veterinary Research, 47, pp. 1–7. https://doi.org/10.1186/s13567-016-0375-4

Berger, L. and Skerratt, L.F. 2013. Where did the frog pandemic come from? Online. https://theconversation.com/where-did-the-frog-pandemic-come-from-14259.

Bezerra-Santos, M.A., Mendoza-Roldan, J.A., Thompson, R.A., Dantas-Torres, F. and Otranto, D., 2021. Illegal wildlife trade: A gateway to zoonotic infectious diseases. Trends in Parasitology, 37(3), pp. 181–184. https://doi.org/10.1016/j.pt.2020.12.005

Bhagirath, A.Y., Li, Y., Somayajula, D., Dadashi, M., Badr, S. and Duan, K., 2016. Cystic fibrosis lung environment and Pseudomonas aeruginosa infection. BMC Pulmonary Medicine, 16, pp. 1–22. https://doi.org/10.1186/s12890-016-0339-5

Blakemore, R.J., 2025. Biodiversity restated:> 99.9% of global species in Soil Biota. ZooKeys, 1224, p. 283. https://doi.org/10.3897/zookeys.1224.131153

Bland, A. 2016. Should we wipe mosquitoes off the face of the Earth? Online. https://www.theguardian.com/global/2016/feb/10/should-we-wipe-mosquitoes-off-the-face-of-the-earth

Blot, S.I., Vandewoude, K.H., Hoste, E.A. and Colardyn, F.A., 2002. Outcome and attributable mortality in critically ill patients with bacteremia involving methicillin-susceptible and methicillin-resistant Staphylococcus aureus. Archives of Internal Medicine, 162(19), pp. 2229–2235. https://doi.org/10.1001/archinte.162.19.2229

Boonman, A., Bumrungsri, S. and Yovel, Y., 2014. Nonecholocating fruit bats produce biosonar clicks with their wings. Current Biology, 24(24), pp. 2962–2967. https://doi.org/10.1016/j.cub.2014.10.077

Boylston, A., 2012. The origins of inoculation. Journal of the Royal Society of Medicine, 105(7), pp. 309–313. https://doi.org/10.1258/jrsm.2012.12k044

Bradley, R. and Wilesmith, J.W., 1993. Epidemiology and control of bovine spongiform encephalopathy (BSE). British Medical Bulletin, 49(4), pp. 932–959. https://doi.org/10.1093/oxfordjournals.bmb.a072654

Brady, M.F., Jamal, Z. and Pervin, N., 2017. Acinetobacter. StatPearls [Internet]. Treasure Island (FL): StatPearls Publishing

Brady, O.J., Osgood-Zimmerman, A., Kassebaum, N.J., Ray, S.E., de Araújo, V.E., da Nóbrega, A.A., Frutuoso, L.C., Lecca, R.C., Stevens, A., Zoca de Oliveira, B. and de Lima Jr, J.M., 2019. The association between Zika virus infection and microcephaly in Brazil 2015–2017: An observational analysis of over 4 million births. PLoS Medicine, 16(3), p. e1002755. https://doi.org/10.1371/journal.pmed.1002755

Brier, B., 2004. Infectious diseases in ancient Egypt. Infectious Disease Clinics, 18(1), pp. 17–27. https://doi.org/10.1016/S0891-5520(03)00097-7

Brown, P., 1998. 1755 and all that: A historical primer of transmissible spongiform encephalopathy. BMJ, 317(7174), pp. 1688–1692. https://doi.org/10.1136/bmj.317.7174.1688

Bullock, D.S., 1956. Vultures as disseminators of anthrax. The Auk, 73(2), p. 21. https://doi.org/10.2307/4081485

Bunnell, T., Hanisch, K., Hardege, J.D. and Breithaupt, T., 2011. The fecal odor of sick hedgehogs (*Erinaceus europaeus*) mediates olfactory attraction of the tick Ixodes hexagonus. Journal of Chemical Ecology, 37, pp. 340–347. https://doi.org/10.1007/s10886-011-9936-1

Burfield, I and Bowden, C. 2022. South Asian Vultures and Diclofenac. Online. https://www.cambridge.org/core/blog/2022/09/28/south-asian-vultures-and-diclofenac/

Burnett, E., Parashar, U.D. and Tate, J.E., 2020. Real-world effectiveness of rotavirus vaccines, 2006–19: A literature review and meta-analysis. The Lancet Global Health, 8(9), pp. e1195–e1202. https://doi.org/10.1016/S2214-109X(20)30262-X

Burton, A., 2014. Vultures: a future foretold. Frontiers in Ecology and the Environment, 12(8), pp. 480–480. https://doi.org/10.1890/1540-9295-12.8.480

Business Standard, 2025. China scientists identify Pangolin as possible host for Coronavirus. Online. https://www.business-standard.com/article/current-affairs/china-scientists-identify-pangolin-as-possible-host-for-coronavirus-120020700815_1.html

Campana, M.G., Kurata, N.P., Foster, J.T., Helgen, L.E., Reeder, D.M., Fleischer, R.C. and Helgen, K.M., 2017. White-nose syndrome fungus in a 1918 bat specimen from France. Emerging Infectious Diseases, 23(9), p. 1611. https://doi.org/10.3201/eid2309.170875

Carías Domínguez, A.M., de Jesús Rosa Salazar, D., Stefanolo, J.P., Cruz Serrano, M.C., Casas, I.C. and Zuluaga Peña, J.R., 2025. Intestinal dysbiosis: Exploring definition, associated symptoms, and perspectives for a comprehensive understanding – a scoping review. Probiotics and Antimicrobial Proteins, 17(1), pp. 440–449. https://doi.org/10.1007/s12602-024-10353-w

Carroll, L. 1865. Alice's Adventures in Wonderland. London: Macmillan and Co.

Carroll, L., 1871. Through the Looking-Glass, and What Alice Found There. The Complete Illustrated Works of Lewis Carroll. London: Chancellor Press.

Carroll, J.F., Klun, J.A. and Schmidtmann, E.T., 1995. Evidence for kairomonal influence on selection of host-ambushing sites by adult Ixodes scapularis (Acari: Ixodidae). Journal of Medical Entomology, 32(2), pp. 119–125. https://doi.org/10.1093/jmedent/32.2.119

Carroll, R. 2001. Online. Did malaria bring Rome to its knees? https://www.theguardian.com/world/2001/feb/21/rorycarroll

Carson, R., 1962. Silent Spring III. New Yorker, 23.

Cassmann, E.D., Frese, R.D. and Greenlee, J.J., 2021. Second passage of chronic wasting disease of mule deer to sheep by intracranial inoculation compared to classical scrapie. Journal of Veterinary Diagnostic Investigation, 33(4), pp. 711–720. https://doi.org/10.1177/10406387211017615

CDC, 2024a. West Nile virus. Online. https://www.cdc.gov/west-nile-virus/about/index.html#:~:text=West%20Nile%20virus%20is%20the,summer%20and%20continues%20through%20fall.

CDC, 2024b. Malaria. Online. https://www.cdc.gov/dpdx/malaria/index.html

CDC. 2024c. Chronic Wasting Disease in Animals. Online. https://www.cdc.gov/chronic-wasting/animals/index.html

Charles, D. 2008. Experts identify fungus suspected in bat die-off. Online. https://www.npr.org/2008/10/30/96342911/experts-identify-fungus-suspected-in-bat-die-off.

Chauhan, R.P., Fogel, R. and Limson, J., 2023. Overview of diagnostic methods, disease prevalence and transmission of MPOX (formerly monkeypox) in humans and animal reservoirs. Microorganisms, 11(5), p. 1186. https://doi.org/10.3390/microorganisms11051186

Chaves-Carballo, E., 2005. Carlos Finlay and yellow fever: Triumph over adversity. Military Medicine, 170(10), pp. 881–885. https://doi.org/10.7205/MILMED.170.10.881

Chippaux, J.P. and Chippaux, A., 2018. Yellow fever in Africa and the Americas: A historical and epidemiological perspective. Journal of Venomous Animals and Toxins including Tropical Diseases, 24, p. 20. https://doi.org/10.1186/s40409-018-0162-y

Choo, S.W., Zhou, J., Tian, X., Zhang, S., Qiang, S., O'Brien, S.J., Tan, K.Y., Platto, S., Koepfli, K.P., Antunes, A. and Sitam, F.T., 2020. Are pangolins scapegoats of the COVID-19 outbreak-CoV transmission and pathology evidence? Conservation Letters, 13(6), p. e12754. https://doi.org/10.1111/conl.12754

Chr, N., 1979. Where have all the species gone? On the nature of extinction and the Red Queen hypothesis. Oikos, 33(2), pp. 196–227. https://doi.org/10.2307/3543998

Chua, K.B., 2003. Nipah virus outbreak in Malaysia. Journal of Clinical Virology, 26(3), pp. 265–275. https://doi.org/10.1016/S1386-6532(02)00268-8

Cillóniz, C., Ewig, S., Menéndez, R., Ferrer, M., Polverino, E., Reyes, S., Gabarrús, A., Marcos, M.A., Cordoba, J., Mensa, J. and Torres, A., 2012. Bacterial co-infection with H1N1 infection in patients admitted with community acquired pneumonia. Journal of Infection, 65(3), pp. 223–230. https://doi.org/10.1016/j.jinf.2012.04.009

Clark, T.W., 1976. The black-footed ferret. Oryx, 13(3), pp. 275–280. https://doi.org/10.1017/S0030605300013727

Clements, A.N. and Harbach, R.E., 2017. History of the discovery of the mode of transmission of yellow fever virus. Journal of Vector Ecology, 42(2), pp. 208–222. https://doi.org/10.1111/jvec.12261

CNN, 2016. Health. Online. https://edition.cnn.com/2007/HEALTH/10/18/mrsa.cases/

Cobb, E. 2023. Disease in Amphibians. Online. https://www.froglife.org/tag/chytrid-fungus/.

Coghlan, A. 2015. Deadly skin eating fungus threatens UK's great crested newts. Online. https://www.newscientist.com/article/dn27461-deadly-skin-eating-fungus-threatens-uks-great-crested-newt/.

Conroy, G. 2023. Tasmanian devil cancer vaccine approved for testing. Online. https://www.nature.com/articles/d41586-023-02124-4

Costa, R.L.D., Lamas, C.D.C., Simvoulidis, L.F.N., Espanha, C.A., Moreira, L.P.M., Bonancim, R.A.B., Weber, J.V.L.A., Ramos, M.R.F., Silva, E.C.D.F. and Oliveira, L.P.D., 2022. Secondary infections in a cohort of patients with COVID-19 admitted to an intensive care unit: Impact of gram-negative bacterial resistance. Revista do Instituto de Medicina Tropical de São Paulo, 64, p. e6. https://doi.org/10.1590/s1678-9946202264006

Couret, J., Dyer, M., Mather, T., Han, S., Tsao, J., Lebrun, R. and Ginsberg, H.S., 2017. Acquisition of Borrelia burgdorferi infection by larval Ixodes scapularis (Acari: Ixodidae) associated with engorgement measures. Journal of Medical Entomology, 54(4), pp. 1055–1060. https://doi.org/10.1093/jme/tjx053

Coutinho-Abreu, I.V., Jamshidi, O., Raban, R., Atabakhsh, K., Merriman, J.A. and Akbari, O.S., 2024. Identification of human skin microbiome odorants that manipulate mosquito landing behavior. Scientific Reports, 14(1), p. 1631. https://doi.org/10.1038/s41598-023-50182-5

Cox, F.E., 2002. History of human parasitology. Clinical Microbiology Reviews, 15(4), pp. 595–612. https://doi.org/10.1128/CMR.15.4.595-612.2002

Cryan, J.F. 2017. Feed Your Microbes: Nurture Your Mind. Online (YouTube). https://www.youtube.com/watch?v=vKxomLM7SVc

Craig, J. 2024. Mpox never stopped spreading in Africa. Now it's an international public health emergency. Again. Online. https://www.vox.com/future-perfect/366903/mpox-monkeypox-africa-continental-emergency-drc-who-clade

Cunha, C.B. and Cunha, B.A., 2008. Brief history of the clinical diagnosis of malaria: from Hippocrates to Osler. Journal of Vector Borne Disease, 45(3), pp. 194–199. http://www.mrcindia.org/journal/issues/453194.pdf

Cunningham, A.A., Smith, F., McKinley, T.J., Perkins, M.W., Fitzpatrick, L.D., Wright, O.N. and Lawson, B., 2019. Apparent absence of Batrachochytrium salamandrivorans in wild urodeles in the United Kingdom. Scientific Reports, 9(1), p. 2831. https://doi.org/10.1038/s41598-019-39338-4

Da Silveira, M.P., da Silva Fagundes, K.K., Bizuti, M.R., Starck, É., Rossi, R.C. and de Resende E Silva, D.T., 2021. Physical exercise as a tool to help the immune system against COVID-19: An integrative review of the current literature. Clinical and Experimental Medicine, 21(1), pp. 15–28. https://doi.org/10.1007/s10238-020-00650-3

Daisley, B.A., Pitek, A.P., Chmiel, J.A., Al, K.F., Chernyshova, A.M., Faragalla, K.M., Burton, J.P., Thompson, G.J. and Reid, G., 2020. Novel probiotic approach to counter Paenibacillus larvae infection in honey bees. The ISME Journal, 14(2), pp. 476–491. https://doi.org/10.1038/s41396-019-0541-6

Dance, A., 2021. The incredible diversity of viruses. Nature, 595(7865), pp. 22–25. https://doi.org/10.1038/d41586-021-01749-7

Dattani, S. and Samborska, V. 2024. HPV vaccination: How the world can eliminate cervical cancer. Online. https://ourworldindata.org/hpv-vaccination-world-can-eliminate-cervical-cancer

Davey Smith, G., 2016. Commentary: Known knowns and known unknowns in medical research: James Mackenzie meets Donald Rumsfeld. International Journal of Epidemiology, 45(6), pp. 1747–1748. https://doi.org/10.1093/ije/dyx029

De Bekker, C., Beckerson, W.C. and Elya, C., 2021. Mechanisms behind the madness: How do zombie-making fungal entomopathogens affect host behavior to increase transmission? MBio, 12(5), pp. 1–15. https://doi.org/10.1128/mBio.01872-21

De la Fuente-Nunez, C., 2019. Toward autonomous antibiotic discovery. Msystems, 4(3), pp. 1–5. https://doi.org/10.1128/mSystems.00151-19

Dean, K.R., Krauer, F., Walløe, L., Lingjærde, O.C., Bramanti, B., Stenseth, N.C. and Schmid, B.V., 2018. Human ectoparasites and the spread of plague in Europe during the Second Pandemic. Proceedings of the National Academy of Sciences, 115(6), pp. 1304–1309. https://doi.org/10.1073/pnas.1715640115

Denning, D.W. 2024. Global deaths from fungal disease have doubled in a decade. Online. https://www.manchester.ac.uk/about/news/global-deaths-from-fungal-disease-have-doubled-in-a-decade--new-study

Deshpande, K., Vanak, A.T., Devy, M.S. and Krishnaswamy, J., 2022. Forbidden fruits? Ecosystem services from seed dispersal by fruit bats in the context of latent zoonotic risk. Oikos, 2022(2), pp. 1–17. https://doi.org/10.1111/oik.08359

Didelot, X., 2016. Heroic sacrifice or tragic mistake? Revisiting the Eyam plague, 350 years on. Significance, 13(5), pp. 20–25. https://doi.org/10.1111/j.1740-9713.2016.00961.x

Dobson, A.P. and Carper, E.R., 1996. Infectious diseases and human population history. Bioscience, 46(2), pp. 115–126. https://doi.org/10.2307/1312814

Doody, J.S., Reid, J.A., Bilali, K., Diaz, J. and Mattheus, N., 2021. In the post-COVID-19 era, is the illegal wildlife trade the most serious form of trafficking? Crime Science, 10, pp. 1–12. https://doi.org/10.1186/s40163-021-00154-9

Drees, K.P., Lorch, J.M., Puechmaille, S.J., Parise, K.L., Wibbelt, G., Hoyt, J.R., Sun, K., Jargalsaikhan, A., Dalannast, M., Palmer, J.M. and Lindner, D.L., 2017. Phylogenetics of a fungal invasion: Origins and widespread dispersal of white-nose syndrome. MBio, 8(6), pp. 1–15. https://doi.org/10.1128/mBio.01941-17

Duffy, S., 2018. Why are RNA virus mutation rates so damn high? PLoS Biology, 16(8), p. e3000003. https://doi.org/10.1371/journal.pbio.3000003

Dunning, H. 2022. Mosquitoes that can't spread malaria engineered. Online. https://www.imperial.ac.uk/news/239931/mosquitoes-that-cant-spread-malaria-engineered/

Ebersole, R. 2023. The condor was saved from extinction. Avian flu threatens to undo that work. Online. https://thebulletin.org/2023/10/the-condor-was-saved-from-extinction-avian-flu-threatens-to-undo-that-work/

Eby, p. 2024. Habitat Restoration Hub. Online. https://restorationdecadealliance.org/habitat-restoration-hub/

ECDC, 2024. Patient stories. Online https://antibiotic.ecdc.europa.eu/en/patient-stories

Eikenaar, C., Ostolani, A., Hessler, S., Ye, E.Y. and Hegemann, A., 2023. Recovery of constitutive immune function after migratory endurance flight in free-living birds. Biology Letters, 19(2), p. 20220518. https://doi.org/10.1098/rsbl.2022.0518

Eisfeld, A.J., Biswas, A., Guan, L., Gu, C., Maemura, T., Trifkovic, S., Wang, T., Babujee, L., Dahn, R., Halfmann, P.J. and Barnhardt, T., 2024. Pathogenicity and transmissibility of bovine H5N1 influenza virus. Nature, 633(8029), pp. 426–432. https://doi.org/10.1038/s41586-024-07766-6

Eklöf, J. and Rydell, J., 2021. Attitudes towards bats in Swedish history. Journal of Ethnobiology, 41(1), pp. 35–52. https://doi.org/10.2993/0278-0771-41.1.35

Epstein, J.H., Anthony, S.J., Islam, A., Kilpatrick, A.M., Ali Khan, S., Balkey, M.D., Ross, N., Smith, I., Zambrana-Torrelio, C., Tao, Y. and Islam, A., 2020. Nipah virus dynamics in bats and implications for spillover to humans. Proceedings of the National Academy of Sciences, 117(46), pp. 29190–29201. https://doi.org/10.1073/pnas.2000429117

Epstein, J.H., Field, H.E., Luby, S., Pulliam, J.R. and Daszak, P., 2006. Nipah virus: Impact, origins, and causes of emergence. Current Infectious Disease Reports, 8(1), pp. 59–65. https://doi.org/10.1007/s11908-006-0036-2

Espunyes, J., Illera, L., Dias-Alves, A., Lobato, L., Ribas, M.P., Manzanares, A., Ayats, T., Marco, I. and Cerdà-Cuéllar, M., 2022. Eurasian griffon vultures carry widespread antimicrobial resistant Salmonella and Campylobacter of public health concern. Science of the Total Environment, 844, p. 157189. https://doi.org/10.1016/j.scitotenv.2022.157189

314   The Nature of Pandemics

Fackelmann, G., Gillingham, M.A., Schmid, J., Heni, A.C., Wilhelm, K., Schwensow, N. and Sommer, S., 2021. Human encroachment into wildlife gut microbiomes. Communications Biology, 4(1), p. 800. https://doi.org/10.1038/s42003-021-02315-7

FAO, 2025. National Ecological Organic Agriculture Strategy (2023–2030). Online. https://www.fao.org/agroecology/database/detail/en/c/1680645/

FANIS, 2025. Investing For Food And Jobs (IFJ): An Agenda For Transforming Ghana's Agriculture (2018-2021). Online. https://www.fanisgh.net/investing-for-food-and-jobs-ifj-an-agenda-for-transforming-ghanas-agriculture-2018-2021

FCCWF, 2024. The Imminent Threat to Condors. Online. https://www.friendsofcondors.org/avianinfluenza

Ferronato, N., Maalouf, A., Mertenat, A., Saini, A., Khanal, A., Copertaro, B., Yeo, D., Jalalipour, H., Raldúa Veuthey, J., Ulloa-Murillo, L.M. and Thottathil, M.S., 2024. A review of plastic waste circular actions in seven developing countries to achieve sustainable development goals. Waste Management & Research, 42(6), pp. 436–458. https://doi.org/10.1177/0734242X231188664

Fischer, W.A., Gong, M., Bhagwanjee, S. and Sevransky, J., 2014. Global burden of influenza: Contributions from resource limited and low-income settings. Global Heart, 9(3), p. 325. https://doi.org/10.1016/j.gheart.2014.08.004

Flies, A.S., Flies, E.J., Fox, S., Gilbert, A., Johnson, S.R., Liu, G.S., Lyons, A.B., Patchett, A.L., Pemberton, D. and Pye, R.J., 2020. An oral bait vaccination approach for the Tasmanian devil facial tumor diseases. Expert Review of Vaccines, 19(1), pp. 1–10. https://doi.org/10.1080/14760584.2020.1711058

Frank, E. and Sudarshan, A., 2024. The social costs of keystone species collapse: Evidence from the decline of vultures in India. American Economic Review, 114(10), pp. 3007–3040. https://doi.org/10.1257/aer.20230016

Frank, E.G., 2024. The economic impacts of ecosystem disruptions: Costs from substituting biological pest control. Science, 385(6713), p. eadg0344. https://doi.org/10.1126/science.adg0344

Frick, W.F., Kingston, T. and Flanders, J., 2020. A review of the major threats and challenges to global bat conservation. Annals of the New York Academy of Sciences, 1469(1), pp. 5–25. https://doi.org/10.1111/nyas.14045

Fu, F., Shao, Q., Zhang, J., Wang, J., Wang, Z., Ma, J., Yan, Y., Sun, J. and Cheng, Y., 2023. Bat STING drives IFN-beta production in anti-RNA virus innate immune response. Frontiers in Microbiology, 14, p. 1232314. https://doi.org/10.3389/fmicb.2023.1232314

Galaway, F., Yu, R., Constantinou, A., Prugnolle, F. and Wright, G.J., 2019. Resurrection of the ancestral RH5 invasion ligand provides a molecular explanation for the origin of P. falciparum malaria in humans. PLoS Biology, 17(10), p. e3000490. https://doi.org/10.1371/journal.pbio.3000490

Gambín, P., Ceacero, F., Garcia, A.J., Landete-Castillejos, T. and Gallego, L., 2017. Patterns of antler consumption reveal osteophagia as a natural mineral resource in key periods for red deer (Cervus elaphus). European Journal of Wildlife Research, 63, pp. 1–7. https://doi.org/10.1007/s10344-017-1095-4

Gates, 2019. Test tube mosquitoes might help us beat malaria. Online. https://www.gatesnotes.com/Test-tube-mosquitoes-might-help-us-beat-malaria

Gautam, A., 2022. Antimicrobial resistance: The next probable pandemic. Journal of the Nepal Medical Association, 60(246), p. 225. https://doi.org/10.31729/jnma.7174

Gelband, H., Panosian, C.B. and Arrow, K.J., eds., 2004. Saving Lives, Buying Time: Economics of Malaria Drugs in an Age of Resistance. Washington, DC: The National Academies Press. https://doi.org/10.17226/11017 https://doi.org/10.17226/11017

Gibson, S., Gunn, p. and Maughan, R.J., 2012. Hydration, water intake and beverage consumption habits among adults. Nutrition Bulletin, 37(3), pp. 182–192. https://doi.org/10.1111/j.1467-3010.2012.01976.x

Giki, 2024. Making it easier to achieve your sustainability goals. Online. https://giki.earth/

Gill Jr, R.E., Piersma, T., Hufford, G., Servranckx, R. and Riegen, A., 2005. Crossing the ultimate ecological barrier: Evidence for an 11,000-km-long nonstop flight from Alaska to New Zealand and eastern Australia by bar-tailed godwits. The Condor, 107(1), pp. 1–20. https://doi.org/10.1093/condor/107.1.1

Gone71, 2020. How to find a reindeer antler. Online. https://www.gone71.com/find-a-reindeer-antler/

Goodfish, 2019. The Good Fish Guide. Online, https://goodfish.org.au/

Goodwin. G. 2021. COVID-19: Symptoms, incubation, prevention, and more. Medical News Today. Online. https://www.medicalnewstoday.com/articles/covid-19

Gonzalez, M.J. and Miranda-Massari, J.R., 2014. Diet and stress. Psychiatric Clinics North America, 37(4), pp. 579–89. https://doi.org/10.1016/j.psc.2014.08.004

Greatorex, Z.F., Olson, S.H., Singhalath, S., Silithammavong, S., Khammavong, K., Fine, A.E., Weisman, W., Douangngeun, B., Theppangna, W., Keatts, L. and Gilbert, M., 2016. Wildlife trade and human health in Lao PDR: An assessment of the zoonotic disease risk in markets. PLOS One, 11(3), p. e0150666. https://doi.org/10.1371/journal.pone.0150666

Green, R.E., Newton, I.A.N., Shultz, S., Cunningham, A.A., Gilbert, M., Pain, D.J. and Prakash, V., 2004. Diclofenac poisoning as a cause of vulture population declines across the Indian subcontinent. Journal of Applied Ecology, 41(5), pp. 793–800. https://doi.org/10.1111/j.0021-8901.2004.00954.x

Greener Practice. 2024. Making your practice greener can save you time and money - and it's easier than you think! Online. https://www.greenerpractice.co.uk/

Greenlee, J.J., Moore, S.J., Cassmann, E.D., Lambert, Z.J., Kokemuller, R.D., Smith, J.D., Kunkle, R.A., Kong, Q. and Greenlee, M.H.W., 2023. Characterization of classical sheep scrapie in white-tailed deer after experimental oronasal exposure. The Journal of Infectious Diseases, 227(12), pp. 1386–1395. https://doi.org/10.1093/infdis/jiac443

Greenspoon, L., Krieger, E., Sender, R., Rosenberg, Y., Bar-On, Y.M., Moran, U., Antman, T., Meiri, S., Roll, U., Noor, E. and Milo, R., 2023. The global biomass of wild mammals. Proceedings of the National Academy of Sciences, 120(10), p. e2204892120. https://doi.org/10.1073/pnas.2204892120

Gregory, X. 2022. Flying fox habitat restoration project designed to lure bats away from Bathurst. Online. https://www.abc.net.au/news/2022-09-16/bathurst-flying-foxes-get-new-home/101443720

Guardian, 2000. BSE Crisis: Timeline. Online. https://www.theguardian.com/uk/2000/oct/26/bse3

Haahtela, T., 2019. A biodiversity hypothesis. Allergy, 74(8), pp. 1445–1456. https://doi.org/10.1111/all.13763

Hall, M. and Tamïr, D., 2022. Mosquitopia: The Place of Pests in a Healthy World. Abingdon: Routledge https://doi.org/10.4324/9781003056034

Halley, A.C., Baldwin, M.K., Cooke, D.F., Englund, M., Pineda, C.R., Schmid, T., Yartsev, M.M. and Krubitzer, L., 2022. Coevolution of motor cortex and behavioral specializations associated with flight and echolocation in bats. Current Biology, 32(13), pp. 2935–2941. https://doi.org/10.1016/j.cub.2022.04.094

Halliday, F.W., Heckman, R.W., Wilfahrt, P.A. and Mitchell, C.E., 2017. A multivariate test of disease risk reveals conditions leading to disease amplification. Proceedings of the Royal Society B: Biological Sciences, 284(1865), p. 20171340. https://doi.org/10.1098/rspb.2017.1340

Hassett, M.O., Fischer, M.W. and Money, N.P., 2015. Mushrooms as rainmakers: How spores act as nuclei for raindrops. PLoS One, 10(10), p. e0140407. https://doi.org/10.1371/journal.pone.0140407

Hawksworth, D.L. and Lücking, R., 2017. Fungal diversity revisited: 2.2 to 3.8 million species. Microbiology Spectrum, 5(4), pp. 1–17. https://doi.org/10.1128/microbiolspec.FUNK-0052-2016

Hayes, C.G., 2001. West Nile virus: Uganda, 1937, to New York City, 1999. Annals of the New York Academy of Sciences, 951(1), pp. 25–37. https://doi.org/10.1111/j.1749-6632.2001.tb02682.x

Hess, I.M., Massey, P.D., Walker, B., Middleton, D.J. and Wright, T.M., 2011. Hendra virus: What do we know? New South Wales Public Health Bulletin, 22(6), pp. 118–122. https://doi.org/10.1071/NB10077

Heymann, D.L. and Rodier, G., 2003. SARS: a global response to an international threat. Brown J. World Aff., 10, p.185. https://doi.org/10.3201/eid1002.031038

Hicks, A.C., Darling, S.R., Flewelling, J.E., von Linden, R., Meteyer, C.U., Redell, D.N., White, J.P., Redell, J., Smith, R., Blehert, D.S. and Rayman-Metcalf, N.L., 2023. Environmental transmission of Pseudogymnoascus destructans to hibernating little brown bats. Scientific Reports, 13(1), p. 4615. https://doi.org/10.1038/s41598-023-31515-w

Hill, E., 2006. The cyanophage molecular mixing bowl of photosynthesis genes. PLOS Biology, 4(8), p. e264. https://doi.org/10.1371/journal.pbio.0040264

Hoang, N.T. and Kanemoto, K., 2021. Mapping the deforestation footprint of nations reveals growing threat to tropical forests. Nature Ecology & Evolution, 5(6), pp. 845–853. https://doi.org/10.1038/s41559-021-01417-z

Honigsbaum, M., 2014. In search of sick parrots: Karl Friedrich Meyer, disease detective. The Lancet, 383(9932), pp. 1880–1881. https://doi.org/10.1016/S0140-6736(14)60905-3

Hook, S.A., Hansen, A.P., Niesobecki, S.A., Meek, J.I., Bjork, J.K., Kough, E.M., Peterson, M.S., Schiffman, E.K., Rutz, H.J., Rowe, A.J. and White, J.L., 2022. Evaluating public acceptability of a potential Lyme disease vaccine using a population-based, cross-sectional survey in high incidence areas of the United States. Vaccine, 40(2), pp. 298–305. https://doi.org/10.1016/j.vaccine.2021.11.065

Horror Show, 2024. Bats in Cinema. Online (YouTube video). https://www.youtube.com/watch?v=qZYwfvthOwg&ab_channel=TheHorrorShow.

Hovmøller, M.S., Sørensen, C.K., Walter, S. and Justesen, A.F., 2011. Diversity of *Puccinia striiformis* on cereals and grasses. Annual Review of Phytopathology, 49(1), pp. 197–217. https://doi.org/10.1146/annurev-phyto-072910-095230

Hoyt, J.R., Langwig, K.E., White, J.P., Kaarakka, H.M., Redell, J.A., Parise, K.L., Frick, W.F., Foster, J.T. and Kilpatrick, A.M., 2019. Field trial of a probiotic bacteria to protect bats from white-nose syndrome. Scientific Reports, 9(1), p. 9158. https://doi.org/10.1038/s41598-019-45453-z

Hughes, A.C., 2021. Wildlife trade. Current Biology, 31(19), pp. R1218–R1224. https://doi.org/10.1016/j.cub.2021.08.056

Hui, D.S., Azhar, E.I., Memish, Z.A. and Zumla, A., 2021. Human coronavirus infections – severe acute respiratory syndrome (SARS), Middle East respiratory syndrome (MERS), and SARS-CoV-2. Encyclopedia of Respiratory Medicine, pp. 146–161. https://doi.org/10.1016/B978-0-12-801238-3.11634-4

Hunink, J.E., Veenstra, T., van der Hoek, W. and Droogers, P., 2010. Q fever transmission to humans and local environmental conditions. FutureWater, Report 90.

Infectious Diseases Society of America (IDSA), 2011. Combating antimicrobial resistance: Policy recommendations to save lives. Clinical Infectious Diseases, 52(suppl_5), pp. S397–S428. https://doi.org/10.1093/cid/cir153

Ingham, H. and Luft, J., 1955. The Johari Window: A graphic model for interpersonal relations. Proceedings of the Western Training Laboratory in Group Development. Los Angeles: University of California.

IPBES, 2020. Biodiversity and Pandemics. Online: https://files.ipbes.net/ipbes-web-prod-public-files/2020-12/IPBES%20Workshop%20on%20Biodiversity%20and%20Pandemics%20Report_0.pdf.

ISID, 2024. Fungal disease awareness week. Online. https://isid.org/fungal-disease-awareness-week/

Jacobi, J., Andres, C., Assaad, F.F., Bellon, S., Coquil, X., Doetterl, S., Esnarriaga, D.N., Ortiz-Vallejo, D., Rigolot, C., Rüegg, J. and Takerkart, S., 2025. Syntropic farming systems for reconciling productivity, ecosystem functions, and restoration. The Lancet Planetary Health, 9(4), pp. e314–e325. https://doi.org/10.1016/S2542-5196(25)00047-6

Jalihal, S., Rana, S. and Sharma, S., 2022. Systematic mapping on the importance of vultures in the Indian public health discourse. Environmental Sustainability, 5(2), pp. 135–143. https://doi.org/10.1007/s42398-022-00224-x

Jin, X., Chua, H.Z., Wang, K., Li, N., Zheng, W., Pang, W., Yang, F., Pang, B., Zhang, M. and Zhang, J., 2021. Evidence for the medicinal value of Squama Manitis (pangolin scale): A systematic review. Integrative Medicine Research, 10(1), p. 100486. https://doi.org/10.1016/j.imr.2020.100486

Johnson, J.S., Reeder, D.M., McMichael III, J.W., Meierhofer, M.B., Stern, D.W., Lumadue, S.S., Sigler, L.E., Winters, H.D., Vodzak, M.E., Kurta, A. and Kath, J.A., 2014. Host, pathogen, and environmental characteristics predict white-nose syndrome mortality in captive little brown myotis (*Myotis lucifugus*). PLOS One, 9(11), p. e112502. https://doi.org/10.1371/journal.pone.0112502

Jones, K.E., Patel, N.G., Levy, M.A., Storeygard, A., Balk, D., Gittleman, J.L. and Daszak, P., 2008a. Global trends in emerging infectious diseases. Nature, 451(7181), pp. 990–993. https://doi.org/10.1038/nature06536

Jones, M.E., Cockburn, A., Hamede, R., Hawkins, C., Hesterman, H., Lachish, S., Mann, D., McCallum, H. and Pemberton, D., 2008b. Life-history change in disease-ravaged Tasmanian devil populations. Proceedings of the National Academy of Sciences, 105(29), pp. 10023–10027. https://doi.org/10.1073/pnas.0711236105

Jones, M.E., Paetkau, D., Geffen, E.L.I. and Moritz, C., 2004. Genetic diversity and population structure of Tasmanian devils, the largest marsupial carnivore. Molecular Ecology, 13(8), pp. 2197–2209. https://doi.org/10.1111/j.1365-294X.2004.02239.x

Jones, W.H.S., 1907. Malaria and history. Annals of Tropical Medicine & Parasitology, 1(1-5), pp. 528–546. https://doi.org/10.1080/00034983.1907.11719270

Judson, O. 2003. A bug's death. Online. http://www.nytimes.com/2003/09/25/opinion/a-bug-s-death.html

Juul, V. and Nordbø, E.C.A., 2023. Examining activity-friendly neighborhoods in the Norwegian context: Green space and walkability in relation to physical activity and the moderating role of perceived safety. BMC Public Health, 23(1), p. 259. https://doi.org/10.1186/s12889-023-15170-4

Kaddumukasa, M.A., Mutebi, J.P., Lutwama, J.J., Masembe, C. and Akol, A.M., 2014. Mosquitoes of Zika Forest, Uganda: Species composition and relative abundance. Journal of Medical Entomology, 51(1), pp. 104–113. https://doi.org/10.1603/ME12269

Kanaujia, R., Bora, I., Ratho, R.K., Thakur, V., Mohi, G.K. and Thakur, P., 2022. Avian influenza revisited: Concerns and constraints. VirusDisease, 33(4), pp. 456–465. https://doi.org/10.1007/s13337-022-00800-z

Karanja, F. 2024. Victorian Poultry Industry. Online. https://agriculture.vic.gov.au/__data/assets/pdf_file/0016/1042612/Poultry-Industry-Fast-Facts_June-2024.pdf

Keesing, F. and Ostfeld, R.S., 2021. Dilution effects in disease ecology. Ecology Letters, 24(11), pp. 2490–2505. https://doi.org/10.1111/ele.13875

Khalil, H., Ecke, F., Evander, M., Magnusson, M. and Hörnfeldt, B., 2016. Declining ecosystem health and the dilution effect. Scientific Reports, 6(1), p. 31314. https://doi.org/10.1038/srep31314

Khan, S., Akbar, S.M.F., Al Mahtab, M., Uddin, M.N., Rashid, M.M., Yahiro, T., Hashimoto, T., Kimitsuki, K. and Nishizono, A., 2024. Twenty-five years of Nipah outbreaks in Southeast Asia: A persistent threat to global health. IJID Regions, 13, p. 100434. https://doi.org/10.1016/j.ijregi.2024.100434

Khonga, E.B. and Sutton, J.C., 1988. Inoculum production and survival of Gibberella zeae in maize and wheat residues. Canadian Journal of Plant Pathology, 10(3), pp. 232–239. https://doi.org/10.1080/07060668809501730

Khoury, C.K., Bjorkman, A.D., Dempewolf, H., Ramirez-Villegas, J., Guarino, L., Jarvis, A., Rieseberg, L.H. and Struik, P.C., 2014. Increasing homogeneity in global food supplies and the implications for food security. Proceedings of the National Academy of Sciences, 111(11), pp. 4001–4006. https://doi.org/10.1073/pnas.1313490111

Kimmerer, R.W. (2013). Braiding Sweetgrass: Indigenous Wisdom, Scientific Knowledge, and the Teachings of Plants. Milkweed Editions.

Kiss, G., Jansen, H., Castaldo, V.L. and Orsi, L., 2015. The 2050 city. Procedia Engineering, 118, pp. 326–355. https://doi.org/10.1016/j.proeng.2015.08.434

Kivinen, S., Berg, A., Moen, J., Östlund, L. and Olofsson, J., 2012. Forest fragmentation and landscape transformation in a reindeer husbandry area in Sweden. Environmental Management, 49, pp. 295–304. https://doi.org/10.1007/s00267-011-9788-z

Klevens, R.M., Morrison, M.A., Nadle, J., Petit, S., Gershman, K., Ray, S., Harrison, L.H., Lynfield, R., Dumyati, G., Townes, J.M. and Craig, A.S., 2007. Invasive methicillin-resistant Staphylococcus aureus infections in the United States. JAMA, 298(15), pp. 1763–1771. https://doi.org/10.1001/jama.298.15.1763

Kouba, M., Bartoš, L., Bartošová, J., Hongisto, K. and Korpimäki, E., 2020. Interactive influences of fluctuations of main food resources and climate change on long-term population decline of Tengmalm's owls in the boreal forest. Scientific Reports, 10(1), p. 20429. https://doi.org/10.1038/s41598-020-77531-y

Larkin, H.D., 2022. Global COVID-19 death toll may be triple the reported deaths. JAMA, 327(15), p. 1438. https://doi.org/10.1001/jama.2022.4767

Learn, J.R., 2023. Birds Stop During Migration to Avoid Disease or Infection. Online. https://www.discovermagazine.com/planet-earth/birds-stop-during-migration-to-avoid-disease-or-infection

Ledford, H., 2025. 154 million lives and counting: 5 charts reveal the power of vaccines. Online. https://www.nature.com/articles/d41586-025-00862-1 https://doi.org/10.1038/d41586-025-00862-1

Lee, D.B. and Hwang, I.S., 2025. Macronutrient balance determines the human gut microbiome eubiosis: Insights from in vitro gastrointestinal digestion and fermentation of eight pulse species. Frontiers in Microbiology, 15, p. 1512217. https://doi.org/10.3389/fmicb.2024.1512217

Leguia, M. and Nelson, M. 2023. Mass die-offs of marine birds and mammals in Peru sound the alarm on the spread of highly pathogenic avian influenza (H5N1) viruses throughout South America. Online. https://communities.springernature.com/posts/

mass-die-offs-of-marine-birds-and-mammals-in-peru-sound-the-alarm-on-the-spread-of-highly-pathogenic-avian-influenza-h5n1-viruses-throughout-south-america#

Leligdowicz, A., Fischer, W.A., Uyeki, T.M., Fletcher, T.E., Adhikari, N.K., Portella, G., Lamontagne, F., Clement, C., Jacob, S.T., Rubinson, L. and Vanderschuren, A., 2016. Ebola virus disease and critical illness. Critical Care, 20, pp. 1–14. https://doi.org/10.1186/s13054-016-1325-2

Leplat, J., Friberg, H., Abid, M. and Steinberg, C., 2013. Survival of *Fusarium graminearum*, the causal agent of Fusarium head blight. A review. Agronomy for Sustainable Development, 33, pp. 97–111. https://doi.org/10.1007/s13593-012-0098-5

Li, Q. (2010). Effect of forest bathing trips on human immune function. Environmental Health and Preventive Medicine, 15, 9–17. https://doi.org/10.1007/s12199-008-0068-3

Liberski, P.P. and Brown, P., 2009. Kuru: Its ramifications after fifty years. Experimental Gerontology, 44(1-2), pp. 63–69. https://doi.org/10.1016/j.exger.2008.05.010

Liu, D.X., Liang, J.Q. and Fung, T.S., 2021. Human coronavirus-229E, -OC43, -NL63, and -HKU1 (Coronaviridae). Encyclopedia of Virology, p. 428–440. https://doi.org/10.1016/B978-0-12-809633-8.21501-X

Liu, W., Li, Y., Shaw, K.S., Learn, G.H., Plenderleith, L.J., Malenke, J.A., Sundararaman, S.A., Ramirez, M.A., Crystal, P.A., Smith, A.G. and Bibollet-Ruche, F., 2014. African origin of the malaria parasite *Plasmodium vivax*. Nature Communications, 5(1), p. 3346. https://doi.org/10.1038/ncomms4346

Liu, X., Chen, L., Liu, M., García-Guzmán, G., Gilbert, G.S. and Zhou, S., 2020. Dilution effect of plant diversity on infectious diseases: Latitudinal trend and biological context dependence. Oikos, 129(4), pp. 457–465. https://doi.org/10.1111/oik.07027

Long, J., Maskell, K., Gries, R., Nayani, S., Gooding, C. and Gries, G., 2023. Synergistic attraction of Western black-legged ticks, *Ixodes pacificus*, to CO2 and odorant emissions from deer-associated microbes. Royal Society Open Science, 10(5), p. 230084. https://doi.org/10.1098/rsos.230084

López-Baucells, A., Revilla-Martín, N., Mas, M., Alonso-Alonso, P., Budinski, I., Fraixedas, S. and Fernández-Llamazares, Á., 2023. Newspaper coverage and framing of bats, and their impact on readership engagement. EcoHealth, 20(1), pp. 18–30. https://doi.org/10.1007/s10393-023-01634-x

Loreto, R.G. and Hughes, D.P., 2019. The metabolic alteration and apparent preservation of the zombie ant brain. Journal of Insect Physiology, 118, p. 103918. https://doi.org/10.1016/j.jinsphys.2019.103918

Low, M.R., Hoong, W.Z., Shen, Z., Murugavel, B., Mariner, N., Paguntalan, L.M., Tanalgo, K., Aung, M.M., Sheherazade, Bansa, L.A. and Sritongchuay, T., 2021. Bane or blessing? Reviewing cultural values of bats across the Asia-Pacific region. Journal of Ethnobiology, 41(1), pp. 18–34. https://doi.org/10.2993/0278-0771-41.1.18

Lu, H., Giordano, F. and Ning, Z., 2016. Oxford Nanopore MinION sequencing and genome assembly. Genomics, Proteomics & Bioinformatics, 14(5), pp. 265–279. https://doi.org/10.1016/j.gpb.2016.05.004

Lunney, D. and Moon, C., 2011. Blind to bats. In The Biology and Conservation of Australasian Bats Edited by Law, B., Eby, P., Lunney, D., and Lumsden, L. pp. 44–63. https://doi.org/10.7882/FS.2011.008

Ma, Z., 2023. Virome comparison (VC): A novel approach to comparing viromes based on virus species specificity and virome specificity diversity. Journal of Medical Virology, 95(4), p. e28682. https://doi.org/10.1002/jmv.28682

Maani, K., 2016. Multi-stakeholder decision making for complex problems: A systems thinking approach with cases. Online. https://ideas.repec.org/b/wsi/wsbook/9294.html

MacIntyre, C.R., Chen, X., Kunasekaran, M., Quigley, A., Lim, S., Stone, H., Paik, H.Y., Yao, L., Heslop, D., Wei, W. and Sarmiento, J., 2023. Artificial intelligence in public health: The potential of epidemic early warning systems. Journal of International Medical Research, 51(3), p. 03000605231159335. https://doi.org/10.1177/03000605231159335

Mackenzieie, D. 2004. Mysterious mass die-off of vultures solved. Online, https://www.newscientist.com/article/dn4617-mysterious-mass-die-off-of-vultures-solved/.

Malembi, E., Escrig-Sarreta, R., Ntumba, J., Beiras, C.G., Shongo, R., Bengehya, J., Nselaka, C., Pukuta, E., Mukadi-Bamuleka, D., Mulopo-Mukanya, N. and Leng, X., 2025. Clinical presentation and epidemiological assessment of confirmed human mpox cases in DR Congo: a surveillance-based observational study. The Lancet, 405(10490), pp. 1666–1675. https://doi.org/10.2139/ssrn.5011551

Madsen, C.L., Kosawang, C., Thomsen, I.M., Hansen, L.N., Nielsen, L.R. and Kjær, E.D., 2021. Combined progress in symptoms caused by *Hymenoscyphus fraxineus* and Armillaria species, and corresponding mortality in young and old ash trees. Forest Ecology and Management, 491, p. 119177. https://doi.org/10.1016/j.foreco.2021.119177

Mansha, M.Z., Aatif, H.M., Ikram, K., Hanif, C.M.S., Sattar, A., Iqbal, R., Zaman, Q.U., Al-Qahtani, S.M., Al-Harbi, N.A., Omar, W.A. and Ibrahim, M.F., 2023. Impact of various salinity levels and Fusarium oxysporum as stress factors on the morpho-physiological and yield attributes of onion. Horticulturae, 9(7), p. 786. https://doi.org/10.3390/horticulturae9070786

Maquart, P.O., Froehlich, Y. and Boyer, S., 2022. Plastic pollution and infectious diseases. The Lancet Planetary Health, 6(10), pp. e842-e845. https://doi.org/10.1016/S2542-5196(22)00198-X

Marani, M., Katul, G.G., Pan, W.K. and Parolari, A.J., 2021. Intensity and frequency of extreme novel epidemics. Proceedings of the National Academy of Sciences, 118(35), p. e2105482118. https://doi.org/10.1073/pnas.2105482118

Marbán-Castro, E., Goncé, A., Fumadó, V., Romero-Acevedo, L. and Bardají, A., 2021. Zika virus infection in pregnant women and their children: A review. European Journal of Obstetrics & Gynecology and Reproductive Biology, 265, pp. 162–168. https://doi.org/10.1016/j.ejogrb.2021.07.012

Markandya, A., Taylor, T., Longo, A., Murty, M.N., Murty, S. and Dhavala, K., 2008. Counting the cost of vulture decline – an appraisal of the human health and other benefits of vultures in India. Ecological Economics, 67(2), pp. 194–204. https://doi.org/10.1016/j.ecolecon.2008.04.020

Martel, A., Spitzen-van der Sluijs, A., Blooi, M., Bert, W., Ducatelle, R., Fisher, M.C., Woeltjes, A., Bosman, W., Chiers, K., Bossuyt, F. and Pasmans, F., 2013. Batrachochytrium salamandrivorans sp. nov. causes lethal chytridiomycosis in amphibians. Proceedings of the National Academy of Sciences, 110(38), pp. 15325–15329. https://doi.org/10.1073/pnas.1307356110

Mason, P.J. and Haddow, A.J., 1957. An epidemic of virus disease in Southern Province, Tanganyika Territory, in 1952-53; an additional note on Chikungunya virus isolations and serum antibodies. Transactions of the Royal Society of Tropical Medicine and Hygiene, 51(3), pp. 238–240. https://doi.org/10.1016/0035-9203(57)90022-6

Mathiesen, S.D., Eira, I.M.G., Turi, E.I., Oskal, A., Pogodaev, M. and Tonkopeeva, M., 2023. Reindeer Husbandry: Adaptation to the Changing Arctic, Volume 1 (p. 278). Springer Nature. https://doi.org/10.1007/978-3-031-17625-8

Maxmen, A. and Mallapaty, S., 2021. The COVID lab-leak hypothesis: What scientists do and don't know. Nature, 594(7863), pp. 313–315. https://doi.org/10.1038/d41586-021-01529-3

Mbala-Kingebeni, P., Rimoin, A.W., Kacita, C., Liesenborghs, L., Nachega, J.B. and Kindrachuk, J., 2024. The time is now (again) for mpox containment and elimination in Democratic Republic of the Congo. PLOS Global Public Health, 4(6), p. e0003171. https://doi.org/10.1371/journal.pgph.0003171

McCallum, H., Jones, M., Hawkins, C., Hamede, R., Lachish, S., Sinn, D.L., Beeton, N. and Lazenby, B., 2009. Transmission dynamics of Tasmanian devil facial tumor disease may lead to disease-induced extinction. Ecology, 90(12), pp. 3379–3392. https://doi.org/10.1890/08-1763.1

McDonald, K.R., Méndez, D., Müller, R., Freeman, A.B. and Speare, R., 2005. Decline in the prevalence of chytridiomycosis in frog populations in North Queensland, Australia. Pacific Conservation Biology, 11(2), pp. 114–120. https://doi.org/10.1071/PC050114

McFarlane, R., Becker, N. and Field, H., 2011. Investigation of the climatic and environmental context of Hendra virus spillover events 1994–2010. PLoS One, 6(12), p. e28374. https://doi.org/10.1371/journal.pone.0028374

McKinney, L.V., Nielsen, L.R., Collinge, D.B., Thomsen, I.M., Hansen, J.K. and Kjær, E.D., 2014. The ash dieback crisis: Genetic variation in resistance can prove a long-term solution. Plant Pathology, 63(3), pp. 485–499. https://doi.org/10.1111/ppa.12196

McMichael, L., Edson, D., Smith, C., Mayer, D., Smith, I., Kopp, S., Meers, J. and Field, H., 2017. Physiological stress and Hendra virus in flying-foxes (Pteropus spp.), Australia. PLOS One, 12(8), p. e0182171. https://doi.org/10.1371/journal.pone.0182171

McMullan, M., Rafiqi, M., Kaithakottil, G., Clavijo, B.J., Bilham, L., Orton, E., Percival-Alwyn, L., Ward, B.J., Edwards, A., Saunders, D.G. and Garcia Accinelli, G., 2018. The ash dieback invasion of Europe was founded by two genetically divergent individuals. Nature Ecology & Evolution, 2(6), pp. 1000–1008. https://doi.org/10.1038/s41559-018-0548-9

McQuiston, J.H., Childs, J.E. and Thompson, H.A., 2002. Q fever. Journal of the American Veterinary Medical Association, 221(6), pp. 796–799. https://doi.org/10.2460/javma.2002.221.796

Meteyer, C.U., Dutheil, J.Y., Keel, M.K., Boyles, J.G. and Stukenbrock, E.H., 2022. Plant pathogens provide clues to the potential origin of bat white-nose syndrome Pseudogymnoascus destructans. Virulence, 13(1), pp. 1020–1031. https://doi.org/10.1080/21505594.2022.2082139

Meyer, K.F., 1942. The ecology of psittacosis and ornithosis. Medicine, 21(2), p. 175. https://doi.org/10.1097/00005792-194205010-00003

Milbank, C. and Vira, B., 2022. Wildmeat consumption and zoonotic spillover: Contextualising disease emergence and policy responses. Lancet Planet Health, 6, pp. e439–48. https://doi.org/10.1016/S2542-5196(22)00064-X

Milisav, I., Šuput, D. and Ribarič, S., 2015. Unfolded protein response and macroautophagy in Alzheimer's, Parkinson's and prion diseases. Molecules, 20(12), pp. 22718–22756. https://doi.org/10.3390/molecules201219865

Minnis, A.M. and Lindner, D.L., 2013. Phylogenetic evaluation of Geomyces and allies reveals no close relatives of Pseudogymnoascus destructans, comb. nov., in bat hibernacula of eastern North America. Fungal Biology, 117(9), pp. 638–649. https://doi.org/10.1016/j.funbio.2013.07.001

Mitchell, R.J., Beaton, J.K., Bellamy, P.E., Broome, A., Chetcuti, J., Eaton, S., Ellis, C.J., Gimona, A., Harmer, R., Hester, A.J. and Hewison, R.L., 2014. Ash dieback in the UK: A review of the ecological and conservation implications and potential management options. Biological Conservation, 175, pp. 95–109. https://doi.org/10.1016/j.biocon.2014.04.019

Monath, T.P., 2001. Yellow fever: An update. The Lancet Infectious Diseases, 1(1), pp. 11–20. https://doi.org/10.1016/S1473-3099(01)00016-0

Morales-Reyes, Z., Pérez-García, J.M., Moleón, M., Botella, F., Carrete, M., Lazcano, C., Moreno-Opo, R., Margalida, A., Donázar, J.A. and Sánchez-Zapata, J.A., 2015. Supplanting ecosystem services provided by scavengers raises greenhouse gas emissions. Scientific Reports, 5(1), p. 7811. https://doi.org/10.1038/srep07811

Morten, J.M., Buchanan, P.J., Egevang, C., Glissenaar, I.A., Maxwell, S.M., Parr, N., Screen, J.A., Vigfúsdóttir, F., Vogt-Vincent, N.S., Williams, D.A. and Williams, N.C., 2023. Global warming and arctic terns: Estimating climate change impacts on the world's longest migration. Global Change Biology, 29(19), pp. 5596–5614. https://doi.org/10.1111/gcb.16891

MSF, 2025. Measles. Online. https://msf.org.au/issue/measles

Mudur, G., 2001. Human anthrax in India may be linked to vulture decline. BMJ, 322(7282), p. 320. https://doi.org/10.1136/bmj.322.7282.320

Mukherjee, P.K., Mendoza-Mendoza, A., Zeilinger, S. and Horwitz, B.A., 2022. Mycoparasitism as a mechanism of Trichoderma-mediated suppression of plant diseases. Fungal Biology Reviews, 39, pp. 15–33. https://doi.org/10.1016/j.fbr.2021.11.004

Murray, C.J., Ikuta, K.S., Sharara, F., Swetschinski, L., Aguilar, G.R., Gray, A., Han, C., Bisignano, C., Rao, P., Wool, E. and Johnson, S.C., 2022. Global burden of bacterial antimicrobial resistance in 2019: A systematic analysis. The Lancet, 399(10325), pp. 629–655.

Musso, D., 2015. Zika virus transmission from French Polynesia to Brazil. Emerging Infectious Diseases, 21(10), p. 1887. https://doi.org/10.3201/eid2010.151125

Mwangilwa, K., Sialubanje, C., Chipoya, M., Mulenga, C., Mwale, M., Chileshe, C., Sinyange, D., Banda, M., Gardner, P.N., Lamba, L. and Kalubula, P., 2025. Attention to COVID 19 pandemic resulted in increased measles cases and deaths in Zambia. Tropical Medicine and Health, 53(1), p. 59. https://doi.org/10.1186/s41182-025-00736-2

Mysterud, A., Tranulis, M.A., Strand, O. and Rolandsen, C.M., 2024. Lessons learned and lingering uncertainties after seven years of chronic wasting disease management in Norway. Wildlife Biology, p. e01255. https://doi.org/10.1002/wlb3.01255

Mysterud, A., Ytrehus, B., Tranulis, M.A., Rauset, G.R., Rolandsen, C.M. and Strand, O., 2020. Antler cannibalism in reindeer. Scientific Reports, 10(1), p. 22168. https://doi.org/10.1038/s41598-020-79050-2

Nalage, D., Sontakke, T., Biradar, A., Jogdand, V., Kale, R., Harke, S., Kale, R. and Dixit, P., 2023. The impact of environmental toxins on the animal gut microbiome and their potential to contribute to disease. Food Chemistry Advances, 3, p. 100497. https://doi.org/10.1016/j.focha.2023.100497

Nambirajan, K., Muralidharan, S., Roy, A.A. and Manonmani, S., 2018. Residues of diclofenac in tissues of vultures in India: A post-ban scenario. Archives of Environmental Contamination and Toxicology, 74, pp. 292–297. https://doi.org/10.1007/s00244-017-0480-z

Neale, M., 2013. In praise of Parsis. Asian Affairs, 44(2), pp. 250–271. https://doi.org/10.1080/03068374.2013.794919

Neill, U.S., 2011. From branch to bedside: Youyou Tu is awarded the 2011 Lasker-DeBakey Clinical Medical Research Award for discovering artemisinin as a treatment for malaria. Journal of Clinical Investigation, 121(10), pp. 3768–3773. https://doi.org/10.1172/JCI60887

Ness, A., Aiken, J. and McKenzie, D., 2023. Sheep scrapie and deer rabies in England prior to 1800. Prion, 17(1), pp. 7–15. https://doi.org/10.1080/19336896.2023.2166749

Newton, J.P., Nevill, P., Bateman, P.W., Campbell, M.A. and Allentoft, M.E., 2024. Spider webs capture environmental DNA from terrestrial vertebrates. Iscience, 27(2), pp. 1–13. https://doi.org/10.1016/j.isci.2024.108904

Nixon, J.V., 2009. Thomas Carlyle's Igdrasil. Carlyle Studies Annual, (25), pp. 49–58.

Nonvignon, J., Aryeetey, G.C., Malm, K.L., Agyemang, S.A., Aubyn, V.N., Peprah, N.Y., Bart-Plange, C.N. and Aikins, M., 2016. Economic burden of malaria on businesses in Ghana: A case for private sector investment in malaria control. Malaria Journal, 15, pp. 1–10. https://doi.org/10.1186/s12936-016-1506-0

Nuding, J. 2024. Troy Swift. Online. https://txsoilsisters.co/podcast/troy-swift.

Nwachukwu, K.C., Ugbogu, O.C., Nwarunma, E. and Nwankpa, C.I., 2023. Eliminating Candida auris: Between ultraviolet-C radiations and medicinal plants, which one is better? Current Clinical Microbiology Reports, 10(3), pp. 131–140. https://doi.org/10.1007/s40588-023-00200-x

O'Reilly, R.L. and Gardner, M.G., 2023. Spatial distribution of the Shingleback nidovirus 1 between Western Australia and South Australia using RT-qPCR. Exploration of Host-Pathogen Interactions in the Australian Skink Tiliqua rugosa, p. 85. Thesis. Online. https://flex.flinders.edu.au/file/02e21485-8486-428c-89a3-4a79d03f5287/1/O%27Reilly2023_MasterCopy.pdf

O'Shea, T.J., Cryan, P.M., Cunningham, A.A., Fooks, A.R., Hayman, D.T., Luis, A.D., Peel, A.J., Plowright, R.K. and Wood, J.L., 2014. Bat flight and zoonotic viruses. Emerging Infectious Diseases, 20(5), p. 741. https://doi.org/10.3201/eid2005.130539

Oladipo, H.J., Tajudeen, Y.A., Oladunjoye, I.O., Yusuff, S.I., Yusuf, R.O., Oluwaseyi, E.M., AbdulBasit, M.O., Adebisi, Y.A. and El-Sherbini, M.S., 2022. Increasing challenges of malaria control in sub-Saharan Africa: Priorities for public health research and policymakers. Annals of Medicine and Surgery, 81, p. 104366. https://doi.org/10.1016/j.amsu.2022.104366

Orrow, G., 2021. Designing an ecological approach to health. BMJ, 375. https://doi.org/10.1136/bmj.n2827

Osdaghi, E., Martins, S.J., Ramos-Sepulveda, L., Vieira, F.R., Pecchia, J.A., Beyer, D.M., Bell, T.H., Yang, Y., Hockett, K.L. and Bull, C.T., 2019. 100 years since Tolaas: Bacterial blotch of mushrooms in the 21st century. Plant Disease, 103(11), pp. 2714–2732. https://doi.org/10.1094/PDIS-03-19-0589-FE

Osterholm, M.T., Anderson, C.J., Zabel, M.D., Scheftel, J.M., Moore, K.A. and Appleby, B.S., 2019. Chronic wasting disease in cervids: implications for prion transmission to humans and other animal species. MBio, 10(4), pp. 10–1128. https://doi.org/10.1128/mBio.01091-19

Ostfeld, R.S. and Keesing, F., 2000. Biodiversity and disease risk: the case of Lyme disease. Conservation biology, 14(3), pp. 722–728. https://doi.org/10.1046/j.1523-1739.2000.99014.x

Otterstatter, M.C. and Thomson, J.D., 2008. Does pathogen spillover from commercially reared bumble bees threaten wild pollinators? PLoS One, 3(7), p. e2771. https://doi.org/10.1371/journal.pone.0002771

Otto, M., 2023. Critical assessment of the prospects of quorum-quenching therapy for Staphylococcus aureus infection. International Journal of Molecular Sciences, 24(4), p. 4025. https://doi.org/10.3390/ijms24044025

Our World in Data, 2022. Per capita meat consumption by type. Online. https://ourworldindata.org/grapher/per-capita-meat-type

Oxitec, 2025. Anopheles programme. Online. https://www.oxitec.com/oxitec-anopheles-program

Pain, D.J., Cunningham, A.A., Donald, P.F., Duckworth, J.W., Houston, D.C., Katzner, T., Parry-Jones, J., Poole, C., Prakash, V., Round, A. and Timmins, R., 2003. Causes and effects of temporospatial declines of Gyps vultures in Asia. Conservation Biology, 17(3), pp. 661–671. https://doi.org/10.1046/j.1523-1739.2003.01740.x

Patel, C., 1991. Breathing. In The Complete Guide to Stress Management (pp. 141–161). Boston, MA: Springer US. https://doi.org/10.1007/978-1-4899-6335-2_10

Peachey, G.C., 1929. John Fewster: An unpublished chapter in the history of vaccination. Annals of Medical History, 1(2), p. 229.

Peeri, N.C., Shrestha, N., Rahman, M.S., Zaki, R., Tan, Z., Bibi, S., Baghbanzadeh, M., Aghamohammadi, N., Zhang, W. and Haque, U., 2020. The SARS, MERS and novel coronavirus (COVID-19) epidemics, the newest and biggest global health threats: What lessons have we learned? International Journal of Epidemiology, 49(3), pp. 717–726. https://doi.org/10.1093/ije/dyaa033

Percival, G.C., 2020. Vaccinating trees? Willow mulch for control of apple scab - Glynn Percival @ Organic Matters. Online (YouTube). https://www.youtube.com/watch?v=-i1RFG8vzkY

Percival, G.C., Graham, S., Percival, C. and Banks, J., 2023. The influence of chitin-and chitosan-based soil amendments on pathogen severity of apple and pear scab. Arboriculture & Urban Forestry, 49(2), pp. 64–74. https://doi.org/10.48044/jauf.2023.006

Plaza, P.I., Gamarra-Toledo, V., Eguí, J.R. and Lambertucci, S.A., 2024. Recent changes in patterns of mammal infection with highly pathogenic avian influenza A (H5N1) virus worldwide. Emerging Infectious Diseases, 30(3), p. 444. https://doi.org/10.3201/eid3003.231098

Plowright, R.K., Ahmed, A.N., Coulson, T., Crowther, T.W., Ejotre, I., Faust, C.L., Frick, W.F., Hudson, P.J., Kingston, T., Nameer, P.O. and O'Mara, M.T., 2024. Ecological countermeasures to prevent pathogen spillover and subsequent pandemics. Nature Communications, 15(1), p. 2577. https://doi.org/10.1038/s41467-024-46151-9

Poinar, G., 2021. Fossil record of viruses, parasitic bacteria and parasitic protozoa. In The Evolution and Fossil Record of Parasitism: Identification and Macroevolution of Parasites (pp. 29–68). De Baets, K., & Huntley, J. W. (Eds.). Cham: Springer International Publishing. https://doi.org/10.1007/978-3-030-42484-8_2

Poland, G.A., 2011. MMR vaccine and autism: Vaccine nihilism and postmodern science. Mayo Clinic Proceedings, 86(9), pp. 869–871. https://doi.org/10.4065/mcp.2011.0467

Pomeroy, R. 2019. Has malaria really killed half of everyone who ever lived? Online: https://www.realclearscience.com/blog/2019/10/03/has_malaria_really_killed_half_of_everyone_who_ever_lived.html.

Powers, A.M., 2011. Genomic evolution and phenotypic distinctions of Chikungunya viruses causing the Indian Ocean outbreak. Experimental Biology and Medicine, 236(8), pp. 909–914. https://doi.org/10.1258/ebm.2011.011078

Prakash, V., 1999. Status of vultures in Keoladeo National Park, Bharatpur, Rajasthan, with special reference to population crash in Gyps species. Journal Bombay Natural History Society, 96, pp. 365–378.

Prasad, V., 2023. Urban sustainability: The way forward. In Urban Environment and Smart Cities in Asian Countries. Insights for Social, Ecological, and Technological Sustainability (p. 1). https://doi.org/10.1007/978-3-031-25914-2_1

Prillaman, J. 2018. Less Water, More Food: Growing the Organic Farming Market in Saudi Arabia. Online.https://thepalladiumgroup.com/news/Organic-Farming-Market-in-Saudi-Arabia

Prist, P.R., Siliansky de Andreazzi, C., Vidal, M.M., Zambrana-Torrelio, C., Daszak, P., Carvalho, R.L. and Tambosi, L.R., 2023. Promoting landscapes with a low zoonotic disease risk through forest restoration: The need for comprehensive guidelines. Journal of Applied Ecology, 60(8), pp. 1510–1521. https://doi.org/10.1111/1365-2664.14442

Pritzkow, S., 2022. Transmission, strain diversity, and zoonotic potential of chronic wasting disease. Viruses, 14(7), p. 1390. https://doi.org/10.3390/v14071390

Puechmaille, S.J., Wibbelt, G., Korn, V., Fuller, H., Forget, F., Mühldorfer, K., Kurth, A., Bogdanowicz, W., Borel, C., Bosch, T. and Cherezy, T., 2011. Pan-European distribution of white-nose syndrome fungus (Geomyces destructans) not associated with mass mortality. PLoS One, 6(4), p. e19167. https://doi.org/10.1371/journal.pone.0019167

Quammen, D. 1996. Natural Acts: A Sidelong View of Science and Nature. Online. https://publicism.info/nature/sidelong/3.html

Queensland Government, 2021. 2018–19 SLATS Report. Online. https://www.qld.gov.au/environment/land/management/mapping/statewide-monitoring/slats/slats-reports/2018-19-report

Quick, J., Loman, N.J., Duraffour, S., Simpson, J.T., Severi, E., Cowley, L., Bore, J.A., Koundouno, R., Dudas, G., Mikhail, A. and Ouédraogo, N., 2016. Real-time, portable genome sequencing for Ebola surveillance. Nature, 530(7589), pp. 228–232. https://doi.org/10.1038/nature16996

Qv, L., Mao, S., Li, Y., Zhang, J. and Li, L., 2021. Roles of gut bacteriophages in the pathogenesis and treatment of inflammatory bowel disease. Frontiers in Cellular and Infection Microbiology, 11, p. 755650. https://doi.org/10.3389/fcimb.2021.755650

Rackham, O. (2016). *Woodlands*. Harper Collins. 9780008156916

Ramadan, N. and Shaib, H., 2019. Middle East respiratory syndrome coronavirus (MERS-CoV): A review. Germs, 9(1), p. 35. https://doi.org/10.18683/germs.2019.1155

Ravikoti, S., Bhatia, V. and Mohanasundari, S.K., 2025. Recent advancements in tuberculosis (TB) treatment regimens. Journal of Family Medicine and Primary Care, 14(2), pp. 521–525. https://doi.org/10.4103/jfmpc.jfmpc_1237_24

Rawal, D., 2020. An overview of natural history of the human malaria. International Journal of Mosquito Research, 7(2), pp. 8–10.

Rebello, F. 2019. Large-scale Syntropic Farming: results and challenges. Online. https://agendagotsch.com/en/large-scale-syntropic-farming-results-and-challenges/

Reuell, p. 2019. Study shows lungless salamanders' skin expresses protein crucial for lung function. Online. https://news.harvard.edu/gazette/story/2019/01/lungless-salamanders-skin-expresses-protein-crucial-for-lung-functionp

Rice, L.B., 2008. Federal funding for the study of antimicrobial resistance in nosocomial pathogens: No ESKAPE. The Journal of Infectious Diseases, 197(8), pp. 1079–1081. https://doi.org/10.1086/533452

Richard, F.J., Gigauri, M., Bellini, G., Rojas, O. and Runde, A., 2021. Warning on nine pollutants and their effects on avian communities. Global Ecology and Conservation, 32, p. e01898. https://doi.org/10.1016/j.gecco.2021.e01898

Ritchie, H. 2023. How many animals are factory farmed? Online. https://ourworldindata.org/how-many-animals-are-factory-farmed

Ritchie, H and Roser, M. 2024. Half of the world's habitable land is used for agriculture. Online. https://ourworldindata.org/global-land-for-agriculture

Ritchie, H. 2025. Almost all livestock in the US is factory farmed. Online. https://ourworldindata.org/data-insights/almost-all-livestock-in-the-united-states-is-factory-farmed

Rizvi, S.A., Einstein, G.P., Tulp, O.L., Sainvil, F. and Branly, R., 2022. Introduction to traditional medicine and their role in prevention and treatment of emerging and re-emerging diseases. Biomolecules, 12(10), p. 1442. https://doi.org/10.3390/biom12101442

RMIT, 2021. Antimicrobial nanotech. Online. https://www.rmit.edu.au/news/all-news/2021/apr/antimicrobial-nanotech

Roberton, S.I., Bell, D.J., Smith, G.J.D., Nicholls, J.M., Chan, K.H., Nguyen, D.T., Tran, P.Q., Streicher, U., Poon, L.L.M., Chen, H. and Horby, P., 2006. Avian influenza H5N1 in viverrids: Implications for wildlife health and conservation. Proceedings of the Royal Society B: Biological Sciences, 273(1595), pp. 1729–1732. https://doi.org/10.1098/rspb.2006.3549

Robinson, J.A., Bowie, R.C., Dudchenko, O., Aiden, E.L., Hendrickson, S.L., Steiner, C.C., Ryder, O.A., Mindell, D.P. and Wall, J.D., 2021. Genome-wide diversity in the California condor tracks its prehistoric abundance and decline. Current Biology, 31(13), pp. 2939–2946. https://doi.org/10.1016/j.cub.2021.04.035

Robinson, J., 2023. Invisible Friends: How Microbes Shape Our Lives and the World Around Us. Pelagic Publishing Ltd.

Robinson, J., 2024. Treewilding: Our Past, Present and Future Relationship with Forests. Pelagic Publishing Ltd. https://doi.org/10.2307/jj.29126477

Robinson, J.M., Aronson, J., Daniels, C.B., Goodwin, N., Liddicoat, C., Orlando, L., Phillips, D., Stanhope, J., Weinstein, P., Cross, A.T. and Breed, M.F., 2022. Ecosystem restoration is integral to humanity's recovery from COVID-19. The Lancet Planetary Health, 6(9), pp. e769–e773. https://doi.org/10.1016/S2542-5196(22)00171-1

Robinson, J.M., Barnes, A.D., Fickling, N., Costin, S., Sun, X. and Breed, M.F., 2024. Food webs in food webs: The micro-macro interplay of multilayered networks. Trends in Ecology & Evolution. 39(10), pp. 913–922 https://doi.org/10.1016/j.tree.2024.06.006

Robinson, J.M. and Breed, M.F., 2019. Green prescriptions and their co-benefits: Integrative strategies for public and environmental health. Challenges, 10(1), p. 9. https://doi.org/10.3390/challe10010009

Robinson, J.M., Breed, M.F. and Beckett, R., 2024. Probiotic cities: Microbiome-integrated design for healthy urban ecosystems. Trends in Biotechnology, 42(8), pp. 942–945. https://doi.org/10.1016/j.tibtech.2024.01.005

Robinson, J.M., Breed, A.C., Camargo, A., Redvers, N. and Breed, M.F., 2024. Biodiversity and human health: A scoping review and examples of underrepresented linkages. Environmental Research, 246, p. 118115. https://doi.org/10.1016/j.envres.2024.118115

Robinson, J.M., Brindley, P., Cameron, R., MacCarthy, D. and Jorgensen, A., 2021. Nature's role in supporting health during the COVID-19 pandemic: A geospatial and socioecological study. International Journal of Environmental Research and Public Health, 18(5), p. 2227. https://doi.org/10.3390/ijerph18052227

Robinson, J.M., Cameron, R. and Jorgensen, A., 2021. Germaphobia! Does our relationship with and knowledge of biodiversity affect our attitudes toward microbes? Frontiers in Psychology, 12, p. 678752. https://doi.org/10.3389/fpsyg.2021.678752

Robinson, J.M., Liddicoat, C., Muñoz-Rojas, M. and Breed, M.F., 2024. Restoring soil biodiversity. Current Biology, 34(9), pp. R393–R398. https://doi.org/10.1016/j.cub.2024.02.035

Rocke, T.E., Kingstad-Bakke, B., Wüthrich, M., Stading, B., Abbott, R.C., Isidoro-Ayza, M., Dobson, H.E., dos Santos Dias, L., Galles, K., Lankton, J.S. and Falendysz, E.A., 2019. Virally-vectored vaccine candidates against white-nose syndrome induce anti-fungal immune response in little brown bats (*Myotis lucifugus*). Scientific Reports, 9(1), p. 6788. https://doi.org/10.1038/s41598-019-43210-w

Roest, H.I.J., Tilburg, J.J.H.C., van der Hoek, W., Vellema, P., van Zijderveld, F.G., Klaassen, C.H.W. and Raoult, D., 2011. The Q fever epidemic in The Netherlands: History, onset, response and reflection. Epidemiology & Infection, 139(1), pp. 1–12. https://doi.org/10.1017/S0950268810002268

Rose, C., 2001. Giants, Monsters and Dragons: An Encyclopedia of Folklore, Legend and Myth. WW Norton & Company. https://doi.org/10.5040/9798400657085

Rosenblum, E.B., James, T.Y., Zamudio, K.R., Poorten, T.J., Ilut, D., Rodriguez, D., Eastman, J.M., Richards-Hrdlicka, K., Joneson, S., Jenkinson, T.S. and Longcore, J.E., 2013. Complex history of the amphibian-killing chytrid fungus revealed with genome resequencing data. Proceedings of the National Academy of Sciences, 110(23), pp. 9385–9390. https://doi.org/10.1073/pnas.1300130110

Roslund, M.I., Nurminen, N., Oikarinen, S., Puhakka, R., Grönroos, M., Puustinen, L., Kummola, L., Parajuli, A., Cinek, O., Laitinen, O.H. and Hyöty, H., 2024. Skin exposure to soil microbiota elicits changes in cell-mediated immunity to pneumococcal vaccine. Scientific Reports, 14(1), p. 18573. https://doi.org/10.1038/s41598-024-68235-8

Roslund, M.I., Puhakka, R., Grönroos, M., Nurminen, N., Oikarinen, S., Gazali, A.M., Cinek, O., Kramná, L., Siter, N., Vari, H.K. and Soininen, L., 2020. Biodiversity intervention enhances immune regulation and health-associated commensal microbiota among daycare children. Science Advances, 6(42), p. eaba2578. https://doi.org/10.1126/sciadv.aba2578

Roslund, M.I., Puhakka, R., Nurminen, N., Oikarinen, S., Siter, N., Grönroos, M., Cinek, O., Kramná, L., Jumpponen, A., Laitinen, O.H. and Rajaniemi, J., 2021. Long-term biodiversity intervention shapes health-associated commensal microbiota among urban day-care children. Environment International, 157, p. 106811. https://doi.org/10.1016/j.envint.2021.106811

Rougeron, V., Boundenga, L., Arnathau, C., Durand, P., Renaud, F. and Prugnolle, F., 2022. A population genetic perspective on the origin, spread and adaptation of the human malaria agents *Plasmodium falciparum* and *Plasmodium vivax*. FEMS Microbiology Reviews, 46(1), p. fuab047. https://doi.org/10.1093/femsre/fuab047

Salomon, J., Hamer, S.A. and Swei, A., 2020. A beginner's guide to collecting questing hard ticks (Acari: Ixodidae): A standardized tick dragging protocol. Journal of Insect Science, 20(6), p. 11. https://doi.org/10.1093/jisesa/ieaa073

Sansonetti, P., 2006. How to define the species barrier to pathogen transmission? Bulletin de L'académie Nationale de Médecine, 190(3), pp. 611–622. https://doi.org/10.1016/S0001-4079(19)33294-7

Santangeli, A., Lambertucci, S.A., Margalida, A., Carucci, T., Botha, A., Whitehouse-Tedd, K. and Cancellario, T., 2024. The global contribution of vultures towards ecosystem services and sustainability: An experts' perspective. iScience, 27(6). 109925. https://doi.org/10.1016/j.isci.2024.109925

Sault, N., 2016. How hummingbird and vulture mediate between life and death in Latin America. Journal of Ethnobiology, 36(4), pp. 783–806. https://doi.org/10.2993/0278-0771-36.4.783

Schonberger, L.B. and Schonberger, R.B., 2012. Etymologia: Prion. Emerging Infectious Diseases, 18(6), p. 1030. https://doi.org/10.3201/eid1806.120271

Science Direct, 2025. Yellow Fever virus. Online. https://www.sciencedirect.com/topics/medicine-and-dentistry/yellow-fever-virus.

Shannon, D.M., Richardson, N., Lahondère, C. and Peach, D., 2024. Mosquito floral visitation and pollination. Current Opinion in Insect Science, p. 101230. https://doi.org/10.1016/j.cois.2024.101230

Sheffield Abundance, 2017. Sheffield Abundance making the best of what we've got. Online. https://sheffieldabundance.wordpress.com/about

Sieradzki, A. and Mikkola, H. 2021. Bats in Folklore and Culture: A Review of Historical Perceptions around the World. Online. https://www.intechopen.com/chapters/80107. https://doi.org/10.5772/intechopen.102368

Simon, M. 2018. Some frogs may be developing a resistance to the disastrous chytrid fungus. Online. https://www.wired.com/story/some-frogs-may-be-developing-a-resistance-to-the-disastrous-chytrid-fungus/

Siomko, S.A., Greenspan, S.E., Barnett, K.M., Neely, W.J., Chtarbanova, S., Woodhams, D.C., McMahon, T.A. and Becker, C.G., 2023. Selection of an anti-pathogen skin microbiome following prophylaxis treatment in an amphibian model system. Philosophical Transactions of the Royal Society B, 378(1882), p. 20220126. https://doi.org/10.1098/rstb.2022.0126

Skerratt, L.F., Berger, L., Clemann, N., Hunter, D.A., Marantelli, G., Newell, D.A., Philips, A., McFadden, M., Hines, H.B., Scheele, B.C. and Brannelly, L.A., 2016. Priorities for management of chytridiomycosis in Australia: Saving frogs from extinction. Wildlife Research, 43(2), pp. 105–120. https://doi.org/10.1071/WR15071

Skill, K., Passero, S. and Farhangi, M., 2022. Cultivating agroecological networks during the pandemic in Argentina: A sociomaterial analysis. Land, 11(10), p.1782. https://doi.org/10.3390/land11101782

Slack, M.P.E., Cripps, A.W., Grimwood, K., Mackenzie, G.A. and Ulanova, M., 2021. Invasive *Haemophilus influenzae* infections after 3 decades of Hib protein conjugate vaccine use. Clinical Microbiology Reviews, 34(3), pp. 1–15. https://doi.org/10.1128/CMR.00028-21

SMH, 2024. The deadly price of a perfect strawberry. Online. https://www.smh.com.au/national/the-deadly-price-of-a-perfect-strawberry-20240513-p5jd78.html

Smith, I., 2003. Mycobacterium tuberculosis pathogenesis and molecular determinants of virulence. Clinical Microbiology Reviews, 16(3), pp. 463–496. https://doi.org/10.1128/CMR.16.3.463-496.2003

Smither, S.J., Piercy, T.J., Eastaugh, L., Steward, J.A. and Lever, M.S., 2011. An alternative method of measuring aerosol survival using spiders' webs and its use for the filoviruses. Journal of Virological Methods, 177(1), pp. 123–127. https://doi.org/10.1016/j.jviromet.2011.06.021

Smulders, F.J.M. & Collins, J.D. (eds), 2002. Food safety assurance: Pre-harvest factors. ECVPH Food Safety Assurance Series, Volume 1. Wageningen Academic Publishers. https://doi.org/10.3920/978-90-8686-508-6

SNF, 2025. Vibhu Prakash profile. Online. https://sanctuarynaturefoundation.org/article/meet-dr.-vibhu-prakash.

Snow, M. 2020. How do biologists fight pandemics in the animal kingdom? Online. https://www.fws.gov/story/2020-07/fighting-chytrid

Sonenshine, D.E., 2018. Range expansion of tick disease vectors in North America: Implications for spread of tick-borne disease. International Journal of Environmental Research and Public Health, 15(3), p. 478. https://doi.org/10.3390/ijerph15030478

Song, H., McComas, K.A. and Schuler, K.L., 2019. Hunters' responses to urine-based scent bans tackling chronic wasting disease. The Journal of Wildlife Management, 83(2), pp. 457–466. https://doi.org/10.1002/jwmg.21593

Soulé, M.E., 1985. What is conservation biology? *BioScience*, 35(11), pp. 727–734. https://doi.org/10.2307/1310054

Spragge, F., Bakkeren, E., Jahn, M.T., BN Araujo, E., Pearson, C.F., Wang, X., Pankhurst, L., Cunrath, O. and Foster, K.R., 2023. Microbiome diversity protects against pathogens by nutrient blocking. Science, 382(6676), p. eadj3502. https://doi.org/10.1126/science.adj3502

Stammnitz, M.R., Gori, K., Kwon, Y.M., Harry, E., Martin, F.J., Billis, K., Cheng, Y., Baez-Ortega, A., Chow, W., Comte, S. and Eggertsson, H., 2023. The evolution of two transmissible cancers in Tasmanian devils. Science, 380(6642), pp. 283–293. https://doi.org/10.1126/science.abq6453

Stapleton, D.H., 2005. A lost chapter in the early history of DDT: The development of anti-typhus technologies by the Rockefeller Foundation's Louse Laboratory, 1942–1944. Technology and Culture, 46(3), pp. 513–540. https://doi.org/10.1353/tech.2005.0148

Steere, A.C., Snydman, D., Murray, P., Mensch, J., Main Jr, A.J., Wallis, R.C., Shope, R.E. and Malawista, S.E., 1986. Historical perspective of Lyme disease. Zentralblatt

Für Bakteriologie, Mikrobiologie Und Hygiene. Series A: Medical Microbiology, Infectious Diseases, Virology, Parasitology, 263(1-2), pp. 3–6. https://doi.org/10.1016/S0176-6724(86)80093-1

Stephens, P.R., Gottdenker, N., Schatz, A.M., Schmidt, J.P. and Drake, J.M., 2021. Characteristics of the 100 largest modern zoonotic disease outbreaks. Philosophical Transactions of the Royal Society B, 376(1837), p. 20200535. https://doi.org/10.1098/rstb.2020.0535

Stoker, B., 1897. Dracula: A Mystery Story. WR Caldwell.

Stokstad, E. 2024. Frog 'saunas' could help endangered species beat a deadly fungus. Online. https://www.science.org/content/article/frog-saunas-could-help-endangered-species-beat-deadly-fungus

Storr, W., 2020. The Science of Storytelling: Why Stories Make Us Human and How to Tell Them Better. Abrams.

Strain, D., 2011. 8.7 million: A new estimate for all the complex species on Earth. 333(6046):1083 https://doi.org/10.1126/science.333.6046.1083

Subramanian, M., 2008. Towering silence. Search: Science, Religion, Culture, p. 144.

Tan, A.F., Yu, S., Wang, C., Yeoh, G.H., Teoh, W.Y. and Yip, A.C., 2024. Reimagining plastics waste as energy solutions: Challenges and opportunities. npj Materials Sustainability, 2(1), p. 2. https://doi.org/10.1038/s44296-024-00007-x

Temmam, S., Vongphayloth, K., Baquero, E., Munier, S., Bonomi, M., Regnault, B., Douangboubpha, B., Karami, Y., Chrétien, D., Sanamxay, D. and Xayaphet, V., 2022. Bat coronaviruses related to SARS-CoV-2 and infectious for human cells. Nature, 604(7905), pp. 330–336. https://doi.org/10.1038/s41586-022-04532-4

Tomás, G., Marandino, A., Panzera, Y., Rodríguez, S., Wallau, G.L., Dezordi, F.Z., Pérez, R., Bassetti, L., Negro, R., Williman, J. and Uriarte, V., 2024. Highly pathogenic avian influenza H5N1 virus infections in pinnipeds and seabirds in Uruguay: Implications for bird-mammal transmission in South America. Virus Evolution, 10(1), p. veae031. https://doi.org/10.1093/ve/veae031

Tranulis, M.A., Tryland, M., Kapperud, G., Skjerve, E., Gudding, R., Grahek-Ogden, D., Eckner, K.F., Lassen, J.F., Narvhus, J., Nesbakken, T. and Robertson, L., 2016. CWD in Norway, European Journal of Nutrition & Food Safety. 9, pp. 301–302 https://doi.org/10.9734/ejnfs/2019/v9i330070

Turunen, M., Soppela, P., Kinnunen, H., Sutinen, M.L. and Martz, F., 2009. Does climate change influence the availability and quality of reindeer forage plants? Polar Biology, 32, pp. 813–832. https://doi.org/10.1007/s00300-009-0609-2

Tyagi, B.K. and Vythilingam, I., 2025. History of mosquito research. In Mosquitoes of India (pp. 3–43). B.K. Tyagi (Ed). Boca Raton: CRC Press. https://doi.org/10.1201/9781003326991-2

UN Decade, 2024. UN Decade on Ecosystem Restoration. Online. https://www.decadeonrestoration.org/

Undark, 2023. The Race to Protect Endangered Condors Against Deadly Bird Flu. Online. https://www.gavi.org/vaccineswork/race-protect-endangered-condors-against-deadly-bird-flu

UNICEF, 2022. UNICEF and WHO warn of a perfect storm of conditions for measles outbreaks, affecting children. Online. https://www.who.int/news/item/27-04-2022-unicef-and-who-warn-of--perfect-storm--of-conditions-for-measles-outbreaks--affecting-children

UNICEF, 2024. Child Health – Malaria. Online. https://data.unicef.org/topic/child-health/malaria

University of Sydney, 2022. First WHO watch list of health threatening fungi released. Online. https://www.sydney.edu.au/news-opinion/news/2022/10/26/first-who-watch-list-of-health-threatening-fungi-released.html

University of Toronto, 2020. Tracking the Coronavirus Pandemic with AI: BlueDot featured on 60 Minutes. Online. https://deptmedicine.utoronto.ca/news/tracking-coronavirus-pandemic-ai-bluedot-featured-60-minutes

UNMC, 2023. Disease forecasters are convinced there's a 27% chance of another COVID-like pandemic within 10 years-but experts believe there's a silver bullet. Online. https://www.unmc.edu/healthsecurity/transmission/2023/04/18/disease-forecasters-are-convinced-theres-a-27-chance-of-another-covid-like-pandemic-within-10-years-but-experts-believe-theres-a-silver-bullet/

USFWS, 2024. Preventing and treating white-nose syndrome. Online. https://www.fws.gov/story/preventing-and-treating-white-nose-syndrome#:~:text=What%20it%20is:%20This%20treatment,of%20white%2Dnose%20syndrome%20decreased.

Van Buskirk, J. and Ostfeld, R.S., 1995. Controlling Lyme disease by modifying the density and species composition of tick hosts. Ecological Applications, 5(4), pp. 1133–1140. https://doi.org/10.2307/2269360

Van den Heever, L., Thompson, L.J., Bowerman, W.W., Smit-Robinson, H., Shaffer, L.J., Harrell, R.M. and Ottinger, M.A., 2021. Reviewing the role of vultures at the human-wildlife-livestock disease interface: An African perspective. Journal of Raptor Research, 55(3), pp. 311–327. https://doi.org/10.3356/JRR-20-22

Van der Hoek, W., Morroy, G., Renders, N.H., Hermans, M.H., Leenders, A.C. and Schneeberger, P.M., 2012. Epidemic Q fever in humans in the Netherlands. Advances in Experimental Medicine and Biology. 984, pp. 329–64. https://doi.org/10.1007/978-94-007-4315-1_17

Van Kerkhove, M.D., Ryan, M.J. and Ghebreyesus, T.A., 2021. Preparing for 'Disease X'. Science, 374(6566), p. 377. https://doi.org/10.1126/science.abm7796

Van Roosmalen, E. and de Bekker, C., 2024. Mechanisms underlying ophiocordyceps infection and behavioral manipulation of ants: Unique or ubiquitous? Annual Review of Microbiology, 78, pp. 575–593. https://doi.org/10.1146/annurev-micro-041522-092522

Voigt, C.C. and Kingston, T., 2016. Bats in the Anthropocene: Conservation of Bats in a Changing World (p. 606). Springer Nature. https://doi.org/10.1007/978-3-319-25220-9

Voloudakis, A.E., Kaldis, A. and Patil, B.L., 2022. RNA-based vaccination of plants for control of viruses. Annual Review of Virology, 9(1), pp. 521–548. https://doi.org/10.1146/annurev-virology-091919-073708

von Mollendorf, C. and Ong-Lim, A.L.T., 2025. How have pneumococcal conjugate vaccines changed the pneumococcal disease landscape? The Lancet Infectious Diseases, 25(4), pp. 367–369. https://doi.org/10.1016/S1473-3099(24)00742-4

Voyles, J., Woodhams, D.C., Saenz, V., Byrne, A.Q., Perez, R., Rios-Sotelo, G., Ryan, M.J., Bletz, M.C., Sobell, F.A., McLetchie, S. and Reinert, L., 2018. Shifts in disease dynamics in a tropical amphibian assemblage are not due to pathogen attenuation. Science, 359(6383), pp. 1517–1519. https://doi.org/10.1126/science.aao4806

Waglechner, N., McArthur, A.G. and Wright, G.D., 2019. Phylogenetic reconciliation reveals the natural history of glycopeptide antibiotic biosynthesis and resistance. Nature Microbiology, 4(11), pp. 1862–1871. https://doi.org/10.1038/s41564-019-0531-5

Wake, D.B. and Koo, M.S., 2018. Amphibians. Current Biology, 28(21), pp. R1237–R1241. https://doi.org/10.1016/j.cub.2018.09.028

Wallace, D. 2023. The real fungal pandemic – amphibian chytrid fungus. Online. https://www.ecolsoc.org.au/blog/the-real-fungal-pandemic/.

Walter, K.S., Carpi, G., Caccone, A. and Diuk-Wasser, M.A., 2017. Genomic insights into the ancient spread of Lyme disease across North America. Nature Ecology & Evolution, 1(10), pp. 1569–1576. https://doi.org/10.1038/s41559-017-0282-8

Walters, J.R., Derrickson, S.R., Michael Fry, D., Haig, S.M., Marzluff, J.M. and Wunderle Jr, J.M., 2010. Status of the California condor (Gymnogyps californianus) and efforts to achieve its recovery. The Auk, 127(4), pp. 969–1001. https://doi.org/10.1525/auk.2010.127.4.969

Washio, M., Kiyohara, C., Hamada, T., Miyake, Y., Arai, Y. and Okayama, M., 1997. The case fatality rate of methicillin-resistant Staphylococcus aureus (MRSA) infection among the elderly in a geriatric hospital and their risk factors. The Tohoku Journal of Experimental Medicine, 183(1), pp. 75–82. https://doi.org/10.1620/tjem.183.75

Weller, T.J., Rodhouse, T.J., Neubaum, D.J., Ormsbee, P.C., Dixon, R.D., Popp, D.L., Williams, J.A., Osborn, S.D., Rogers, B.W., Beard, L.O. and McIntire, A.M., 2018. A review of bat hibernacula across the western United States: Implications for white-nose syndrome surveillance and management. PLoS One, 13(10), p. e0205647. https://doi.org/10.1371/journal.pone.0205647

Wells, K., Hamede, R.K., Kerlin, D.H., Storfer, A., Hohenlohe, P.A., Jones, M.E. and McCallum, H.I., 2017. Infection of the fittest: Devil facial tumour disease has greatest effect on individuals with highest reproductive output. Ecology Letters, 20(6), pp. 770–778. https://doi.org/10.1111/ele.12776

Whitfield, J., 2002. Portrait of a serial killer. Nature. https://doi.org/10.1038/news021001-6 https://doi.org/10.1038/news021001-6

WHO, 2024. Tuberculosis. Online. https://www.who.int/news/item/29-10-2024-tuberculosis-resurges-as-top-infectious-disease-killer

WHO, 2025. Poliomyelitis (polio). Online. https://www.who.int/health-topics/poliomyelitis#tab=tab_1

Widmaier, E.P., Raff, H. and Strang, K.T., 2019. Vander's Human Physiology: The Mechanisms of Body Function (15th edn). McGraw-Hill Education.

Wilbur, S.R. and Kiff, L.F., 1980. The California condor in Baja California, Mexico. American Birds, 34(6), p. 8.

Wilderness Society, 2024. The stats that expose Australia's hidden deforestation crisis. Online. https://wilderness.org.au/protecting-nature/deforestation/the-stats-that-expose-australias-hidden-deforestation-crisis

Wilder-Smith, A. and Massad, E., 2018. Estimating the number of unvaccinated Chinese workers against yellow fever in Angola. BMC Infectious Diseases, 18, pp. 1–4. https://doi.org/10.1186/s12879-018-3084-y

Wilkinson, G.S. and South, J.M., 2002. Life history, ecology and longevity in bats. Aging Cell, 1(2), pp. 124–131. https://doi.org/10.1046/j.1474-9728.2002.00020.x

Will, I., Attardo, G.M. and de Bekker, C., 2023. Multiomic interpretation of fungus-infected ant metabolomes during manipulated summit disease. Scientific Reports, 13(1), p. 14363. https://doi.org/10.1038/s41598-023-40065-0

Wille, M. 2024. Chickens, ducks, seals and cows: a dangerous bird flu strain is knocking on Australia's door. Online. https://theconversation.com/chickens-ducks-seals-and-cows-a-dangerous-bird-flu-strain-is-knocking-on-australias-door-230013.

Williams, B.A., Jones, C.H., Welch, V. and True, J.M., 2023. Outlook of pandemic preparedness in a post-COVID-19 world. npj Vaccines, 8(1), p. 178. https://doi.org/10.1038/s41541-023-00773-0

Williams, A., 2025. Expected ban on Mexican GM corn fetches praise – and worry over imports. Online. https://news.mongabay.com/2025/02/expected-ban-on-mexican-gm-corn-fetches-praise-and-worry-over-imports/

Wirsenius, S., Azar, C. and Berndes, G., 2010. How much land is needed for global food production under scenarios of dietary changes and livestock productivity increases in 2030? Agricultural Systems, 103(9), pp. 621–638. https://doi.org/10.1016/j.agsy.2010.07.005

Wise, I.J. and Borry, P., 2022. An ethical overview of the CRISPR-based elimination of Anopheles gambiae to combat malaria. Journal of Bioethical Inquiry, 19(3), pp. 371–380. https://doi.org/10.1007/s11673-022-10172-0

WNS, 2004. White-nose syndrome: where is it now? Online. https://www.whitenosesyndrome.org/about/where-is-it-now

Woodroffe, R., Astley, K., Barnecut, R., Brotherton, P.N., Donnelly, C.A., Grub, H.M., Ham, C., Howe, C., Jones, C., Marriott, C. and Miles, V., 2024. Farmer-led badger vaccination in Cornwall: Epidemiological patterns and social perspectives. People and Nature, 6(5), pp. 1960–1973. https://doi.org/10.1002/pan3.10691

WUR, 2016. Vaccine stops Dutch elm disease. Online. https://edepot.wur.nl/431509

WWF, 2025. Newly patented technology helps save endangered black-footed ferrets. Online. https://www.worldwildlife.org/stories/newly-patented-technology-helps-save-endangered-black-footed-ferrets

Yao, W., Zhang, X. and Gong, Q., 2021. The effect of exposure to the natural environment on stress reduction: A meta-analysis. Urban Forestry & Urban Greening, 57, p. 126932. https://doi.org/10.1016/j.ufug.2020.126932

Yokoyama, T. and Mohri, S., 2008. Prion diseases and emerging prion diseases. Current medicinal chemistry, 15(9), pp. 912–916. https://doi.org/10.2174/092986708783955437

Zhang, Y., Li, Y., Zeng, J., Chang, Y., Han, S., Zhao, J., Fan, Y., Xiong, Z., Zou, X., Wang, C. and Li, B., 2020. Risk factors for mortality of inpatients with Pseudomonas aeruginosa bacteremia in China: Impact of resistance profile in the mortality. Infection and Drug Resistance, pp. 4115–4123. https://doi.org/10.2147/IDR.S268744

ZSL, 2024. 7 things you can do to help prevent the Illegal Wildlife Trade. Online. https://www.zsl.org/what-we-do/conservation/protecting-species/7-things-you-can-do-help-prevent-illegal-wildlife-tradeZSL

Zukal, J., Bandouchova, H., Brichta, J., Cmokova, A., Jaron, K.S., Kolarik, M., Kovacova, V., Kubátová, A., Nováková, A., Orlov, O. and Pikula, J., 2016. White-nose syndrome without borders: Pseudogymnoascus destructans infection tolerated in Europe and Palearctic Asia but not in North America. Scientific Reports, 6(1), pp. 1–17. https://doi.org/10.1038/srep19829

Index

Page numbers in *italics* refer to illustrations; those with n refer to notes.

Waddle, Anthony 268
walkability in Norwegian
    neighbourhoods 306n13
walking ecosystems 64–5, 97, 216,
    227, 260
wandering behaviour 117–8
wastewater systems 275
'water absorption response' posture
    299n7
West Nile Virus 19–20, 128, 283n9
Western fence lizard (*Sceloporus
    occidentalis*) 59
Western medicine 259
wheat (*Triticum* spp.) 205, 207, 249
wheat stem rust (*Puccinia
    graminis*) 205
wheat stripe rust (*Puccinia
    striiformis*) 205
whistling tree frog (*Litoria
    verreauxii*) 165
White Cliffs of Dover 81
white-footed mouse (*Peromyscus
    leucopus*) 56, 58, 60, 289n15
white-nose syndrome (WNS) 144
    cave-dwelling bat with *145*
    fungus 148
    long-wave UV illuminating
        lesions *149*
    to North America 149–50, *151*
    sylvatic plague 152–5
    wildlife pandemic 150–2
white-tailed deer (*Odocoileus
    virginianus*) 192, 288n5,
    302n16
wildlife habitat encroachment 18
wildlife markets 236
    illegal 241
    in Lao PDR 306n5

wildlife meat markets 236
wildlife pandemics, realms of 11–2
wildlife reserves 275
wildlife trafficking 10, 241, 243–4
Wired (Rowley) 164
Wodonga restoration project 221
wolf (*Canis lupus*) 55, 213
World Health Organization (WHO) 7,
    68, 272, 274
World Wildlife Fund 265
Wuhan Institute of Virology in
    China 93

'X-shredder' skew offspring 297–8n26
Xylella 237
Xylella fastidiosa 237, 306n6

yellow fever 17, 88–9
    disruptions in supply or distribution
        of 91
    vector for 136
Yersin, Alexandre 2
Yersinia pestis 2, 8–9, 11, 109, 152
'yes' camp 135–8

Zigas, Vincent 189
Zika virus 17–9, 128, 282n5, 282n6,
    283n7
'zombie ant' fungus 121, 295n7
zombie-causing fungi 116–7, 279–80
zooanthroponosis 109–10
Zoological Society of London (ZSL)
    243, 264
zoonotic diseases 3–4, 305n15
zoonotic outbreaks 281n2, 283n13
zoonotic Puumala hantavirus 62
zoonotic times 3–4
Zoroastrian sky burial practices 285n21